NEW DIMENSIONS IN AGRICULTURAL GEOGRAPHY

*VOLUME FOUR*

# LANDUSE AND AGRICULTURAL PLANNING

*Concept's International Series in Geography*

*Volume 1* **Historical Dimensions of Agriculture**
*Volume 2* **The Ecology of Agricultural System**
*Volume 3* **Socio-economic Dimensions of Agriculture**
*Volume 4* **Landuse and Agricultural Planning**
*Volume 5* **Spatial Dimensions of Agriculture**
*Volume 6* **Human Perception and Technological Change in Agriculture**
*Volume 7* **Dynamics of Agricultural Development**
*Volume 8* **Anthropogenic Dimensions in Agriculture**

*Concept's International Series in Geography No. 4*

NEW DIMENSIONS IN AGRICULTURAL GEOGRAPHY

VOLUME FOUR

# LANDUSE AND AGRICULTURAL PLANNING

**NOOR MOHAMMAD**
*Editor*

**CONCEPT PUBLISHING COMPANY, NEW DELHI 110059**

ISBN 81-7022-400-4 (Set)
ISBN 81-7022-404-7 (Vol. 4)

First Published 1992



*Published by*
Ashok Kumar Mittal
Concept Publishing Company
A/15-16, Commercial Block, Mohan Garden
NEW DELHI 110059 (India)

*Lasertypeset by*
Microtech Advance Printing Systems (Pvt.) Ltd.
H-13, Bali Nagar,
NEW DELHI 110015

*Printed at*
Sunil Printers
CB 386, Ring Road, Naraina
NEW DELHI 110028

# PREFACE

AGRICULTURE CONTINUES to be the key industry. It is not only the dominant form of employment but also forms the basis for a way of life for, perhaps, two-thirds of the world's households. Agriculture being the function of physical, socio-institutional and techno-economic factors which are dynamic in nature, keeps on changing with the basic objective of increasing production and generation of foodgrains surplus. The traditional system of farming is in the process of transformation to modern agriculture which itself is undergoing a set of equally and generally even in more rapid changes. So it has added a new dimension in agriculture which needs to be studied and analysed.

A project was, therefore, taken up to accomplish this task. Original research papers dealing with various facets of agriculture were invited from scholars of different discipline from all over the world. The response was very encouraging and a large number of original papers were received. Besides these, some papers which have classical and everlasting value have also been picked up from my earlier publication – *Perspectives in Agricultural Geography*. The papers have been classified according to their basic themes and have been put into following eight volumes so as to give a systematic picture of various themes. The articles dealing with concepts, theories and models have been kept first followed by those dealing with the same theme at macro, meso and micro levels respectively.

The First Volume – *Historical Dimensions of Agriculture* – containing 19 articles deals with the origin, diffusion and development of agriculture, research trends in agricultural geography, world agricultural systems and methods and techniques in agricultural geography.

The Second Volume – *The Ecology of Agricultural System* – deals with the selected parameters of physical environment specially the climatic elements which affect the agricultural landscape in spite of technological advancement in agriculture.

The Third Volume – *Socio-economic Dimensions of Agriculture* –

containing 19 articles deals with socio-economic environment which have bearings on agricultural systems. The first 8 articles deal with social parameters specially labour, caste, women etc., while the last 11 articles examine the impact of techno-economic factors specially the various facets of irrigation which is a basic agricultural input.

The Fourth Volume – *Landuse and Agricultural Planning* – contains 19 articles dealing with concepts, theories, models of landuse, land capability and suitability, wastelands and dynamics of landuse.

The Fifth Volume – *Spatial Dimensions of Agriculture* – includes articles on agricultural systems of specific areas such as Himalayas, Bastar, Uttara Khand etc. It also deals with dynamics of crop landuse, cropping pattern and its diversification, crop ecology and crop-combination analysis.

The Sixth Volume – *Human Perception and Technological Change in Agriculture* – contains only 15 articles and covers four basic themes. The first set of 3 articles deals with human perception and his consequential response, while the second set includes articles on correlates of technological change and adoption of agricultural innovations. The third set of articles deals with the assessment of socio-economic transformation in rural India. The last set deals with agriculture and human settlements.

The Seventh Volume – *Dynamics of Agricultural Development* – includes articles on concept of development and its various facets. Besides assessing the trends of levels of agricultural development, it also examines and evaluates its impact on environment and ecological parameters.

The Eighth Volume – *Anthropogenic Dimensions of Agriculture* – containing 17 articles deals with regional imbalances in agricultural productivity, methods of its measurement, and also its correlates and determinant. It also focuses on food availability and food security at regional and continental level.

The collection of papers in eight volumes of *New Dimensions in Agricultural Geography* is designed to encapsule the major themes and recent trends and concepts in the field of agricultural geography with particular reference to conceptual revolution in the discipline. All the volumes contain a brief introduction which gives a glimpse into the topics dealt with in each of them.

The variety of articles contributed by the renowned scholars of different disciplines has made the character of these volumes multi-disciplinary and has enhanced their value. These volumes have been designed to help not only the students of a particular discipline but also to general students who wish to keep abreast of the latest developments in the field of agricultural geography. From these points of view these volumes hold a promise of rich fare and in addition to being heuristic, will provide a mine of information for discerning scholars. It is hoped that this unique collection of research papers will not only be valuable to the researchers in the fields of agricultural geography, economics, sociology, history, and agricultural science, but will also provide guidelines to decision-makers and planners.

The completion of this project has been possible due to the constant inspiration and unstinting co-operation of a host of people too numerous to acknowledge individually. However, I owe a debt of gratitude to my mentor — Dr. Mohammad Shafi, Professor Emeritus, Department of Geography, A.M.U., Aligarh who gave me valuable comments and suggestions ever since the inception of this project. Prof. R. Ramachandran and my other colleagues who offered sincere comments on the title of volumes and the collection of articles deserve special thanks. I would also like to thank my father, wife and children who provided a favourable working atmosphere to me for successful completion of this project. Last but not the least, I would like to put a special word of appreciation, love and affection to Mohammad Imran Ali Khan who is a source of hope and inspiration.

Department of Geography
Delhi School of Economics
University of Delhi
Delhi-110007

NOOR MOHAMMAD

*January 26, 1992*

# PUBLISHER'S NOTE

When Concept's International Series in Geography No. 1 — *Perspectives in Agricultural Geography* edited by Dr. Noor Mohammad in five volumes was issued over a decade back, we expressed our apprehensions that 'Posterity will be the best judge of our efforts in this direction. The response to this set of volumes will provide the impetus for hastening the publication of further series that are already on the anvil'.

*Perspectives in Agricultural Geography* received such an overwhelming response that it became out of print soon after its publication. Its success provided the necessary encouragement to issue thereafter Concept's International Series in Geography No. 2 — *Perspectives in Urban Geography* edited by Dr. C.S. Yadav in 16 volumes and No. 3 — *Perspectives in Geomorphology* by Dr. H.S. Sharma in four volumes.

We are glad to offer now Concept's International Series in Geography No. 4 — *New Dimensions in Agricultural Geography* edited by Dr. Noor Mohammad in eight volumes. In view of the new emerging trends, concept, theories and models in the field of agricultural geography, the set presents a wide panoramic view of the latest advances in this field. Hope this set too would be equally welcomed by the students, researchers, scholars and planners in various disciplines.

*February 11, 1992* NAURANG RAI

# CONTENTS

# LIST OF FIGURES

# LIST OF TABLES

# LIST OF CONTRIBUTORS

**Bandooni, Suresh Kumar**

Department of Geography, Bhagat Singh College (Ev.), University of Delhi, Delhi

**Bhattacharya, Rabindranath**

Department of Geography, University of Calcutta, Calcutta

**Casetti, Emilio**

Department of Geography, The Ohio State University, Ohio, USA

**Chaturvedi, Arun**

National Bureau of Soil Survey and Land Use Planning, Indian Agricultural Institute Campus, New Delhi

**Coleman, Alice**

Director, Second Land Utilization Survey of Britain and Head of the School of Human Environment Studies, King's College, Univesity of London, London

**Khan, A. R.**

Department of Geography, Shivaji College, University of Delhi, Delhi

**Kumar, A.**

P.G. Department of Geography, University of Bhagalpur, Bhagalpur

**Lahiri, Sulekha**

Department of Geography, Loreto College, Darjeeling

**Majeed, Abdul**

Department of Geogaphy, S.S. School, Jamia Millia Islamia, New Delhi

**Mohammad, Noor**

Department of Geography, Delhi School of Economics, University of Delhi, Delhi

**Pandit, Kavita**

Department of Geography, Franklin College of Arts and Science, University of Georgia, Athens, Georgia 30602, USA

**Pofali, R. M.**
National Bureau of Soil Survey and Land Use Planning, Seminar Hills, Nagpur

**Sen, Amal Kumar**
Central Arid Zone Research Institute, Jodhpur

**Sharma, G. P.**
Geography, T.N.B. College, Bhagalpur University, Bhagalpur

**Sharma, S. C.**
Department of Geography, M.L.K. Degree College, Balrampur, Gonda

**Siddiqi, M. Farooq**
Department of Geography, Aligarh Muslim University, Aligarh

**Siddiqi, N. A.**
Professor of Geography (Retd), Department of Geography, Aligarh Muslim University, Aligarh, India

**Singh, R. B.**
Department of Geography, Delhi School of Economics, University of Delhi, Delhi

**Singh, R. P.**
Professor of Geography (Retd), Department of Geography, Magadh University, Bodh Gaya

**Singh, Surendra**
Central Arid Zone Research Institute, Jodhpur

**Stola, Wladyslawa**
Institute of Geography and Spatial Organization, Polish Academy of Sciences, Warsaw, Poland

**Tyszkiewicz, Wieslawa**
Institute of Geography and Spatial Organization, Polish Academy of Sciences, Warsaw, Poland

**Vats, P. C.**
Central Arid Zone Research Institute, Jodhpur

# INTRODUCTION

THE PHYSICAL environment of man consists of land forms, climate, water, soil, natural vegetation and animal life. This resource complex is necessary for human existence, since it provides him living space, food, and a vast variety of raw materials used in the satisfaction of human needs (Chauhan, 1966). Land resources play a strategic role in the determination of man's economic, social and cultural progress which could be easily seen and clearly understood through reference to economic history of different countries. But unfortunately, the development and use of such an important resource was left in the past entirely to trial and error methods which has naturally led to its improper use in many parts of the world. Consequently, this basic resource has gradually deteriorated, although in a world short of food and other basic raw materials, its conservation is one of the basic necessities. In countries where efforts are being made to modernise agriculture through the adoption of scientific methods, there is an excellent opportunity for correcting past errors in landuse and to get rid of future errors through detailed mapping and classification of land. This is possible only when an exhaustive landuse survey of every piece of land is properly done, and based on these information, land is suitably classified in different categories (Stamp, 1954). The value of such landuse survey lies in detailed information it provides with regard to existing use and misuse of land. This information, when analysed, leads to a better understanding of the present use, the land has been put to and the possibilities of its further development (Mohammad, 1978). However, on the basis of the information, thus collected, land should properly be classified in different categories.

The classification of land is quite essential and rather a prerequisite in any landuse planning and development. Unless it is determined as to what is the optimum use of a particular piece of land, what are its problems and their adequate solutions, no planning can successfully be proposed. This is now an integral part of landuse planning and is receiving attention of the planners in many countries of the world. In fact land classification is a method of grouping land on the basis of their inherent characteristics, existing use, yield capacity and their classes as assessed in relation to their suitability to different uses envisaged in the plan. It relates to climatic factors, soil characteristics, slope and degree of erosion, water supply and drainage and similar environmental conditions affecting landuse and productivity.

Land classification may be general or of specific nature. The essence of the former is to give a summary of the inherent physical characteristics not only of the tract under consideration, but of other areas, which could provide alternative site for development. Such a classification had the advantage that once it is completed, it can be referred to without delay to assist in the solution of many planning problems. The specific classification attempts to put the land into different groups with certain specific theme, such as land capability or land suitability etc. The former, devised by the U.S. Soil Conservation Service is mainly concerned with erosion hazards and is based on the inherent soil characteristics, external land features and environmental complex that limit the use of land. In this system soils are grouped at three levels – capability classes, capability sub-classes and capability units – in descending hierarchical order. Land suitability classification is a broad assessment of the usefulness of the land for certain specific kind of agricultural enterprise. The capability is the inherent capacity of land to perform at given level for general landuse, while suitability is a statement of the adaptability of a given area for a specific landuse type (F.A.O., 1974, 1976).

However, realising the importance of land classification different countries have classified their land on various bases depending upon their requirements, resources, time, and socio-economic conditions. Various methods have been employed for rating the land into capability classes in different parts of the world. For example Morgon's (1939) method based on degree of adaptation of land types of various crops, Storie's (1933) method of land rating, Bennett's (1939) method based on site and physical properties of land, Marbut's (1935) method based on crop yield, Stamp's (1954) potential production unit method, Raychaudhuri and Shome (1960) soil rating index and so on.

The articles included in this volume deal with concepts, theories and methods of land classification on the one hand, while on the other, certain papers give an analytical review of land classification done in different parts of the world. In the third category of articles attempts have been made to classify the land into general, capability, and suitability categories at macro, meso and micro levels.

However, this volume opens up with **Chaturvedi's** paper which tries to explain the concept of land capability and land suitability which is very complex and intricate. Actually the quality of land is a complex attribute which acts in a distinct fashion in its influence on the suitability of land for a specific kind of use. But in brief capability is an inherent capacity of land to perform at a given level for a general landuse, while suitability is a statement of adaptability of a given area for a specified landuse type. Land capability is, by and large, ascertained by inherent soil characteristics, external land features, and environmental factors limiting landuse. The capability classification provides three major categories of soil groupings viz., capability unit, capability sub-class, and capability class. The author, after discussing these capability classes, has evaluated three major factors which affect them viz., inherent soil characteristics, landscape features and climatic factors. Although suitability studies are old one, a systematic framework for land suitability classification was given first by the F.A.O., and four types of land suitability classification were recognised.

For a rational planning of landuse and development the significance of proper classification of land hardly needs any emphasis. In fact land classification is an attempt to integrate land according to their characteristics. Due to ever increasing pressure of population on land resources, different countries of the world have categorised the land with the recommendation of their optimum use. **Sharma** in his paper has given a resume of land classification in some selected countries of the world. In Great Britain, on the basis of land use surveys conducted in thirties under the guidance of Stamp, the soil is classified into three categories on the basis of landuse — past and present, type of farming, fertility and productivity. These three categories viz. good quality land, medium quality land, and poor quality land have further been divided, keeping their minor details into account. To have a comparative view of these soil types, Stamp introduced the concept of Potential Production Unit (P.P.U.). In U.S.A. the land is classified on the basis of its capability and suitability. In this method the relevant factors of both physical and human environment are tabulated and numbered. The number of each unit area of land are expressed in fractional code. The author has also reviewed the inductive method of land classification and has discussed Storie's index which is based on soil profile expressed in three factors. He has also mentioned briefly the criteria suggested by Zoorykin for classification of land. In India attention on scientific classification of land has been given much later. But, however, various attempts have been made from time to time in this direction, and the author has given reference to work of Raychaudhuri, Chaudhary, Sharma, Shafi, and others. However, the fact remains that land classification is of immense importance and of practical value in landuse planning and development.

Land reform is one of the prominent measures of agricultural development as it provides the base for agricultural operations. **Jyszkiewicz** has studied the issue of land lease in Poland during the past three decades. The study examines in detail the trend and patterns of lease in Poland together with nature of lease such as long term lease, short-term lease, or land given on cash or kind basis, and has assessed its merits and limitations. The study finally concludes that land lease in Poland has been recognised by many farmers as a convenient form of enlarging the size of the farms. It plays an important role in agricultural production, since it offers a chance to some farmers to enlarge their farms and to others a chance of diminishing it, or of resigning from running a farm when there is a shortage of labour or for other reasons.

The concept of landuse is a wide and complex one, it is a function of land, water, air and man, and has been used by man since time immemorial to fulfil his changing demand with the passage of time. Land with its varied topography, slope, field pattern, soil, temperature, precipitation, natural cover and countless creatures, have to be planned to engrave an economy where man can maintain a standard of existence. Man has played his part on land to portray the different phases of his ties with it. However, **Singh** has presented in his paper the concept of landuse from historical view point highlighting the modern problem of landuse. The key to most important aspect of landuse lies in relation of population of land. Hence, the crux of the review refers to the study of the problems in use of land by man. The author has discussed in detail rural landuse with more emphasis on agricultural landuse and its related problems. A brief review of urban landuse and various landuse studies done in India and abroad is also presented. The author goes in further depth and discusses the problem of land conservation, land classification and landuse planning.

Land resources play a strategic role in the socio-economic development of a nation. Among the land resources, agricultural land resource has played a vital role since time immemorial engaging the largest percentage of the inhabitants of the world. Therefore, in any plan which aims at utilising the resources in scientific, rational and economic manners, the landuse surveys are of fundamental importance. Hence, the different countries of the world have conducted the landuse surveys on different lines keeping their physical, social and economic conditions in view. **Majeed** in his paper has given a critical review of the major landuse surveys conducted in the world, and has ascertained that there are mainly three approaches to landuse surveys, viz., census approach, unit area approach and sampling approach.

The census approach was firstly adopted by Stamp in thirties in Great Britain to take a national stock of land resources for its proper utilisation in order to increase the food production. British Ordnance Survey maps were used for primary survey carried out by volunteers of different organisations. The survey was revised by Miss Coleman in 1960 on almost the same line with minor alterations.

In unit area approach which was basically adopted in U.S.A., the area of survey is divided into small units and a comprehensive survey of each unit is conducted. The author has reviewed Michigan Land Economic Survey, Tennessee Valley Authority, and Rural Land Classification Programme in Puerto Rico. In the former the experts from different disciplines were engaged and a complete landuse survey was done. The author has given more emphasis on T.V.A. which was lineal descendent of Michigan project. In this survey a fractional code system was adopted for survey in which numerator represents human conditions and denominator expresses physical environment. The complete symbol of a unit was in three parts viz., roman numerals, short fraction and long fraction. The land classification programme of Puerto Rico is a more complicated method of recording landuse.

In sampling approach either random or systematic purposive sampling approach is adopted keeping time, resources and other facilities in view. Buck conducted the landuse survey in China in 1929 on random sampling basis with a view of collecting accurate information about China's agriculture which could be used as basis for national agricultural policies. Shafi and subsequently his students followed purposive sampling approach for conducting landuse surveys in different parts of the country. In this approach field to field survey of selected villages is carried out. Data are analysed, mapped and suitably interpreted. Finally recommendations are made for the agricultural development of the region.

**Siddiqui** in his article on "Land classification for agricultural planning—A Study in Methodology" has given a review of various land classification studies conducted in different countries of the world. It is well known fact that in the past due to several reasons, the development of land resources was left to trial and error method leading to its improper use and its deterioration. But due to development of science and technology, and fast increasing pressure of population on land resources, people have realised its significance and are adopting various measures to protect it. For this the classification of land, which attempts at integrating land according to their inherent characteristics, existing uses, yield capacity and their suitability for producing plant of economic significance, is a pre-requisite. Hence, keeping the requirements, resources, time and socio-economic conditions in view, various countries have adopted different methods of land classification. The author has given a critical review of land classification in U.S.A., Great Britain and Soviet Union. Besides examining various techniques of land classification, the author has also given a critical review of land classification done in India. However, the study reveals that a uniform method and criteria cannot be applied for land classification all over the world because of the varied and complex conditions, but for the overall development of the nation proper land classification is pre-requisite.

The concept of landuse has been developed to serve practical purpose and implies interaction between land and environment. The basic aim of landuse survey is to record the distribution of land under various uses in different socio-economic and environmental conditions and to put the resources to improved uses based on scientific principles, and without any degradation. The landuse planning, is thus, needed with changing needs and pressure. The function of the landuse planning is to guide decisions on landuse in such a way that all the available resources of the environment are put to the most beneficial use together with their conservation for future. It is the primary duty of the landuse planners to study the various land characteristics in respect of different natural aspect, and suggest best use under defined management practices for sustained utilisation. Any alternative use of land suggested should be without disturbing the ecological as well as socio-economic balance of the area (F.A.O., 1976). **Pofali** in his paper has given a critical outlook of landuse survey and planning. After discussing briefly the nature, concept and aim of landuse planning, the author has suggested the scale of maps to be used and levels of landuse classification.

Landuse requires multi-disciplinary approach to arrive at sustained landuse planning. Landuse survey information should include details on physical, social and economic aspects but Pofali has dealt with environmental aspect in detail to find out suitability, productivity, capability and potentiality for suitable landuse planning specially in the field of agriculture in Vidarbha region in Maharashtra. The author has examined the Storie's index, land capability classification, nine factor method for evaluating productivity and potentiality and irrigability class. The author has also highlighted the use of aerial photograph and LANDSAT imageries for landuse survey and planning so that data could be collected quickly and simultaneously, and did not become out-dated by the time it is completely collected, processed, analysed and mapped.

Geomorphological studies conducted in different parts of western Rajasthan reveal that identification of different landforms, their genetic classification, mapping and assessment is an essential prerequisite for rational landuse planning and its development. **Singh** and **Vats** have identified 10 landform units on the basis of morphological, slope, drainage and landsat spectral characteristics in the Barmer district of western Rajasthan, and have highlighted the salient features having relevance with landuse planning of each region. They have assessed physical potentials and limitations of major landform units and sub-units, and have examined the possibilities of their development into croplands, woodlands, pastures, silvipastures and desert catchment depending upon prevailing climatic conditions and landuse capability. The study concludes that fluvial and aeolian landforms have been created by various processes under wet and dry phases, and have different morphological and landsat spectral characteristics. Due to variations in physical potentials and limitations, these landforms have direct bearing on landuse patterns and planning. Fluvial landforms, characterised by almost level slope, fertile sediments, limited hazards, good groundwater potential and better rainfall conditions are fit for rainfed and irrigated crops, while aeolian landforms due to undulating topography, poor fertility of aeolian sediments, severe to very severe wind erosion/deposition hazards, poor rainfall and moisture regime, could be developed for cropland, pastures, silvipastures and hortipastures. On the whole animal husbandry and agro-industries will be the main occupation in the western part of the district, while in the eastern part crop husbandry, animal husbandry and agro-industries will become the main source of livelihood. The changing landuse pattern based on proper land management will play a significant role in the economic development of the region.

For the scientific planning of land resources, the assessment of its use and misuse is pre-requisite which is possible with land utilisation survey and mapping. The ultimate purpose of land utilisation survey is to arrive at systems of landuse and management best suited to the trends or resources and capabilities of the land composing it. **Sen** has critically assessed the existing landuse, and has suggested rational landuse for Rajasthan desert. The author has studied the present landuse and its rotation to agricultural aspects including crop yields, cropping pattern during kharif and rabi seasons in each of the agro-ecological regions i.e., arid and transitional between arid and semi-arid of Rajasthan Desert. He has also made inter-crop and inter-regional comparisons at various levels. Based primarily on the interpretation of five Landsat imageries, landuse data have been generated, discussed in detail and mapped for the three rainfall zones i.e., < 300 mm rainfall, 300-400 mm rainfall and >400 mm rainfall. It, however, does not exactly corresponds with ground truth of the area. The author has finally suggested the landuse plan for the three rainfall zones under desert ecosystem. Only one-third of the physical area will be under crop cultivation, and the remaining will be occupied by grassland and pastures, forests, and fruits and vegetables. The author has also described the resources required to implement the suggested landuse plan and has finally accounted its benefits in detail in monetary value as well as in terms of employment.

World War II, by and large, is considered to be a turning point in the overall growth and development of a nation. In Poland also considerable changes have taken place in agricultural landuse since World War II, as a result of socio-economic development. **Stola** has very systematically studied these changes in historical perspective. Changes in the proportion of the separate forms of landuse i.e. between arable land, permanent grass land and perennial crops and their inter-structure are systematically represented by orientation of agricultural landuse, which determines the system of agricultural use and types of agricultural production. For the identification of the orientation, the method of successive quotient is used with number of quotients being established at six. The study has investigated the orientation of agricultural landuse by powitas in seven times cross section between 1947 and 1988, both for agricultural as a whole and separately for private, state and collective farming.

For the overall development of a region in general and that of agricultural in particular, the significance of intensive landuse survey at micro levels hardly needs any emphasis. **Lahri**'s paper is an attempt in this direction. She has selected three sub-divisions — Darjeeling, Kurseong and Kalimpong — of Darjeeling district of Himalayan region. The character of landuse of this particular area is most closely related to its physical environment including altitude, relative relief, degree of slope, soil qualities, drainage conditions, availability of water, length of growing season, insolation in hill villages, difficulties in transportation, thickness of soil, and ruggedness of terrain. All these factors are reflected in the landuse — both at macro and micro levels. Besides this, political and economic factors have also exerted considerable influence. However, forest under proper management is the ideal form of landuse in the rugged regions with high degree of slope. High forest and plantation cover a large part of the total area of the hill tract. The plantation was introduced by the British in the last century. It may not be very ideal form of landuse but it is very firmly established feature of the cultural landscape. Indigenous cultivation with stress on maize and millet in the rural areas and vegetables near the market centres is other major form of landuse. Any way the current landuse practices are not very economical everywhere and steps are being taken for correct appraisal and utilisation of resources of this area. Mixed farming with more emphasis on fruit growing and animal husbandry, bee-keeping and sericulture would be more suitable form of landuse.

The constant interference and nature's course of yielding did not for all the time howered around the perfect balance but the insatiable need often ushered man to the statues where it had to pay heavy toll for the misdeed. In the areas of fragile natural ecosystem where human pressure and interaction go on increasing unabated, the activities turn counter productive. Besides the nature, calamities like landslides, flood, and famine, the slow depletion of vital productive nutrients from the soil, is going to affect the productivity in the long run. Hence, the monitoring of activities is essential. It is more in landuse and agriculture because land is finite and its uses are restricted. The correct status, mode, rate, direction and resultant ramifications is possible when the watch on the activity undertaken in detailed one. With the level of statistics available, the monitoring of larger areas in this regard is rather difficult. **Raza** and **Singh** have, therefore, taken up a small area Mandi tahsil of Mandi district located in the fragile ecosystem of Himalayas for monitoring landuse and agriculture. Here considerable agricultural development has taken place due to massive induction of modern technology with the help of Federal Republic of Germany. On the basis of field survey and personal field observations, the authors highlight the changing pattern of landuse during 1984-85 and 1986-87 in detail at district and selected tahsil levels and discuss in detail the agricultural characteristics — terracing, gradient irrigation, nutrient type and spatial pattern of crop landuse, in the *katil*, *upraon* and *talaon* areas. The monitoring of crop system is described and explained at meso-tahsil, and micro-village level where, a comparative study of total cropped area and production of food crops — cereals, pulses and others, and non-food crops, is done together with changing productivity of principal crops. A brief but explanatory account of horticulture and livestock and poultry is also given. The study concludes that the sub-systems of landuse agricultural system such as historisity, interactive mechanism, different activities involved, popular strategies, input-output analysis, human involvement, livestock and fisheries development etc. hold great responsive role in the elaboration and extension of their monitoring to have a balanced and sound strategy formation for the effective utilisation of the resources.

The movement of population is inherent in the process of socio-economic development. The causes of population movement are varied but its basic objectives have primarily been to seek better economic opportunities in order to enhance the individuals income leading to material advancement and enhanced social esteem. Out-migration has, by and large, brought prosperity by way of flow of funds, technology and cultural contacts in the region of origin. Eventually it affects various sectors of the economy of the region. **Mohammad** and **Bandooni** have studied the impact of out-migration on patterns of agricultural landuse in the Pauri Garhwal district of U.P. Himalayas. The study is predominantly based on primary data but secondary data have also been used for meso-level study. The impact of out-migration on patterns of agricultural landuse has been assessed at two levels i.e., the district level and the respondent level. The former is based on secondary data taking tahsil as unit of analysis, while the later is based on primary data collected from 14 selected villages located in different physiographic zones of the district. From these villages 168 households have been selected for in depth micro-level study. The study reveals that there has been a lot of inter- and intra-sectoral changes in the uses of land. The different categories of landuse excluding forest have shown a positive change with 0.2 to 3.18 index of volume of change. The maximum increase is recorded in case of area under groves and fruits which indicates a structural shift in cropping pattern from crop farming to horticulture.

The micro-level study also indicates the changes that have occurred in agricultural landuse. The level of change differs from village to village and respondent to respondent. The degree of change is more in villages with more out-migration as compared to the villages with less out-migration. An other striking feature is that the area under horticulture and cash crop is increasing fast, while the area under millet has shown a declining trend. The areal strength of wheat and rice has remained almost the same.

During the process of economic development of a region, the proportion of its labour force changes from one sector to the other sector. Generally, it tends to decline in agriculture and increase in manufacturing first, and then it declines in it also with gain in services. However, manufacturing growth was stronger and materialised earlier in some countries, while in others it remained low and was associated with greater growth of service sector. **Casette** and **Pandit** have modelled these two historical development patterns using a non-linear dynamic frame of reference, and have interpreted as reflecting favourable and perverting dynamic equilibria. The sectoral shift model consisting of differential equations relating rates and levels of sectoral labour force is arrived at using the Expansion Method due to changing prices, capital stock and technological levels. The manufacturing and service dominated sectoral shifts are explained in terms of multiple systematic equilibria that materialise at intermediate development level. The study reveals that conventional economic presuppositions imply that the single stable equilibrium path in existence at low development levels is eventually complemented by a second stable dynamic equilibrium. The latter is manufacturing dominated and is favourable, while the former is service dominated and perversive. The paper provides a theoretical rationale for many government interventions aimed at promoting the growth of the manufacturing sector. It may be construed as attempt to pay a country loose from a "perverse attractor" and bring it into the domain of the "positive attractor" represented by the manufacturing dominated development path.

During the past few decades particularly after adopting various strategies for agricultural development, landuse and cropping pattern has considerably changed almost all over the country. **Bhattacharya** in his paper has studied these changes at meso level in the state of Bihar. These changes in landuse under five-fold categories — land not available for cultivation; other uncultivated land excluding fallow; fallow land; net area sown; and net cropped lands, are studied during 1958-59 and 1970-71 both for the state as a whole and district-wise, as they existed before bifurcation. The cropping pattern has also been studied in the same pattern during the four major agricultural seasons (*Bhadoi, Aghani, Rabi* and *Garma*) of the state. Lastly the cropping pattern in the said two periods are compared at district level. It has been found that the changes are significant, though the intensification of agricultural land has not very much advanced from 1958-59 level. The study reveals that the primary landuses remain very much the same but there is an appreciable change in cropping pattern and crop-combination.

Since Independence, our country has decided to induct massive technology in agriculture which has began to bear the fruits as early as mid-fifties but an outstanding breakthrough in production could be achieved in late sixties when several package schemes and new strategies were adopted for the development of agriculture. But neither the induction of technology nor the agricultural development has been uniform all over the country and there are variations at all levels. **Siddiqi** in his paper has described and explained the agricultural practices and agricultural changes during 1950-51 and 1985-86 in Bundelkhand region of Uttar Pradesh. The author has given a reasoned account of agricultural development, general landuse and its irrigation system in the region where canal is dominant source of irrigation accounting 75 per cent of the irrigated area. Kharif and Rabi are the main agricultural seasons. Bajra, rice, arhar and small millet are the major crops of the former, while wheat, gram, lentil, barley, oilseeds are principal crops of the latter season. These crops are, by and large, well adjusted to the prevailing environmental conditions during these seasons. The region has a peculiar system of farming-mixed farming, in which farmers sow two/three crops requiring different amount of rainfall. This system is adopted due to typical environmental conditions specially in case of soil, and climate. The system also improves the soil fertility. Due to poor quality of soil, the double cropped area is insignificant and the chance of its increase is also very meagre.

The concept and definition of wasteland is highly controversial and debatable. In spite of the continuous efforts of the individual scholars and research institutions together with several voluntary and government organisations, a precise definition of wasteland has not yet emerged. Similar is the case with areal strength of wasteland and their spatial distribution. **Mohammad** in his paper has described and explained some of these issues and problems. The author has given a critical comment on definitions of wasteland as given in *Chambers Dictionary, Webster's International Dictionary* and the *Dictionary of Geographical Terms*. He also comments on definition given by National Commission on Agriculture, Bhumle and Khare, Yadav, Technical Task Group of National Wasteland Development Board (its original and revised definitions) and concludes that firstly, it is difficult to frame a concise definition of wasteland, secondly, wastelands should be defined on the basis of its productive and functional utility, and finally, it should be defined with reference to its area of occurrence because a particular type of land may be waste in one region but productive in other region. The areal strength of wasteland is also controversial and it ranges from 51 million hectares to 175 million hectares. The author makes a critical assessment of estimates of wasteland given by National Commission on Agriculture, Bhumle and Khare, Institute of Coastal and Offshore Research and National Remote Sensing Agency (NRSA) and recommends that NRSA's estimate is much closer to reality. Likewise there is lack of maps showing spatial distribution of wastelands. However, recently NRSA has come up with the wasteland map of each state and the country as a whole which are comparatively authentic and reliable.

With increasing pressure, urbanisation and industrialisation, the area under agricultural uses is diminishing very fast and it has taken a serious turn in U.K. The findings of Second Landuse Survey indicate that if the contemporary rate of loss continued, all improved land would vanish within 400 years, while the subsequent Sample Survey (1970) results show an acceleration which makes 200 years a more reasonable estimate. Moreover, there is a high positive correlation between land grab and land take which means that loss in productivity is even greater than loss in area. **Coleman** has studied this problem highlighting the causes of land losses, land uses compatible with agriculture, and possibility of reversing the present trend. In fact the problem started with the idea of converting the high density inner city area into low density area by shifting the population towards the rurban fringe. Consequently, the farmers impinge upon wildlife habitat with the urge to provide "More Food From Our Own Resources". Thus, the chain reaction of outward pressure extends from the inner city to the wildlife habitat with the farmland in between. And as such three interests clash and none of them really benefiting from it. The authoress has tried to assess in detail the landuse which are compatible with agriculture. She has examined the case of wildlife habitat, landscape beauty, historical and archaeological sites and recreational amenities which are compatible with farming. She has finally examined as to whether the present trend can be reversed or not, and concludes that the implication of turning the tide are enormous, and the inner city is source of great chain-reaction of landuse pressure.

Among several methods of measurement of agricultural productivity, caloric output per unit of land/man is an important one. This is a more effective measure in the areas of foodgrains subsistence economy. **Kumar** and **Sharma** in their paper have adopted this technique at micro-level in Haspura Anchal of Bihar state. The authors have briefly described population growth and distribution, landuse pattern, food crops production and average yield, availability of cereals, food and dietary standards and availability of calories from cereals, and have finally delineated the surplus and deficit areas. On the basis of standard requirement of calories (2,000 per head per day) they have converted the calories into Standard Nutrition Unit and have correlated with total sown area in order to measure agricultural productivity. However, the study reveals that Haspura Anchal has adequate agricultural potential to ensure a minimum level of nutrition to the people. At present a proper planning is required to change the traditional outlook of farmers with a proper combination and intensity of cropping with a view to increase the yield of pulses, cereals and the cultivation of high yielding cash crops.

## REFERENCES


Bennett, H. H., 1939 : *Soil Conservation*, New York.

Chauhan, D. S., 1966 : *Studies in Utilization of Agricultural Land.*

F.A.O., 1974 : Approaches to Hand Classification, *Soil Bulletin* No. 22.

F.A.O., 1974 : A Framework for Land Evaluation, *Soil Bulletin* No. 32.

F.A.O., 1976 : A Framework for Land Evaluation, *Soil Bulletin* No. 1.

Marbut, C., 1935 : Soils of the United States, *Atlas of American Agriculture*, U.S. Department of Agriculture.

Mohammad, N., 1978 : *Agricultural Landuse in India*, Inter-India Publication, Delhi.

Morgan, M. F., 1939 : *Bulletin Connected with Agriculture Experiment Station.*

Raychaudhuri, S. P. and Shome, K. B., 1960 : 'Rating of Soils of India', *Proceedings of the National Institute of Science of India*, 25, Sept. I.

Stamp, L. D., 1954 : *The Under Developed Land of Britain.*

Storie, T. E., 1933 : *An Index for Rating the Agricultural Values of Soil*, California Experiment Station Bulletin No. 556.

# 1

# THE CONCEPT OF LAND CAPABILITY AND SUITABILITY

ARUN CHATURVEDI

**"Land is an area of the earth's surface, the characteristics of which embrace all reasonably stable, or predictably cyclic, attributes of the biosphere vertically above and below this area including those of the atmosphere, the soil and the underlying geology, the hydrology, the plant and animal populations, and the results of past and present human activity, to the extent that these attributes exert a significant influence on present and future uses of the land by man."**

— FAO, *Soils Bulletin No. 32*, 1976

THE MOST important aspects of land are its role in providing anchorage space to all resources, and the fact that most human activities take place on land.[1] Land is limited in supply and there is competition for its use. In the light of increasing population, the demand is also increasing, thus optimum use of land has become a necessity. Land is also unevenly distributed in terms of its qualities. It has its limitations for different uses. Land can be improved for a particular use by certain measures, it can also be improved by a certain kind of landuse or at least sustained production can be assured. Land can deteriorate by its mismanagement, wrong use or by certain wrong cultivation practices. Thus if misuse and wastage of land have to be avoided, it should be used judiciously considering its capability and suitability for a particular use.

"Land quality is a complex attribute of land which acts in a distinct manner in its influence on the suitability of land for a specific kind of use". (FAO, *Soils Bulletin No. 32*, 1976). It includes the ecological qualities for crop growth such as availability of nutrients, water and radiation energy, erosion resistance, flooding hazard and temperature regime. Land qualities related to management and inputs include possibility of mechanisation and accessibility and size of land holdings. The concept of land quality is relevant only for suitability studies for a particular landuse. It can be used for estimating potentials only in relation to defined land utilisation types.

### Soil-mapping Unit

A soil-mapping unit is a portion of the landscape that has similar characteristics and qualities and whose limits are fixed by precise definitions. Within the cartographic limitations and considering the purpose for which a map is made, the soil-mapping unit is the unit about which the greatest number of precise statements and predictions can be made. Likewise, a term being used frequently now is land-mapping unit which is a mapped area of land with specified characteristics. These units are defined and mapped by natural resource surveys, e.g. soil survey, forest inventory. Their degree of homogeneity or of internal variation varies with the scale and intensity of the study.

The basic mapping units are the basis for all interpretative groupings of soils. They furnish information needed for developing capability units, crop suitability groupings, range site groupings and other interpretative groupings. The most specific management practices and estimated yields are related to the individual mapping unit, which can be called the building stones of the system.

### Land Capability

Capability is the inherent capacity of land to perform at a given level for a general landuse. Land can be classified on the basis of capability indices according to its potential and limitations for sustained production. The doyen of such a classification system is the land capability classification of the USDA, Conservation Service (Klingebiel and Montgomery, 1961). This system is one of a number of interpretative groupings of soil mainly based on: (i) the inherent soil characteristics, (ii) external land features, and (iii) environmental factors that limit the use of land.

The capability classification provides three major categories of soil groupings: (i) capability unit, (ii) capability sub-classes, (iii) capability classes.

According to this system the first category, capability unit, is a grouping of soils that have about the same responses to systems of management of common cultivated crops and pasture plants. This is the smallest category of the land capability classification system, and considers the long-time estimated yields, which should not vary more than about 25 per cent in one land capability unit. Yields are one of the criteria used in establishing capability units within a capability class and are estimated under the common management practices. The soils in a capability unit are sufficiently uniform to: (a) produce similar kinds of cultivated crops and pasture plants with similar management practices, (b) require similar conservation treatment and management under the same kind and condition of vegetative cover, (c) have comparable potential productivity. Information provided in a capability unit can be used for planning individual tracts of land, field by field.

The second category, the capability sub-class, is a grouping of capability units having similar kinds of limitations and hazards. Four general kinds of hazards and limitations are recognised and the symbol given with each is used as a suffix to the concerned land capability class. These are: (1) erosion and run-off (including risk of erosion and post-erosion damage)—*e*; (2) excess of water (wetness, high waterable, problem of drainage, overflow)—*w*; (3) root zone limitations (shallow depth, low water holding capacity, salinity or alkalinity)—*s*; and (4) climatic limitations—*c*.

Where soils have two kinds of limitations, both can be indicated, the dominant one being used first. Where two kinds of limitations are essentially equal, priority is to be given to *e*, followed by *w* and then *s*. Land capability sub-classes have been used widely in many parts of the world with slight modifications mostly incorporating local hazards.

The third and broadest category in the capability classification is the capability classes, eight of which are recognised. "The risk of soil damage or limitations in use become progressively greater from Class I to VIII. Soils in Class I to IV are capable of producing adapted plants, such as forest trees or range plants, and the common cultivated field crops and pasture plants. Soils in Classes V, VI and VII are suited to the use of adapted native plants. Soils in Class VIII do not return on-site benefits of inputs of management for crops, grasses, or trees without major reclamation" (Klingebiel and Montgomery, 1961). In other words, Class I to IV lands are regarded as suitable for cultivation, Class V to VII as suitable for pasture and forestry and Class VIII as suitable for wildlife and watershed.

## Factors Influencing Land Capability Classification

There are three factors taken into consideration for placing land into capability groupings: (1) inherent soil characteristics that have a definite role to play in the behaviour of soils and its management; (2) landscape features like slope and erosion conditions which may limit the safe and productive use of soil; and (3) climatic factors that limit the use of land.

### *Inherent Soil Characteristics*

The inherent soil characteristics considered are given below with the importance attached to each:

*Nature of Parent Material*: Gives an idea of the potential nutrient status of the soil.

*Soil Colour*: Speaks of organic matter content and the state of leaching and hydration.

*Texture and Structure*: Influences the air-water balance in the root zone and movement of water.

*Types of Clay Minerals*: Modify the exchange capacity of the soil and the condition of maintaining a supply of plant nutrients, available water and other related factors.

*Permeability*: Slow permeability of the subsurface layers such as claypan, fragipan, etc., can influence root development and cause continuous of periodic waterlogging, a high water table or flooding.

*Root Distribution*: Determines the depth of soil material to layers inhibiting root penetration.

*Soil Reaction*: Determines the state of base saturation as also the balance of available plant nutrients or their toxicity or deficiency.

*Soil Depth*: Limitations control root development and affect moisture retention and infiltration resulting in enhanced runoff and soil loss. It modifies the rooting system of plants which is ultimately reflected in crop growth and yield.

A classification of depth ranges, related depth classes, their description and suggested land capability classes are given below:

| Depth Class Symbol | Description | Land Capability Classes |
|---|---|---|
| $d_5$ | Very deep | I |
| $d_4$ | Deep | II |
| $d_3$ | Moderately deep | III |
| $d_2$ | Shallow | IV |
| $d_1$ | Very shallow | VI, VII |

### *Landscape Features*

Landscape features taken into consideration for placing areas in land capability groupings are slope and erosion condition.

*Slope*: Gradient and length of slope profoundly affect the rates of run-off and soil removal and indirectly the amount of moisture absorbed by the soil and thus influence the landuse capability. The slope percentage determines the erosion susceptibility of the soil depending on its nature and helps in placing land in suitable capability classes. The erstwhile All India Soil and Landuse Survey Organisation, IARI, has modified slope classes and ranges to suit a variety of soil conditions in India, and published in their *Soil Survey Manual* in 1960.

| Slope Classes | Slope Range | Description |
|---|---|---|
| A | 0– 1% | Nearly level |
| B | 1– 3% | Very gently sloping |
| C | 3– 5% | Gently sloping |
| D | 5–10% | Moderately sloping |
| E | 10–15% | Strongly sloping |
| F | 15–25% | Moderately steep to Steep |
| G | 25–33% | Steep |
| H | 33–50% | Very steep |
| I | over 50% | Very very steep |

The above-given classes are to be considered in the light of the nature of soil, including the kind of soil, depth, permeability, etc. The Slope Class A will normally represent a condition of least erosion in moist kinds of soils (alluvial, red soil, etc.) and other conditions being satisfactory the Capability Class I will be linked with lands falling within this slope class. In the case of other slope classes the erosion and consequent soil loss will become pronounced, and the need to put land under a group higher than Class I will depend upon the soil characteristics. Thus in the case of black soil, the land on Slope Class B and C has to be put anywhere between Class II and IV, depending upon the intensity of management practices, because black soils generally lie in gently undulating land with a long continuous slope and are more susceptible to erosion. In the case of black soils of low permeability, lands falling under even Slope Class A will need to be put under a group higher than Class I.

*Erosion Condition*: Four erosion classes are recognised, their description and associated land capability are given below:

| Erosion Class Symbol | Description | Capability Class |
|---|---|---|
| $e_1$ | No erosion or slight erosion | I and II |
| $e_2$ | Moderate erosion | III |
| $e_3$ | Severe erosion | IV |
| $e_4$ | Very severe erosion | VI and VII |

Slight erosion ($e_1$) may be associated with mild sheet erosion, and moderate erosion ($e_2$) with sheet and rill erosion. Severe erosion ($e_3$) to be associated with excessive surface erosion, exposing the subsoil and recognised by a tendency to form small gullies. Very severe erosion ($e_4$) may be associated with extensive loss of surface and subsurface soil (25-75 per cent B horizon). It generally includes a few shallow gullies and occasional deep gullies (All India Soil and Land Use Survey, *Soil Survey Manual*, 1960).

## Climatic Factors

Climatic limitations (temperature and moisture) affect capability. Extremely low temperatures and short growing seasons may limit plant growth. Limited moisture supply affects capability in sub-humid, semi-arid and arid climates. In an arid climate the moisture from rain and snow is not enough to support crops. Arid land can be classed as suited to cultivation (Classes I to IV) only if the moisture limitation is removed by irrigation. Wherever the moisture limitation is removed in this way, the soil is classified considering the effects of other permanent features and hazards, without losing sight of practical requirements of irrigation farming. Provided other soil conditions are favourable the influence of climate in land classification can be summarised in the following way (*Soil Survey Manual*, IARI, 1960).

| Characteristics of Climate | Influences on Capability Classes |
|---|---|
| (1) Humid Climate with well-distributed rainfall | May start from Class I |
| (2) Humid Climate with occasional dry spells | May start from class II |
| (3) Sub-humid; crop yield frequently reduced by droughts | May start from Class II to III |
| (4) Semi-arid | May start from Class III to IV |
| (5) Arid | May start from Class IV |

The capability of soil decreases with the decrease in effective rainfall.

**Assumptions**

Certain assumptions are to be made in assigning soils to various capability groupings. Klingebiel and Montgomery have listed 14 such assumptions in their handbook *Land Capability Classification*. Some of the most important assumptions are:

(1) A favourable ratio of output to input exists, based on long-time economic trends for average farms and farmers using moderately high-level management.
(2) Within a capability subclass or unit, soils have similar limitations or hazards.
(3) Response that can be expected from applying a given set of alternatives is approximately uniform for the soils in a capability unit.
(4) The capability classification of soils in an area may be changed when major reclamation projects are implemented that permanently change the limitation in use or reduce the hazards.
(5) Soils suited to cultivation are also suited to other uses such as pasture, range, forest and wildlife.
(6) Accessibility, size and shape of the soil area, location within field, skill or resources of individual operators and other characteristics of land ownership patterns are no criteria for capability groupings.

The land capability classification system has been applied in nearly all countries with some variations, and have been adapted to a greater or lesser extent to meet local conditions particularly in respect of limitations diagnostic of sub-classes.[2] The argument against this classification system is that despite its name, it is a classification of soils rather than land.

The soil capability system was designed initially as a tool to curb the spreading soil erosion in the United States but it has had a wider application. Lately it has become a part of all soil surveys, up to the sub-class level. It has also been applied for suggesting landuse plan for agriculture, especially in regard to limitations for particular crops and cropping system.

The USDA system is based on the influence of crops, grasses and trees adapted to American conditions mainly on the performance of maize and such aerated crops. There is no adequate provision for locally important crops such as paddy which differs in optimum requirement from wheat, maize and other row crops. In the Indian context even the good paddy land cannot be put under Class I. Paddy needs intensive management measures to retain water even to raise a normal crop and thus the land has to be put in a higher class.

Due to these problems major variations in the class system have been attempted making it more specific and are being used in some countries. "For example, in Canada the soil capability classification relates solely to agriculture. Forestry has also been excluded from consideration in Ghana, where the capability classification focuses on mechanised and hand cultivation for crop and livestock production. In Ireland, in addition to a general soil suitability classification that has only six classes but is otherwise similar in principle to the U.S. system, a separate classification is used, which groups soils into five use ranges. These ranges refer to the uses to which the soils are suited under normal management and fertiliser practice and extent from wide to extremely limited (Gardiner in FAO, 1974). This extract from the general system is of considerable interest since its meaning is unencumbered by other ideas. The use range could be especially valuable in planning situations that involve competition for the use of agricultural land" (Smyth, 1978).

Another kind of variation of the USDA system relates to adding other sets of assumptions, or "groupings of soils under two sets of management; with low technology and low capital investment and the modern level where intensive use is made of capital and the level of technology is high. Such groupings have been made for soils in Brazil and East Africa. In Guyana, South America, capability groupings contain five classes, at the most general level some of the basic assumptions also differ, for example, the need for major drainage works was not considered in classifying the wet soils. Also more sub-classes were recognised and the capability units are not numbered or named by any uniform system from one area to another but are numbered consecutively within sub-classes in each soil survey area and are given descriptive names that express some of the key soil characteristics" (FAO, 1966).

**Land Suitability**

The biggest advantage and usefulness of land capability classification lies in the context of broad planning, particularly in making broad assessments of the usefulness of land in terms of agricultural versatility. The suitability of land for specific crops or for specific kinds of agricultural enterprise, such as mechanised arable farming or market gardening, is beyond the domain of land capability classification system and is covered by land suitability studies. "Capability is the inherent capacity of land to perform at a given level for a general landuse while suitability is a statement of the adaptability of a given area for a specific landuse type" (FAO, 1976). Two areas of same capability can be quite different in suitability.

Suitability studies are as old as soil survey. Earlier suitability studies were confined to suitability of soils for particular crops. A systematic framework for land suitability classification was first given by FAO after an expert consultation on land evaluation held at the International Agricultural Centre, Wageningen, Netherlands, in October, 1972. The document, "A framework for land evaluation" was published by FAO in 1974 in the form of a *Soils Bulletin (32)* and was circulated to experts for review and additional comment. One of the outstanding features of this classification system was the inclusion of economic aspects of land as consideration for land evaluation. The relationship between resource surveys and economic and social analysis was discussed and two approaches to land evaluation were suggested:

1. Two-stage approach to land evaluation.
2. Parallel approach to land evaluation.

"In the two-stage approach the first stage is mainly concerned with qualitative land evaluation, later (although not necessarily) followed by a second stage consisting of economic and social analysis. In the parallel approach the analysis of the relationships between land and landuse proceeds concurrently with economic and social analysis" (FAO, *Soils Bulletin 32*, 1976).

Four main kinds of land suitability classifications have been recognised:

1. Qualitative Classification,

2. Quantitative Classification,
3. Current Suitability (land in its present condition),
4. Potential Suitability (after specified major improvements have been completed).

A Qualitative Classification is one in which relative suitability is expressed in qualitative terms only, without precise calculation of costs and returns. They are based mainly on the physical productive potential of the land, with economics only present as a background.

A Quantitative Classification is one in which the distinctions between classes are defined in common numerical terms, which permits objective comparison between classes relating to different kinds of landuse. Quantitative classifications are generally not used in the undeveloped and developing countries as they involve statistics of costs and prices, inputs and outputs, which are generally not available in these countries.

According to FAO *Soils Bulletin No. 32* the land suitability classification is divided into four categories : Orders, Classes, Sub-classes and Units. Land suitability orders indicate whether land is assessed as suitable or not suitable for the use under consideration. There are two *orders* represented by the symbol S and N which stand for *suitable* and *not suitable* respectively. Land may not be suitable for a particular landuse because of its ecological limitations or the investment may not be justified considering the lesser output. Land suitability *classes* refer to the degrees of suitability, three classes being recognised for the order *suitable* (S): Class $S_1$ for highly suitable, $S_2$ for moderately suitable and $S_3$ for marginally suitable land. Within the order *not suitable* (N) two classes have been suggested: Class $N_1$ for the land which is currently not suitable, but where limitations can be removed by new technology and Class $N_2$ for land which is permanently not suitable, the limitations being irremovable to make the land suitable for the given landuse. Land suitability Sub-classes reflect the limitations in the same way as land capability sub-classes and are indicated by small letters, e.g. $S_2$w, $S_2$m. Class $S_1$ land has no sub-classes as it is regarded as having no limitations.

Land suitability *units* are sub-divisions of a sub-class and reflect minor differences in required management within them, showing the differences in detail of their limitations. They can be used for planning at the farm level. Suitability units are represented by Arabic numerals following a hyphen, e.g. $S_2$w-1, $S_2$w-2. Any number of units can be recognised within a sub-class.

"Suitability evaluation does not identify a single farm-use as 'best' on each land unit. Suitability classes for different uses cannot be compared in a routine automatic manner" (FAO 1976). The physically and economically viable alternatives have to be considered for planning decisions.

The importance of Land Capability and Land Suitability studies lies in its application for the purpose of Agricultural Landuse Planning. The planning could be done broadly for large areas using classes and sub-classes or for smaller areas using units as categories.

## NOTES

1. Some human activities do take place on water, on ocean beds, but there are economic limitations to its use.
2. Diagnostic Criterion is a variable, which may be a land quality, a land characteristic or a function of several land characteristics that has an understood influence on the output from, or the required inputs to, a specified kind of landuse, and which serves as a basis of assessing the suitability of a given type of land for that use. (FAO, *Soils Bulletin No. 32* — A framework for land evaluation, 1976).

## REFERENCES

All India Soil and Landuse Survey Organisation, 1960 : *Soil Survey Manual*, IARI.

Beek, K.J. and Bennema, J., 1972 : *Land Evaluation for Agricultural Landuse Planning, An Ecological Methodology*, Agriculture University, Wageningen, The Netherlands.

Bennema, J., 1978 : *Land Evaluation in Agricultural Landuse Planning: Soil Resource Data for Agricultural Development*, Ed. L.D. Swindale, College of Tropical Agriculture, University of Hawaii, Honolulu.

FAO, 1966 : *Special Fund Final Report on the Soil Survey of British Guyana*. FAO ST/19 BRG.

FAO, 1976 : "A framework for land evaluation", *Soils Bulletin No. 32*, Rome.

FAO, 1974 : "Approaches to Land Classification", *Soils Bulletin No. 22*, Rome.

Klingebiel, A.A. and Montgomery, P.H., 1961 : "Land Capability Classification", *Agriculture Handbook No. 210*, USDA.

Riquier, J., 1974 : "A Summary of Parametric Methods of Soil and Land Evaluation, Approaches to Land Classification", FAO *Soils Bulletin No. 22*, Rome.

Smyth, A. J., 1978 : "Contribution of Soil Survey Interpretation in Land Appraisal", *Soil Resource Data for Agricultural Development*, (ed.), L.D. Swindale, University of Hawaii, Honolulu.

Steele, J. G., 1967 : "Soil Survey Interpretation and Its Use", FAO *Soils Bulletin No. 8*, Rome.

# 2

# LAND CLASSIFICATION

S. C. SHARMA

DEVELOPMENT OF land resources has very often been left in the past entirely to trial-and-error methods. This has naturally led to improper use of land in many parts of the world, with the results that an important basic resource has gradually deteriorated (Siddiqi, 1971). In the world of ours confronted with the incessantly growing pressure of population on limited land resources, we cannot afford to allow this state of affairs to persist any longer, and it is essential that optimum use of every acre of land be made in the national interest and no piece of land should remain as waste-land. This necessitates the determination of every piece of land to the optimum use. There is consequently need of some system of land classification (Stamp, 1960). The importance of land classification into types based on the quality and intensity of its use hardly needs any elaboration in landuse studies (Mohammad, 1978).

Classification of land means assigning each tract, or piece of land within a specific area (or area under reference), its proper class within a system of classes. The system relates to a particular quality or characteristics of land or distinctions of the quality or characteristics. In a simple way land classification means dividing the land into different categories or classes according to a single factor or a particular interpretation (Chauhan 1966). It defines the value or quality of land for any one type of use (Sharma 1975).

Land classification attempts at integrating lands according to their native characteristics, pre-existing use, yield capacity, and these classes are then assessed in relation to the suitability for different uses envisaged in the plan (Siddiqi, 1971). It relates climatic factors, soil characteristics, slope and degree of erosion, water supply and drainage and similar physical environmental conditions to the landuse and productivity (Arunachalam, 1959). According to Gray and Regan (1940) land classification is aimed at the determination of the best use from both the entrepreneurial and social standpoints.

It is thus clear that classification is the categorisation or grading and grouping of land according to its quality, which is based on certain factors that are taken into consideration in such a classification. The first requirement for the successful development and application of any technical system of land classification is a clear understanding of the problems for which the classification is needed (Kellog, 1940). Generally, two sets of factors are important in land classification, viz. the physical on the one hand and the social-economic on the other; the former expressing the inherent or natural properties of the soil which are of permanent nature. Therefore, a physical classification of land is thus permanent and once made can serve the basis on which social and economic classification can be superimposed. A change in the socio-economic set-up may necessitate changes in the classification based on the second set of factors. According to Arunachalam (1959) a classification of land can be developed on the basis of a productivity index calculated from the yield and quality of crops, "under physically defined systems of management". This treatment will obviously apply to arable land only on the basis of the present landuse and it will also reflect the influence of socio-economic factors such as the skill or intensity of labour and investment applied to any piece of land. Any land classification for planning landuse must be based on natural characteristics rather than upon the skill of an individual (Arunachalam, 1959).

In recent years various attempts have been made in different countries of the world to classify land from different points of view and for different purposes using a variety of methods. Stamp (1960) has aptly remarked that it is not surprising that there is a considerable divergence of points of view on the classification of land. Some important classifications may be discussed here.

### Land Classification in Great Britain

The Utilisation Survey of Britain which was established in the fall of 1930 and which worked under the guidance of Stamp, after about two years of discussion with experts in various fields, drew up a simple classification of land into 10 types, to be used for a broad national policy of landuse planning and conservation of land resources. This was later embodied in the Land Classification Map (scale 1 : 625000) published by the Ordinance Survey for the Ministry of Town and Country Planning, now the Ministry of Housing and Local Government (Stamp, 1960). The Land Utilisation Survey took into account all the relevant factors, viz., present utilisation, past utilisation, results of soil, vegetation and fertility, productivity surveys, types of farming, etc. The classification is as follows:

*Major Category I — Good Quality Land*

This is highly productive when under good management. Land in this category has the following characteristics:

Site 1. Not too elevated.
2. Level, gently sloping or undulating.
3. Favourable aspect.

Soil 1. Deep.
2. Favourable water conditions (actual or potential).
3. Texture mostly loamy, but including some peats, sands, silts and clays.

The three criteria under Site apply to all the types of good agricultural land, but a simple differentiation of soil conditions gives the following four types :

1. *First Class Land Capable of Intensive Cultivation:* This land is especially good for the cultivation of foodstuffs for human consumption [and hence designated 1(A) where A indicates arable]. The soils are deep and in texture mainly loamy, the drainage is free but not excessive, and the soils must not be excessively stony and must work easily in all seasons.
2. *Good General Purpose Farmland:* This is similar to No.1, but has a restricted range of usefulness due to: (a) less depth of soil, or (b) presence of stones, or (c) occasional liability to drought, or wetness, or (d) some limitation of seasons when the soils work easily. It is possible to distinguish (A) suitable for ploughing, and (AG) suitable for crops or permanent grass (G indicates grass).
3. *First Class with Water Conditions Especially Favouring Grass:* This is also similar to 1(A) but owing to high permanent water table or liability to winter or occasional flooding or somewhat

heavier or less tractable soils, it is less suitable to arable cultivation than to grass. Such land may be converted into 1(A) by drainage operations. The designation is 3(G).

4. *Good but Heavy Land:* Although such land has soils of a good depth and the natural fertility is often high, the soils are heavy, with the result that both the period of working and the range of possible crops are restricted. Designation is 4(AG).

*Major Category II — Medium Quality Land*

This land is only of medium productivity even when under good management, productivity being limited by reason of the unfavourable operation of one or more of the factors of site or soil character, e.g.

| | | |
|---|---|---|
| Site | 1. | High elevation, |
| | 2. | Steepness, |
| | 3. | Unfavourable aspect, |
| Soil | 1. | Shallowness of soil, |
| | 2. | Defective water conditions. |

This includes two types of land, viz.:

1. *Medium Quality Light Land:* This is land defective by reason of lightness or usually shallowness of soil. Some of it can be ploughed, but in other cases it affords excellent permanent grass, hence the designation 5(AG).
2. *Medium Quality General Purpose Farmland:* This is land defective primarily by reason of relief-land broken up by steep slopes with patches of considerable elevation with varied aspects and varied water conditions. Most of the land is usually equally suitable for crops or grass and designated as 6(AG).

*Category III — Poor Quality Land*

This is land of low productivity by reason of the extreme operation of one or more factors of site and soil and includes the following:

1. *Poor Quality Heavy Land:* This is land suffering from extreme heavyness or wetness of soils and needs extensive drainage work before it can be rendered agriculturally useful. The heavy clay

lands tend to be in grass often rush-infested, hence designated as 7(G).

2. *Poor Quality Mountain Moorland:* This is land rendered poor by extreme elevation or ruggedness of relief, usually combined with shallowness of soil, or occurrence of peat. General designation is 8(H), where H indicates heathland or moorland.
3. *Poor Quality Light Land:* This group includes so-called "hungry" lands, rendered poor by extreme lightness of soil, with attendant overdrainage resulting in drought and poverty in nutrient materials. Designation is 9(H).
4. *Poorest Land:* This is land where several factors combine to render the land agriculturally useless or almost so. In some cases reclamation is possible.

Regional classification of land in Great Britain has generally been based on the three major categories. The Land Utilisation Survey of Northern Ireland found that four rather than three major categories were desirable for conditions in Ireland and evolved a scheme which was similar to that used by the Department of Agriculture for Scotland (Symons, 1970). These schemes are as follows:

| *N.I. System* | | *Scottish System* |
|---|---|---|
| *Category A* | : High quality land | A |
| *Category B* | : Medium quality land | B |
| Class $B_1$ | : Medium quality, light | B+ |
| Class $B_2$ | : Medium quality, but rather heavy or with water table | B/B+ |
| Class $B_3$ | : Medium, but with adverse factors of slope, shallow soil, etc., and unevenness of quality | B-/B |
| Class $B_4$ | : Medium, with adverse factors more adverse than $B_3$ and/or unevenness of quality more marked | B-/C |
| *Category C* | : Poor quality land | C |
| Class $C_1$ | : Poor-marginal hill land, rocky, or steep lowland | C |
| Class $C_2$ | : Poor-peaty and ill-drained lowland | C |
| *Category D* | : Very poor quality land | D |
| Class $D_1$ | : Very poor, mountain | D |

Class $D_2$ : Very poor, lowland bog, marsh, sand-dunes, etc. D

For measuring the relative productivity of different types of land mentioned earlier, Stamp introduced the concept of "Potential Production Unit" (PPU) which may be defined as the potential production of one acre of good average farmland under good management. As a standard, a very rough equation is as follows (Stamp, 1960 a):

One acre of types 1 and 3 = 2 P.P.U.
One acre of types 2 and 4 = 1 P.P.U.
One acre of types 5 and 6 = 0.5 P.P.U.
One acre of types 7, 8 and 9 = 0.1 P.P.U.
One acre of type 10 = 0.0 P.P.U.

**Land Classification in the USA**

Dr. Vink in a paper which he presented to the International Seminar on Land Development for Agricultural Uses, August 25th–September 12, 1958, at Wageningen, succinctly described the many varied schemes of land classification. He noted that we may distinguish at least six categories of classification (Stamp, 1960b):

(1) In terms of inherent characteristics (Soils and Relief),
(2) In terms of inherent qualities (Soil Quality Classification),
(3) In terms of present use (Soil Use),
(4) In terms of use capabilities (Soil Suitability),
(5) In terms of recommended use,
(6) In terms of programme effectuation.

American workers have stressed Category 4—land capability classes—and one scheme widely used is the eight-fold scheme of the U.S. Soil Conservation Service (USAD, 1954), with a primary division into land suited and not suited to cultivation. This is concerned mainly with the erosion hazard, but which introduces other aspects of land potential and aims at a qualitative assessment of land for agricultural and watershed protection purposes.

This is obviously a "common-sense" classification, but it lacks any sure scientific background and is very much a matter of individual subjective judgement (Stamp, 1960b). It has little value where there are no erosion problems, and moreover, the assessment of productivity is done in most general terms (Shafi, 1969).

### *Unit-Area Land Classification*

A unit-area method of land classification was developed for the Tennessee Valley Authority by the U.S. National Resource Planning Board (USA, 1942). In this method, the relevant factors of both physical and human environments are tabulated and numbered, and the numbers for each unit area of land are expressed as a fractional code. The denominator contains the digits representing the physical conditions, while the numerator shows the human data. Whenever a change occurs in one or more of the items recorded, a new unit area is delimited. The complete symbol for each land unit is in three parts—a Roman numeral, a short fraction and a long fraction,

for example IV $\frac{3}{4}$ $\frac{1A212}{2221123}$

The Roman numeral shows the severity or absence of problems in the area, as revealed by the details that follow, in five classes: I. Problems insignificant; II. Problems not critical; III. Moderately critical; IV. Very critical; V. So critical that a radical change in landuse appears desirable. The short fraction is the summary of the long fraction. In the long fraction, numerator relates to items which are grouped under the heads: major landuse (1 to 10 types); agricultural emphasis (A to F); field size (1—4); amount of idle land (1—4); and quality of farmsteads and equipment (1—5); The denominator goes up to seven digits and their subsequent orders relate to slope (1—5); drainage (1—5); erosion (1—5); stoniness (1—4); rock exposure (1—5); soil depth (1—4); and soil fertility (1—5). (*See* Appendix I).

The classification is as follows:

**Numerator**

Areas classified on the basis of:

*Class 1. Excellent*—Medium to large fields; little or no idle land; complete agricultural utilisation of the land under a regime especially well suited to the land.

*Class 2. Good*—Similar to Class 1; efficient agricultural use under a regime well suited to the land.

*Class 3. Medium*—Small-to-medium fields, limited idle land; moderately efficient agricultural use under a regime moderately well suited to the land.

*Class 4. Poor*—Small-to-medium fields, in many cases interrupted; considerable idle land; low grade, inefficient, or destructive agricultural use under a regime ill-suited to the land.

*Class 5. Very Poor*—Very small, interrupted fields; excessive idle land; very low grade, inefficient, or destructive agricultural use under a regime in most cases entirely unsuited to the land.

**Denominator**

Areas classified on the basis of the present quality of the land for arable farming:

*Class 1. Excellent*—Lands exceptionally well-suited to intensive arable farming.

*Class 2. Good*—Lands well-suited to arable farming and moderately suited to intensive forms of arable farming under proper land management.

*Class 3. Medium*—Lands suited to less intensive forms of arable farming or to general farming under proper land management.

*Class 4. Poor*—Lands for the most part poorly suited to all forms of arable farming even under proper land management; in many cases best suited to grazing.

*Class 5. Very Poor*—Lands not suited to either arable farming or grazing; in most cases suited to forest production, recreation, etc.; in some cases, waste land.

The suitability of this method of classification and mapping depends on the availability of facilities for detailed field work, aerial photographs, etc. It has been used with necessary modifications of the digits in some areas in New Zealand (Cumberland, 1944). The classification is not, however, well-suited to distinguishing potential from existing productivity and the varying requirements for different uses are not recognised adequately (Symons, 1969).

**Quantitative Classification**

In the examples of land classification discussed above, there is no attempt to relate one class to another quantitatively but many such attempts have been made to devise an objective and quantitative form of land classification. The methods used may be classed generally as inductive by adding or integrating "marks" awarded to certain properties of the land or soil that influence productivity or deductive (e.g. deduced from the yield data), but many methods in actual use are a combination of both types (Symons. 1970).

**Deductive Method of Land Classification**

The U.S. Soil Survey "productivity ratings" illustrate the deductive type of classification (Jacks, 1946). An example is given below:

**Soil type: Miami Loam**

| Corn | | Wheat | | Rye | | Alfalfa | | Sugar Beet | | Productivity grade | |
|---|---|---|---|---|---|---|---|---|---|---|---|
| A | B | A | B | A | B | A | B | A | B | A | B |
| 70 | 90 | 70 | 100 | 80 | 100 | 70 | 90 | 60 | 70 | 3 | 1 |

Columns A show the yield of crops with the common practices of management in the area, and B those with the best practices, as percentages of standard yields. Standard yields represent average yields on the more extensive and widely developed soils of the regions in the United States in which the crop is a principal product (Ableiter, 1940). Suppose the standard for the corn is 50 bushels per acre and the average yield under average management is 35 bushels, and under the best management 45 bushels. The "productivity grade" is obtained by a simple percentage weighting of the crop ratings according to each crop's local importance, Soils with a weighted average 100 and 90 are graded 1, between 90 and 80, 2, and so on. In the above example "productivity rating", of the soil would be :

$$A: \frac{35 \times 100}{50} = 70, \text{ while } B: \frac{45 \times 100}{50} = 90$$

In this manner, the "productivity ratings" of the soil with respect to all the crops grown in the area under best management are obtained. The ratings are added and divided by the number of crops.

It should be noted that productivity grading is an arbitrary procedure, depending upon the validity of the system of assessment.

## Inductive Methods of Land Classification

The "Storie Index" is an inductive method which is based on the soil profile (Storie, 1933). It has been used in the USA, as well as in other countries. The main characteristics of the profile are expressed in three "factors", one of which is surface texture. Each factor is quoted as an estimated percentage of the optimum conditions of that factor for plant growth and the factors are then multiplied to give the rating as a percentage of the possible "score". One of the defects of this method is that multiplying instead of adding the figures for the characteristics enables any one very detrimental factor to bring down the index for the soil as a whole, for example, 100 × 100 × 20 = ratio 20, i.e.

$$\frac{200{,}000}{1000{,}000} = .2, \text{ whereas } 100+100+20=\text{ratio } 73, \text{ i.e.}$$

$$\frac{220}{300} = .73 \text{ by addition.}$$

It will be seen that although quantitative, such schemes have been unable to avoid a high degree of subjectivity in allocating the points, or in deciding what constitutes a normal management, optimum conditions, etc.

## Land Classification in the Soviet Union

In the USSR, there has been a growing realisation that in order to obtain increased farm production it is essential to pay greater attention to yields and costs on different classes of land. K.V. Zvorykin (1963) has suggested five stages for assessing the worth of individual areas of land in monetary terms. The first stage is the mapping of the land according to its physical attributes and quality of existing cultivation. The suggested indicators are: (1) relief and slope, (2) soil variety (with indication of bedrock and degree of erosion), (3) landuse type, (4) vegetation (cultivated crops where appropriate), (5) tillage conditions of the land (especially presence of weeds in cropped land), and depth of water table.

In the second stage, the established landuse is mapped according to the properties of the previous few years, recording features such as type of cropping, actual rotation of the last five to 10 years, application of fertilisers, irrigation, drainage and terracing.

In the third stage, data on crop yields, and on the labour and cost required to grow crops on various kinds of lands with unlike natural properties are collected and processed, which helps towards selecting the most economical landuse methods for particular lands.

The fourth stage consists of research and description of the properties of the land with a view to explaining differences in fertility and workability. Climatic, water, chemical, etc., properties of the soil are important subjects of this stage of work.

It is in the fifth stage that the actual land classification is worked out. All the land units of "modifications" found in the first stage are combined on the basis of information from the later stage of the work into higher-rank classification units. Modifications with similar observed properties and fertility are grouped into a land class. Land classes suitable for identical uses but differing in fertility and workability are grouped together as land types. Land classes that favour different combinations of crops would thus be assigned to different land types. Where parts of the farm territory contain a frequent alteration of small plots of land belonging to different types and classes, the purely typological scheme is replaced by composite groups and sub-groups called agro-production land groups.

The principal characteristic of all these land categories, whether types and classes or agro-production groups or sub-groups, is their relative suitability for a given type of landuse and for growing various crops or groups of crops. Zvorykin argues that a solution of the problems of classification and land appraisal is hindered by undue emphasis on soil types, but if reliable values can be attached to each unit of the soil classification, the soil map will be of great assistance to the land classification.

## Land Classification in Tropical Regions

Since very different conditions prevail in the tropical regions from those in the temperate regions, the problem of land classification here should be considered on a different footing. Systems used in temperate lands can give guidance in the tropics in general principles only. In view of the conditions prevailing in the tropical regions, it is not easy to arrive at any definite land classification scheme and hence a multiple land classification system may be recommended which envisages a long-term optimum and a more immediate aim for a given area. This view is well supported in the report of the landuse survey team in British Honduras.

**Land Classification in India**

In India, land classification on a scientific basis is quite recent. The importance of soil classification, which forms mainly the basis for land classification, was recognised in the recommendations of different Famine Inquiry Commissions in the years between 1866 and 1901 (Choudhury, 1968). According to S.P. Raychaudhri (1962) who has tried to trace the evolution of Indian agriculture from 2500 B.C. to the recent days, in the early days soils were divided into two broad groups, i.e. "urvara" (fertile) and "anurvara" (sterile), the former were classified on the basis of different crops grown and the latter on the basis of "ushara" (salt in land) and "maru" (desert).

In the history of soil survey of India, the revenue soil classification is probably the earliest (Rao, 1956). It dates back to the periods when the lands were assessed for revenue in the 16th century (Raychaudhuri, 1966). This classification was based upon fairly scientific principles, such as texture, productive ratings, colour and other physical properties of soils (Satyanarayana and Sankaran, 1968). "Assessment Statements of the Villages" reveal that the lands in the vicinity of settlement sites, which are locally known as "gauhan" or "goind", etc., since they receive most of the village refuse as well as a good quantity of night soil and are frequently manured, are the most fertile and fetch the highest rent. Adjacent to this zone are the lands of the Second Circle, locally known as Manjha or Manjhar (Singh, 1976), which are the second best or medium quality lands. Situated at the periphery of the villages are the inferior or poor quality lands which are locally known as "uparhar" or "palo", etc. As such, as a matter of fact, the best soil is merely always in the Gauhan, the second best in the Manjha and the worst in the Har (Liddiard, 1911-14).

It is commonly realised that the revenue classification is outdated and it does not conform to the modern methods of soil classification. But it should be mentioned, since no system is perfect, and it takes time before a scheme of scientific classification of soils becomes available, the revenue classification supplemented by local classification and modified, to the extent possible, by field observation, can be relied upon in the preliminary studies (Rao, 1956).

In the earlier part of the present century more stress was laid on improving the productivity of the land and as such fertility classification of soils was undertaken based on the content of available nitrogen, phosphorus, potash, etc., which helped a lot in classifying the land on scientific basis. The regional soil laboratories located in different parts of the country are entrusted with the task of preparing soil maps which are of immense value in the work of land classification.

In recent years, the problem of land classification has been tackled by many researchers. The pioneer work in the State of Uttar Pradesh is done by Prof. M. Shafi who, in his renowned study entitled ***Land Utilisation in Eastern Utter Pradesh*** (Shafi, 1960), has classified the village fields according to their fertility and productivity into the following categories:

*Good Quality Land "A"*

This generally includes all fields which on account of their high fertility and productivity, with the addition of manures and irrigation facilities, yield two crops a year.

*Medium Quality Land "B"*

This is of medium fertility and productivity and has further been subdivided into BI and BII lands. BI are superior to BII, although not as productive as A.

*Poor Quality Land "C"*

This is the land which generally is left unutilised either due to the presence of injurious salts at or near the surface or owing to the sandy nature of the soil or some other limitations.

It may be mentioned here that no quantitative assessment of fertility could be given, but the writer collected information on soil characteristics, on the availability of irrigation water, on the supply of manures and the crops grown in each field, and on the basis of this classified the land into these classes. A number of researchers have followed this system of land classification in their research studies, some of whom have also adopted the P.P.U. concept introduced by Prof. Stamp mentioned earlier for comparing the potential production of land of different types (Sharma, 1968; Amam, 1977; Jalal, 1976; Mohammad, N., 1978; Mohammad,1978).

Arunachalam, after deriving the land grades by taking both soil and water supply characteristics into account, has classified the land of the village Vanamadevi into five major land types, viz. "A"—Best, "B"—Good, "C"—Average, "D"—Poor and "E"—Very Poor (Arunachalam, 1959). Parush R. Sharma has attempted the land classification in the Chhattisgarh region by measuring the quality of land by the physical and chemical characteristics of soil as well as socio-economic considerations of the land. He has integrated these three groups of variables, and classified the land into four major categories, i.e. Good quality land, Medium quality land, Poor quality land and Very Poor quality land (Sharma, 1975).

In recent years, schemes for combined soil-land classification have been attempted. The land capability classification is the best illustration, very useful work on which has been done in some areas commanded under irrigation projects. (Soil Survey Reports). The procedure for soil surveys and the scheme of land classification is laid down in the *Soil Survey Manual* issued by the All India Soil and Landuse Survey Organisation, IARI, New Delhi. The all India standard capability classification pattern with some modifications is mainly based on the land capability classification as devised by the U.S. Soil Conservation Service and takes into account the inherent characteristics of soils, external land factors and the degree of limitations in the use of land. Eight such classes have been recognised with four sub-groups which further have been sub-divided into capability units.

It may be remarked here that this type of land classification is of immense importance and practical utility in landuse planning and development. It is desirable that this scheme should be adopted on a uniform basis for the whole country. Although it would need enormous labour and technical knowhow together with huge funds, it would help greatly in predicting the suitability and adaptability of soils to various farming practices, pasture, grazing, forestry, land reclamation and soil conservation practices.



Turnham, D., 1971 : *The Employment Problem in Less Development Countries,* Paris: OECD.

United Nations Economic Commission for Europe, 1977 : *Structure and Change in European Industry,* New York: United Nations.

Wan, H. Y., 1971 : *Economic Growth,* New York: Harcourt Brace Jovanovich Inc.

APPENDIX-I

## Unit-Area Land Classification

CRITERIA FOR DELIMITING HOMOGENEOUS LANDUSES*

**A. Major landuses shown by the numerator of the long fraction**

| First Digit Major Landuse | Second Digit Agricultural Emphasis | Third Digit Field Size | Fourth Digit Amount of Idle Land | Fifth Digit Quality of Farm-steads and Equipment |
|---|---|---|---|---|
| 1. General Farming | A. Corn | 1. Large | 1. Little | 1. Excellent |
| 2. Animal Industry | G. Grain (small) | 2. Medium | 2. Limited | 2. Good |
| 3. Cash-crop Farming | B. Beef cattle | 3. Small | 3. Considerable | 3. Medium |
| 4. Part-time Farming | D. Dairying | 4. Very Small | 4. Excessive | 4. Poor |
| 5. Subsistence Farming | S. Sheep | | | 5. Very Poor |
| 6. Forest Land | H. Hogs | | | |
| 7. Recreational Area | M. Mules and/or Horses | | | |
| 8. Rural-Village Area | P. Poultry | | | |
| 9. Urban Area | T. Tobacco | | | |
| 10. Manufacturing and Mining Areas | C. Cotton | | | |
| | W. Truck | | | |
| | O. Orchard | | | |
| | N. No emphasis | | | |
| | F. Forage | | | |

*APPENDIX-I (concld.)*

**B. Major physical conditions, shown by the denominator of the long fraction**

| First Digit<br>Slope | Second Digit<br>Drainage | Third Digit<br>Erosion | Fourth Digit<br>Stoniness |
|---|---|---|---|
| 1. Relatively Level | 1. Thorough | 1. Little or no observable erosion | 1. Free from stones |
| 2. Relatively Level to Undulating | 2. Adequate | 2. Little denudation by erosion | 2. Moderately stony |
| 3. Undulating to Moderately Hilly | 3. Poor | 3. Sheet erosion and ephemeral gullies | 3. Stony |
| 4. Hilly | 4. Very Poor | 4. Excessive sheet erosion and gullying | 4. Very stony |
| 5. Steep | 5. Excessive | 5. Excessive gully erosion | |

| Fifth Digit<br>Rock Exposure | Sixth Digit<br>Soil Depth | Seventh Digit<br>Soil Fertility |
|---|---|---|
| 1. Little or no rock exposure | 1. Deep (6 feet or more) | 1. Exceptionally fertile |
| 2. Limited rock exposure | 2. Moderately deep (3 to 6 feet) | 2. Fertile |
| 3. Considerable rock exposure | 3. Shallow (1 to 3 feet) | 3. Moderately fertile |
| 4. Excessive rock exposure | 4. Very shallow (less than 1 foot) | 4. Low in fertility |
| 5. Rock exposure dominant | | 5. Very low in fertility |

**Source:* Jacks, G.V. (1946), *Land Classification for Landuse Planning*, Imperial Bureau of Soil Science, Harpenden. For details *see also* U.S.A. National Resources Planning Board (1941), p. 121.

## REFERENCES

Ableiter, J. K., 1940 : "Productivity ratings of soil types", *Missouri Agri. Expt. Sta. Bull.*, p. 421.

Arunachalam, B., 1959 : "Land Classification and Possibilities of Re-Planning the Landuse in Vanamadevi Village", *Bombay Geographical Magazine*, Vols. VI-VII, September, No. 1, p. 19.

Chauhan, D. S., 1966 : *Studies in Utilisation of Agricultural Land*, p. 34.

Choudhury, R. N., 1968 : An Approach to the Scientific Classification of Soils of Arambagh Subdivision, District Hooghly, West Bengal", *Geographical Review of India*, Vol. XXX, Number 4, p. 73.

Chumberland, K. B., 1944 : "The survey and classification of land in New Zealand: A basis for planning, *Transaction of Royal Society of New Zealand*, Vol. 47, pp. 185-95.

Kellog, C. E., 1940 : "Theory of Land Classification", *Missouri Bulletin*, No. 421, pp. 164-73.

Liddiard, E. S., 1911-14 : *Final Settlement Report of Etawah District*, p. 65.

Mohammad, N., 1978 : *Agricultural Landuse in India*, p. 93.

(i) Patiala-Punjab Soil Survey Report, 1972 (Regional Pilot Project for Soil and Water Management). A Cooperative Soil Survey by the Punjab Deptt. of Soil Conservation and Water Management Division, Ministry of Agr., Govt. of India and USAID; (ii) Bellary, Distt. Mysore, Soil Survey Report, 1972 (Regional Pilot Project for Soil and Water Management). A Cooperative Soil Survey by the Water Management Division, Ministry of Agr., Govt. of India, Mysore Deptt. of Agr. and USAID; (iii) Soil Survey Reports of Bhathat Block (Distt. Gorakhpur), 1976, Padrauna Block (Distt. Deoria) 1976, Pipraich Block (Distt. Gorakhpur) 1977, Sardarnagar Block (Distt. Gorakhpur) and Block Captainganj (Distt. Deoria) 1978 — all under the Gandak Command Area Development Authority. Similar work has been taken up in the Sharda Sahayak and Ramganga Command Areas.

Rao V. L. S. P., 1956 : "A Note on Soil Classification", *Bombay Geographical Magazine*, Vol. IV, No. 1, December, p. 24.

Raychaudhuri, S. P., 1966 : *Land and Soil*, National Book Trust, New Delhi, p. 5.

Satyanarayana, P. and Sankaran, A., 1970 : "The Soil Map of Andhra State", *The Andhra Agricultural Journal*, March, p. 70.

Sharma, P. R., 1975 : Land Classification in Chhattisgarh Region: An Appraisal, *National Geographical Journal of India*, Vol. XXI, Part 2, June, p. 108.

Sharma, S. C., 1968 : *Land Utilisation in Etawah District of Uttar Pradesh*, Ph.D. Thesis — (Unpublished), p. 69.

(i) Sharma, S. C., op.cit.; (ii) Jalal, D. S.: *Land Utilisation in the District of Pithoragarh*, 1976, pp. 79-82; (iii) Amani, K. Z., *Agricultural Landuse in Aligarh District*, Aligarh, 1977; (iv) Noor Mohammad, *Agricultural Landuse in India*, 1978, pp. 93-94; (v) Ali Mohammad, *Situation of Agriculture, Food and Nutrition in Rural India*, 1978, pp. 111-209, and a host of other such studies.

Siddiqi, N. A., 1971 : "Land Classification for Agricultural Planning, A Study in Methodology", *The Geographer*, Vol. XVIII, p. 102.

Singh, H. H., 1976 : *Land Utilisation Survey: District Gonda*, p. 26.

Stamp, L. D., 1960b : *Our Developing World*, pp. 113-14.

Storie, T. E., 1933 : An Index for Rating the Agricultural Value of Soils, *California Experiment Stations Bulletin*, p. 556.

Symons Leslie, 1970 : *Agricultural Geography*, London, p. 245.

U.S.A., 1951 : Department of Agriculture, *Soil Survey Manual*, (1954), *A Manual on Conservation of Soil and Water.*

U.S.A., National Resources Planning Board (1942), *Land Classification in the United States*, p. 119.

Zvorykin, K. V., 1963 : "Scientific Principles for an Agro-production Classification of Lands", *Geografiya i Khozyaistvo, transaction in Soviet Geography, Review and Translation*, Vol. 4, pp. 3-10.

# 3

# LAND LEASE IN POLAND

WIESLAWA TYSZKIEWICZ

AN ANALYSIS of the changes of agrarian structure in Poland between 1960 and 1980 has shown that land lease played an important role in these processes. The system of land lease in Poland has developed along the lines different from those in other European countries.[1] However, it has been a prominent factor in the economic life in the countryside and in the transformations of agriculture.

A certain phenomenon can be observed in agriculture. Some farms cannot cultivate their whole land because of the lack of labour force and on the other hand, in some farms there is an idle labour since too small arable land in the farms renders the employment of labour surplus impossible. In such circumstances a better utilisation of the factors of agricultural production can be achieved by means of an extra employment or land lease.

Land lease in Poland has never been too popular and the land reform of 1945 made it less popular than ever. However, after 1956 as people were outflowing from the countryside and those who remained were ageing, many abandoned farms were taken over by individual farmers on lease. The available data show that in 1960 as many as 1395 hectares were given on lease, out of which 52.7 per cent leased from the State. In 1970, about 1230 hectares, that is, 7.5 per cent of land cultivated by individual farmers were on lease. According to A. Szemberg (1973) the land leased in 1972 already reached 1880 ha. which constituted 12 per cent of the land owned by individual farmers and the number of farms holding land on lease constituted 27 per cent of the total number of farms. Although the correctness of the above data can arise some reservations, there certainly was a tendency between 1960 and 1972 to expand land lease.

In the northern and western territories the land lease was the most common (about 19 per cent of the land owned by individual farmers, which constituted 54 per cent of the number of farms), and in central Poland it was popular too (11% and about 30% of the farms). In southern and eastern Poland it gained the least popularity (9% and 20%, respectively). In central and south-eastern Poland the average number of farms holding land on lease constituted 5 per cent of the total number of farms and the land leased by them — about 1.5 per cent of the total arable area.

In Great Poland, Lower Silesia and in the northern and western territories land lease has become more and more popular. In 1970, as many as 50 per cent of individual farms rented about 15 per cent of the arable land. On the scale of poviats (districts) the oscillations were still greater: in the Limanowa district 12 per cent of individual farms leased 0.6 per cent of arable land; in the Swidin district — 58 per cent of the farms leased 28 per cent of land and in the Szczytno district 63 per cent and 22.7 per cent, respectively.

The dynamics of this phenomenon was investigated by Szemberg (1973) but only for the years 1967-1972. She claims that in 1972 in the western and northern territories the proportion of the land leased amounted to nearly 20 per cent of Poland's arable land and in the central voivodships it was more than 11 per cent, while in the eastern voivodship it reached about 2 per cent but showed a tendency to increase.

In Poland the number of farms leased as a whole is small. They constitute 3 per cent of the total number of farms and usually they are small-sized farms up to 2 ha. However, renting some land additionally by individual land owners has been more and more prevalent all over Poland.

In 1970s, in northern and western Poland, 30-40 per cent of the total land given on lease fell to bigger farms of 10–15 ha; in eastern and central Poland 50 per cent of the leased land were rented additionally by owners of the farms of 7 ha; in northern and south-eastern Poland by owners of the farms of 3–5 per cent. An average area of land rented additionally was 2.3 ha and its size was almost identical in all parts of Poland. Generally, the bigger the farm, the more land it rents additionally. However, the share of land rented additionally is greater in smaller farms (Szemberg, 1973).

Out of the total land rented in 1970, about 49 per cent were taken on lease from the State and in 1972 35 per cent. These proportions were different depending on the part of Poland. While in 1972 in the western and northern territories, the land was predominately leased from the State (more than 62%), in central Poland it constituted 36 per cent and in southern and eastern Poland only about 11 per cent.

Short-term lease from people emigrating from the countryside or ageing was prevalent from of tenure. These short-term lease contracts were not conducive to the proper utilisation of the land.

The land leased from the state came from the so-called State Land Fund (SLF) which has been established to regulate the turnover of land between the farmers who abandoned their land or relinquished its cultivation to those who are eager and able to do it. The State Land Fund transfers this land to individual farmers, agricultural circles or socialised farms in the form of lease or sale. Between 1960-1970, the State Land Fund took over more than 2 million ha of land in which the land of individual land owners constituted above 40 per cent, the land owned by the State Treasury and found out during the various censuses — 55 per cent, and the land belonging to socialised farms — 5 per cent. The greatest amount of land was taken over in the northern and western voivodships, the Poznan and Bydgoszcz including.

The above-mentioned amount of land was transferred as follows: 30 per cent to state farm, about 9 per cent to production cooperatives, a little more then one per cent to other institutions and 15 per cent were sold to individual farmers. There still remained more than one million ha of land — the greater proportion of which was leased to individual farmers and this proportion rose from 55-60 per cent in 1959 to 65-68 per cent in 1966-1971 and to more than 80 per cent in 1972.

The questionnaire conducted in the 1960s (Olesiak, 1968) showed that among the farmers who gave land on lease the greatest group constituted those who had earlier abandoned the village. The remaining proportions were those who lived in the village, were lonely or ageing, and thus, were not able to cultivate their whole land. Among the farmers who rented land additionally the most numerous were those who own their own farms (51.2%). The next group constituted lease holders who continually worked outside agriculture — 48.8 per cent of the total. As many as 35 per cent of them did not possess any land of their own (full lease holders).

In 1986, the total amount of land owned by the State Land Fund was 775,000 ha. The structure of land utilised by the SLF has not been principally changed for many years. Lease holding is the most common form of land utilisation (55% of the land owned by SLF). Among the lease holders individual farmers are the most numerous (98% of the

SLF is leased by 400 thousands lease holders).

Private turnover of land testifies to a steady demand of farmers for land. The number of purchase-sale contracts in 1986 has grown by 19 per cent as compared to 1985 (33,996 and 28,659 respectively). The number of donations has not significantly changed — they have grown by 20 per cent only (15,525 and 15,213 respectively). An average price of agricultural land in 1986 was, according to the Central Statistical Office, 195 thousand zlotys per one ha and was by 14.5 per cent lower than in 1985 (Szemberg, 1987).

In the private turnover of land, land lease played an essential role. In many regions of the country it is a common form of enlarging the size of the farm. Similarly to the former periods land is most often leased out by elder people (in exchange for care, agricultural products) or by people abandoning the village.

The available data show that in 1983 the share of the land given on lease in the percentage of the total area cultivated by individual farmers was decreased as compared to the earlier periods and constituted 5.5 per cent (850 thousand ha).

The total land on lease in 1983 included the share of the land rented from the State Land Fund (3.2%), the share of the land rented from private persons on the contract or non-contract basis as well as the share of the land rented from socialised farms (0.5%).[2]

In Poland in 1983, the average area of leased land per one farm was 2.9 ha. The area of leased land per one farm was enlarging as the size of farms was growing.

The developmental degree of land lease in farms grouped according to the size is not the same. Like in 1972 now too farms of 5-10 ha and of 10–15 ha are the most interested in tenancy. This concerns central, northern and western Poland. In eastern Poland farms above 5 ha rent the greatest amount of land, while in southern and south-eastern parts of the country farms of 3.5 ha rent land most often. In industrial centres and around great urban-industrial agglomerations land lease attracts farms in the groups of 1-2 ha and 7-10 ha. Spatial differentiation of the total tenancy and of the land rented from the State Land Fund as well as of the land rented from private persons in terms of percentage of the individual holdings to total area in Poland in 1983 is shown in Figs. 3.1 to 3.3.

As far as the degree of development of land lease in Poland in 1983 is concerned, we can distinguish the following areas:

1. The western and northern territories — the highest land lease coefficients;
2. Some voivodships in western, southern and eastern Poland — an average land lease coefficient;
3. Central and southern Poland — the lowest land lease coefficients.

In western and northern territories the proportion of the leased land was the highest and in many communes it surpassed 20 per cent or even 30 per cent of the total area of the land cultivated by agricultural farms.

The highest share of the leased land in the total percentage of the area utilised by individual farmers was found in the communes, for instance: Przybierow — 66.7 per cent; Wegorzyno — 48.4 per cent; Drawno — 35.6 per cent; Drezdenko — 33.2 per cent; Tuplice — 32.3 per cent; Wymiarki — 33.4 per cent; and Zaborow — 32.9 per cent.

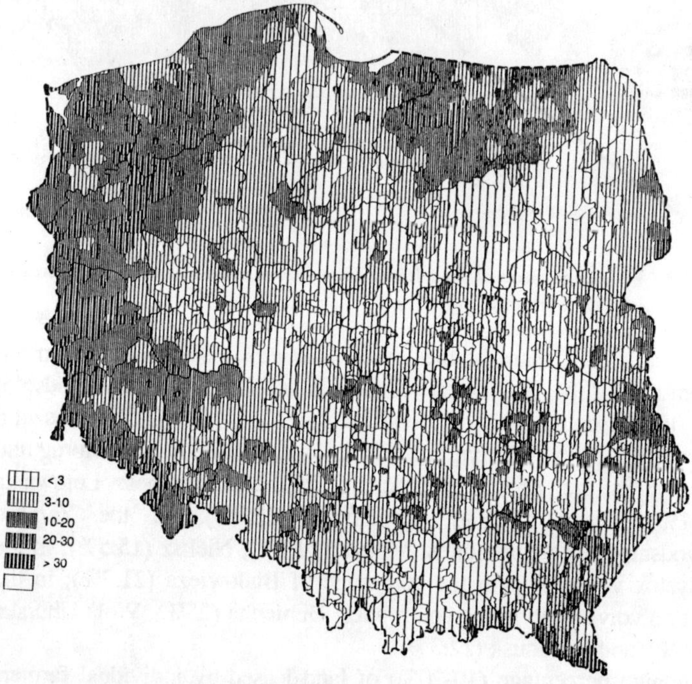


**FIG. 3.1: Percentage share of total land leased in the total area of individual farming.**

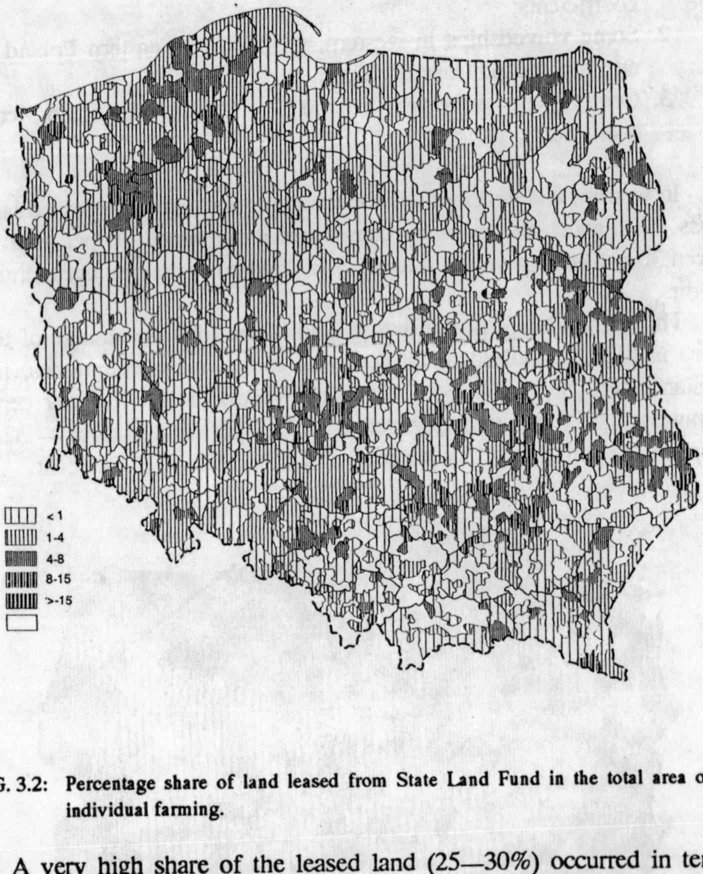


**FIG. 3.2: Percentage share of land leased from State Land Fund in the total area of individual farming.**

A very high share of the leased land (25—30%) occurred in ten communes of the Walbrzych voivodship and Jelenia Gora voivodship; also in the communes of Bobolice and Grzmiaca in the Koszalin voivodship. In the Olsztyn voivodship in the communes of Morag and Dywity; in the Krosno voivodship in the communes: Lutowiska (56.1%), Cisna (29.5%), Komancza (22.9%); in the Zamosc voivodship in the communes: Mircze (24.7%), Nielisz (15.5%); in the Bialystok voivodship in the commune of Bialowieza (21.7%); in the Chelmn voivodship in the communes: Dubienka (25%), Wola Uhurska (22.6%), and Dorohusk (22.3%).

A high percentage (10-20%) of land leased by individual farmers was found in the communes in western Poland, Western Pomerania,

Great Poland, on the Lower Vistula, in many communes of Upper Cieszyn Silesia, in some communes of Silesia Loweland and the Sudeten Foothills.

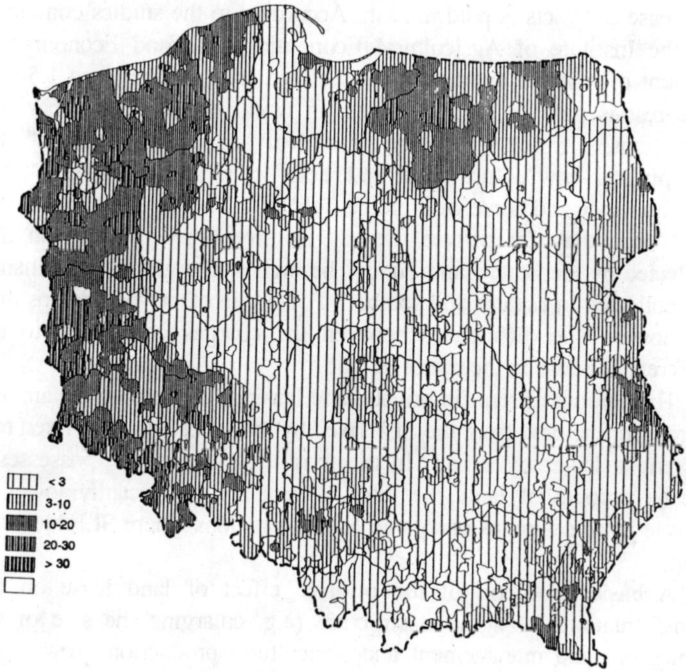


**FIG. 3.3: Percentage share of land leased from private owners in the total area of individual farming.**

While in the western and northern territories land lease from the State Land Fund was prevalent (more then 76%), its share in central Poland did not exceed 25 per cent and in the south and east of Poland it was only 12 per cent (Fig. 3.2).

In 1983 land lease was the least developed in central and southern Poland. The average percentage of the leased land was 2.5 per cent of the total. On the scale of communes the oscillations were greater, from below 1 per cent to above 5 per cent (Figs. 3.1 to 3.3). On the national scale in 68 communes land lease have never occurred at all.

It follows from the studies conducted by the Institute of Agricultural Economics and Food Economy that land lease as a means of enlarging the farm size has been recognised by some farmers as convenient.

The principal form of the rent is payment in cash. In all parts of Poland hire money is the most common. As many as 70-80 per cent of the lease contracts is paid in cash. According to the studies conducted by the Institute of Agricultural Economics and Food Economy the amount of rent per one ha in terms of rye oscillates between 1.5-4 q. The rest is paid in farm products, in work, or there are other tied conditions, e.g. for "discharge of a debt and renting a room", for "keeping a cow", for "care and help in running the farm", for "maintenance and horse work in the field", for "money and board".

Hire money for private tenancy is three times higher than that collected by the State Land Fund. The highest charges for the tenancy are collected in eastern and southern Poland, in other voivodships they are lower. The differences between the charges correspond to the differences between the prices of land.

The amount of rent (in private land leasing) depends on the amount of rent for the land from the SLF fixed by the state. The state-fixed rate of rent is lower by a half. In the territories where SLF possesses a greater amount of land, the rates of rent are significantly lower in private land leasing as compared to those regions where SLF has little land.

A basic condition of the positive effect of land lease on the transformations of agrarian structure (e.g. enlarging the size of the farm), on land management and agricultural production growth, are long-term lease contracts containing a guarantee of the endurance of the terms of contract. As it has been shown by the Institute of Agricultural Economics and Food Economy, in Poland short-term contracts (up to 5 years) constitute 27 per cent, medium-term contracts (6-19 years) — 23.7 per cent and long-term and non-term contracts the remaining per cent.

A lack of guarantee of a long-term tenancy, a constant fear of termination of a lease contract or even taking the land back as was the case with renting land from SLF (Grochowski, 1975, Olesiak, 1974), — all this is not conducive to a rational land management and prevent land holders from making outlays on land fertilisation, investments, etc. At present, the longest period of tenancy is 16 years. The law does not sanction the non-term contracts. What the legislator aims at is preventing the sale of land as a result of non-term lease contracts (Adamowski, 1988).

Summing up, we may conclude that land lease in Poland has been recognised by many farmers as a convenient form of enlarging the size of the farms. It plays an important role in agricultural production, since it offers a chance to some farmers to enlarge their farms and to others a chance of diminishing it, or of resigning from running a farm when there is a shortage of labour or for other reasons.

## NOTES

1. The level of development of the system of land lease in Europe is differentiated. Great Britain has a highly developed system of land lease since about 54 per cent of the land is cultivated by lease-holders. In Holland there were 40,000 lease holders (22.5%) per 181,000 farms and the land held by them on lease constituted 26.1 per cent of arable land. These figures do not comprise partial lease-holders (71.3 thous.) who being land owners themselves rented some land additionally. In France lease holders together with share tenants cultivated more than 29.3 per cent of the total arable land. *Cf.* W. Kwiecien, Wspolczesne rolnictwo Holandii, Holland's Contemporary Agriculture, Warszawa 1975, LSW, pp. 184, 186.
2. The data coming from the General Census of 1983 pertaining to the rented land (on the scale of communes and voivodships) given by the Department of Agriculture and Food, Central Statistical Office.

## REFERENCES

Adamowski, Z., 1988 : Przemiany strukuralne w rolnictwie polskim (Structural Transformations in Polish Agriculture), *SGGW-AR,* Warszawa, p. 192.

Grochowski Z., 1975 : Kierunki i tendencje przekształcen rolnictwa polskiego (Directions and tendencies of transformations of Polish agriculture), *IER,* Warszawa.

Olesiak K., 1968 : Obrot ziemia w gospodarstwach chlopskich (Turnover of land in peasant farms), *KIW,* Warszawa, p. 135.

———, 1974 : Obrot i gospodarka ziemia w Polsce Ludowej (Turnover and Land Management in People's Poland), *IER,* p. 79.

Szemberg A., 1987 : Gospodarka ziemia i przemiany agrarne — tendencje (Land Management and Changes in Agrarian Structure), *Nowe Rolnictwo,* 10, pp. 1-3.

Tyszkiewicz W., 1978 : Struktura agrarna Polski 1945-1975. Analiza przestrzenno-czasowa (Some Changes in the Agrarian Structure in Poland 1945-1976. A spatial and temporal analysis), *Dokumentacja Geograficzna,* 1, p. 87.

———, 1979 : Recent changes in the agrarian structure in Poland, in: *Rural Transformations in Hungary and Poland.* A Polish-Hungarian Seminar, Budapest, pp. 62-73.

———, 1983 : Long-term changes in size structure of the Polish individual farm, in: Geographical transformation of rural areas. *Proceedings of the 3rd Yugoslav-Polish Geographical Seminar,* Ljublana-Maribor 19-21. X. 1983, pp. 161-170.

———, 1986 : Agrarian structure in Poland, *Geographia Polonic 52,* pp. 221-233.

farms, [illegible] processing the sale of land as a result of [illegible] these sources (Adamowski, 1983).

Summing up, we may conclude that land lease in Poland has been [illegible] by many [illegible] of enlarging the size of the farms. It plays an important role in agricultural production, since it offers a chance to some farmers to enlarge their farms and to others a chance of [illegible] of resigning from running a farm when there is a shortage of labour or for other reasons.

NOTES

1. The level of [illegible] of the system of land lease in [illegible] is differentiated. Great Britain has a highly developed system of land [illegible] about 54 per cent of [illegible] land [illegible] 40,000 [illegible] 38 per cent of [illegible] These figures do not comprise partial [illegible] [illegible] land owners themselves [illegible] [illegible] cultivated more than [illegible] per cent of the total arable land [illegible] Warsaw [illegible] Agriculture, Warszawa 1970, [illegible] pp. 181–186.

2. The data obtained from the [illegible] Census of 1970 pertaining to the leased land [illegible] by the Department of Agriculture and [illegible] Central Statistical Office.

REFERENCES

Adamowski, Z., 1983. [illegible] Structural [illegible] Warsaw.

[illegible] (Changes and tendencies of transformations of Polish agriculture), PSE, Warszawa.

[illegible] (Turnover of land in [illegible]), Warszawa, [illegible]

[illegible] (Turnover and Land Management in Poland), Warszawa, [illegible]

[illegible] Management and changes in Australian [illegible]

[illegible], A., 1978, [illegible] Polsce 1945–1976, [illegible] Changes in the Agrarian Structure in Poland 1945–1976, [illegible]

[illegible] 1979, Agrarian changes in the agrarian structure in Poland, In: [illegible] Property [illegible] London, [illegible]

[illegible] Changes in the structure of the Polish individual farm in [illegible] Geographical [illegible], 1983, pp. [illegible]

# 4

# CONCEPT OF LANDUSE

R. P. SINGH

LANDUSE IS a function of four variables—land, water, air and man. Each plays its own role in composing its life history. Land constitutes its body, water runs through its veins life blood, air gives it oxygen and man acts as the dynamic actor to reflect its types, pattern and distribution. In fact, man—the user of land—is himself the product of atmospheric behaviour, hydrological action and lithospheric expressions. Man can survive without air for a few minutes, without water for a few hours and without food for a few days.

The earth, with its invisible girdle of atmosphere and multi-pattern water bodies, sets the stage for the tree of life to germinate, to grow and to bear a variety of vertibrates and invertibrates. Land spans over 57 to 58 million sq. miles or 37 billion acres, covering two-and-a-half of the earth's area whereas water engirdles one-and-a-half of the earth's area. Polar regions share 6 million sq. miles of the total land area. Of the remaining land area of 52 million sq. miles, forests, deserts and grasslands share two-fifths, one-third and one-fourth of the total remaining land area, respectively (Zimmer Mann, 1933). Land is, therefore, a vast theatre containing multi-coloured curtains defining various natural landscapes from the Equator—the land of heat and humidity to the poles—the land of cold and ice sheets, where man enacts his drama of activities.

## The Earth

The deciphered gaseous form from the parent body cooled and attained form making atmosphere, hydrosphere and lithosphere. The earth since then has registered several chapters of effusions, earth movements, denudation and sedimentation. This dynamic behaviour of our planet presents two architectural plans:

(i) Shield stable, compact, immobile gneiss nuclei; and
(ii) mobile welts sedimentary wrappers with woven wrinkles attaining altitudes out of geosynclines in different geological periods.

The shields have been denuded to form lands of ancient relief constituting the cell of the earth's anatomy. The mobile welts of the caledonian, hercynian and tertiary have reached respective stages of uplands, highlands, plateaux and mountains, chains and ranges. The recent mobile welt is formed by the young fold mountains running from west to east in the old world and north to south in the new world.

The land expresses itself in the form of plains, peneplains, plateaux, hills, highlands, valleys and basins, mountains and ranges. They are found in humid, arid and cold regions to be sculptured by varying processes on different climatic anvils.

Land, therefore, is clad with luxuriant vegetation in humid regions, grasslands in the interior of the continents having rain below 20 inches and scrubs, bare lands and sandy lands in arid regions and meadows in the arctic lands (Singh, 1967). Recent studies in the fields of anthropology, archaeology, geography, plant geography, meteorology, sociology and zoo-geography reflect that the cradle land of mankind ranges from the arid heartland of the Eurasian continent to the hot-wet forest land of Africa. Man has spread all over the corners of the globe from the cradle land, using and misusing the land.

**Changing Climate**

The face of the land has changed and so has changed the climate of the world in different geological periods. The prevailing patterns of climate during pre-carboniferous, carboniferous and post-carboniferous were different from the present climatic pattern. Triassic dryness made its mark on the world climate. Wandering of the continental plates from the ancient anchorage to their present site coupled with the pole's fluctuation and changing equator explains the present distribution of climate. Man has probably set up his shelter with the use of land when the world attained the present expression with well-carved climatic regions and vegetable zones.

The Pleistocene period registered the waxing and waning of the ice-sheets and almost shut these segments from the approach of the *Homo-sapiens*. The glacial epoch of Gunz, Min del, Riss and Wurm and the inter-glacial epochs mark the cold and warm periods respectively. The relics of the Neanderthal in the Alpine regions indicate an advance of glaciers engulfing their cave-shelters. The Ramapithicus, the Sivapithicus and the Siwalik man also felt the waves of glacial epochs. The climate of Central Asia changed. The humid forested heartland was transformed into semi-arid and arid land. The Pithecanthropus became Pithecanthropus erectus in this transitional phase. The *Homo-sapiens* not only had to migrate from the cradleland but had also to learn the use of land by trial-and-error.

**The Man**

Man is the finest fruit of the tree of life, for he can think, understand and execute. Where was he born? In what climatic pattern was he shaped? Was he a hunter, a shepherd or a tiller? Several chapters of the story of land and man can only respond to one or many hundreds of such questions. Land—the primary source of food and water—the essential ingredient to nourish the food plants—have to be treated to balance the study of landuse. The meaning of landuse is restricted to various uses that are given to land, but in this use we find sheds of overuse, underuse and misuse which assist us to make a diagnosis for administering the proper dose to attain optimum landuse.

The Biblical and Babylonian akin chronology of the creation of light, firmament and dryland to man, the epic concept of creation due to cosmic struggle, mythical origin of the world in half a dozen days, the garden of Eden and the story of the offerings of the very best "Of the first-lings of his flock" and "Of the fruit of the ground" by Eve's sons, Abel and Cain, respectively and subsequent killing of Abel by Cain in a feat of jealousy, do reflect that man learnt the use of land in forest, grassland and river valley. The rivalry between two brothers, Cain and Abel does symbolise the age-old struggle between the pastoralist and the agriculturist in the ancient Middle East; for Abel was "A keeper of sheep and Cain a tiller of the ground" (*Life*, Editor's Note, 1965).

Every Hindu in day-to-day sacred rites and *Poojas* uses cow-dung, grasses, Akshat (rice) which indicate his age-old pastoral tie and his later wedding to soil as a tiller. He is a farmer of many centuries. A scientific interpretation of numerous relics of ancient India not only mirrors the genesis of landuse but they show the changing landuse in man's migratory, transitory and sedentary phases (Singh, 1967).

## Relics of the Socio-economic Pattern

Man in the secluded areas of the Peninsular India has many miniature relics of the age-old socio-economic pattern. He carved them out in the changing environment. His tie with the forestland is seen in his food, dress and shelter. He loves to hunt. His children still go to collect fruits and flowers in the nearby forest tract. He himself takes to hill cultivation or slash-and-burn cultivation on the desiccated highlands and plateaux. His sedentary habits compel him to take to culture in valleys and uplands of gently rolling countryside lying at different levels. The Todas of the undulating countryside of the Nilgiri near Ootacamund have not as yet changed their age-old pastoral occupation. Taking a few examples from Chota Nagpur, it is clear that the changing use of land in the changing environment suggests the concept of landuse. The incursions of the Mundas, Oraon and Santhals into this forest land of considerable dimensions reveal how man forced by circumstances unlearns and learns the use of land and its resources. The Birhor (Bir=forest, hor=man) every day goes to the forest for hunting and food-gathering. His entire economy is forest economy. But the changing face of the land and the adoption of sedentary life are a slow change in their occupation. The Paharias in the Rajmahal Highlands love to live on the top of hills. Slash-and-burn cultivation has squeezed his land. The misuse of land is teaching them to use it properly. The Manjhis living on the summit of the Baghmundi bear the distinct scar of seclusion and practices of landuse. The hill tribes of the Garo, Khasi, Jaintia, Patkoi and Lushai Hills are the finest examples of the evolution of primitive economy and its subsequent modification in the light of their social and political make-up. The development schemes, projects and plans in the country seem to reorient the uses of land even in the secluded segments.

## Concept of Landuse

Land varies in altitudes, forms and expressions. Man has played his part on land to portray the different phases of his ties with it. The *Homo-sapiens* moved from one topography to another where climate, flora and fauna also changed. He used land, flora and fauna to fit his limited wants. Men were not many and wants were limited. The use of land likewise was limited. Men multiplied, their wants increased and became complex. The uses of land also increased. Methods and technology also changed. Man was making his own map on the face of the earth to portray his link, adaptation, creation and destruction. This cultural map forms the appendix to the natural landscape. Man's task was tremendous for he had to draw outlines from scratch. He had to experiment himself to add any expression and colour to his use of land. His primeval shelter was the branches of trees where he wandered and struggled for his existence. He ate leaves, fruits and flowers. He was gatherer, hunter, woodcutter, herder, tiller, miner, builder, engineer, planner, etc. He has been using the land from one primary use to another general use. Man has cleared the forest for shifting (Jhum) cultivation. He then used the land for large-scale farming, small-scale farming, intensive farming, mixed farming, dry farming, etc. He has used the land for one crop or another is a minor landuse problem, but to use each plot of land for the right cultivation under optimum conditions to obtain optimum yield is a significant problem. Man has learnt the use of grasslands, semi-arid and arid lands to his own advantage by applying improved methodology and utilisation of his accomplishments.

Landuse primarily relates to public problems, whereas farm management involves private problems as faced by an entrepreneur of a firm, who in this case is farmer.

The key to the most important aspect of landuse lies in the relation of population to land. The crux of the review, therefore, refers to the study of the problems in use of land by man.

Landuse particularly is linked with problems initiated in the process of deciding upon and translating into action the optimum use of limited land between the alternative major types of land utilisation.

The major types of rural landuse are:

(1) *Agricultural Land:*

(a) Dry farming areas,

(b) Irrigated lands,

(c) Grazing areas.

(2) *Village Forest Lands.*

(3) *Forest Land:*
   (a) Forests,
   (b) Cut over (land on which forests have been recently depleted),
   (c) Environmental and recreational land.

Landuse finds solutions of the problems in conversion of land from one major use to another general use. The study of landuse shows that whether a particular forest land should be cut for use as farmland. In turn it also aims at solving problems arising from blind cutting of forest trees either through aforestation or other suitable measures. There is little doubt that efficient use of land depends as much upon the capacity of man to treat the land and manage it, as upon the systems of farming, systems of tenures and units of enterprise (size of farms). The level of production or the efficiency in the use of land is, to that extent, dependent upon the nature of institutions and institutional framework within which the production function is carried on by the farmer.

Strictly speaking, the study of institutions in relation to landuse may be outside the scope of landuse but divorce between production forces and production relationship will be irrational in a scientific analysis of landuse.

The meaning and scope of landuse is explained in outline by Zimmermann in his *World Resources and Industries.* In fact "land" is used by the economist in a broader sense to denote natural resources in general. The problem, in that case, becomes one of resource appraisal. But restricting the review to agricultural land, factors determining the limits of the agricultural uses of land are being reflected.

## Physical Frontiers of Agriculture

In fact physical frontiers of agriculture may be termed as limiting factors for agriculture also. There are four major physical frontiers, i.e. temperature, moisture environs, topography and soil. Temperature conditions particularly in growing and harvesting seasons are predominating bases. A slight tilt either this way or that can turn the pages of its growth and yield. Temperature and dates of occurrence of spring and fall, frosts are a conspicuous base for agriculture growth. Moisture environs tied up with temperature environs play a significant role as physical frontiers of agriculture. Varied forms of condensation like rainfall, snowfall, hail, fog, mist, sleet, humidity, rate of evaporation, etc., portray their well-defined roles as the physical frontiers of agriculture.

Varied topographical expressions along with their degree and direction of slope and ruggedness or smoothness form the significant base for agricultural growth. Soils—the binder and feeder of agricultural plants—define their physical, chemical and bacteriological characteristics to portray their roles as physical frontiers. It is for the geographers to determine the physical frontiers particularly in the spheres of climate and pedology (soil science). Take wheat—the staff of life--for illustration. Only one acre in 10 of the land of the world is physically set free for wheat production. Adverse climatic conditions prevent production on nearly four-fifths of the land, and of that climatically available land about 11 million square miles, over one-third is tor, hilly or rough for wheat cultivation, shortening the area to 7 million square miles. Unfavourable soils still further reduce by a fifth the area climatically and topographically available, leaving only five-and-a-half million square miles physically fit for wheat. In fact, less than one per cent of the land surface of the earth is under wheat cultivation at present.

Physical frontiers outlined above portray the limits of agricultural potentiality. The extent to which agriculture is actually practised depends upon purely relative factors of human behaviour such as the arts, population pressure, standard of living, etc.

Man has achieved a lot to limit physical limits to increase the capability of land. The following table gives physical frontiers and their corrective arts:

| *Physical Frontiers* | *Corrective Arts* |
|---|---|
| 1. Temperature | 1. Plant-breeding producing quick maturing varieties such as Canadian wheat; ensilage which turns immature corn into useful feed; mechanical devices which speed up farm work and thus make agriculture possible in regions where the growing period otherwise would be too short. |
| 2. Moisture | 2. Dry farming, irrigation, drainage, terracing, tree culture; selection of |

| | |
|---|---|
| | drought-resisting varieties such as Kaffir corn and Kaoling. |
| 3. Topography | 3. Terracing; contour-ploughing; tree culture; goat and donkey pasture. |
| 4. Soils | 4. Improved methods of ploughing with improved implements; other progressive forms of cultivation; improved understanding of the bacteriological aspects of soil behaviour. |

Punjab, the ace State in the field of agriculture in India, has given the best example of how corrective arts have helped them to defeat the physical frontiers of agriculture. Punjab of India was weaker in agricultural status than the Punjab of Pakistan, but the Punjabi farmer intensely devoted to his land and willing to try new methods coupled with investment assisted him to be called the best farmer not only of India but also of the world. Punjab has achieved a foodgrain production level unparalleled in the world—from 3.4 million tonnes in 1965-66 it now produces nearly 11 million tonnes. Punjab grows wheat, paddy, potato, sugarcane, etc., where the farmers have established their control over the physical frontiers by corrective arts.

**Settlements**

Settlements, though they appear as a dot, a circle, a vertically or horizontally stretched human abode of great variety, share a fraction of land used by man for several other purposes. Man has erected his hut and hamlet to form an isolated, scattered, sparsed, compact and complex settlement. He has spread horizontally to build imposing giant cities like London, Paris, Moscow, Tokyo, etc., or to spread vertically to erect skyscrapers of New York and Chicago.

His movements mark his footprints to formulate the concept of communication. Man was the first beast of burden and he, therefore, was the originator of transport. His feet symbolise the vehicles of any type and his footprints personify lines of communication. His footpaths were transformed into cart tracks, roads—unmetalled, metalled and highly macadamised and railways. They form lines, nets and webs of any imaginable pattern. These prints constitute the internal and external structures of the settlements of any size and form.

He evolved many patterns of waterbodies, varying in size and form, to serve the growing needs. Numerous waterbodies like ditches, wells, ponds, tanks, ahars, pynes, tunnels, canals, irrigation channels and karez were constructed to suit the clime, country and countrymen.

Man not only scratched or ploughed the land, but has punctured the earth to make use of the hidden treasure within the vaults of the earth. Many mining camps and mining centres were pitched and opened in any nook and cranny of the world to tap the mineral wealth. The mining camps of nitrates and copper in Chile, copper camps in Rhodesia, Katanga, western USA, Gold camps in Witwatersrand and other mining camps of the world are also examples of landuse. Immense varied wealth of the vaults of the earth helped man to change the use of land by erecting industrial plants to convert the land into industrial centres, districts and belts.

The industrial and highly urbanised sectors of the globe not only replaced the former use of land but also created complex circumstances to re-orient the use of land to suit the needs of such complex regions with an intricate make-up and highly dynamic wants.

## Landuse and Pasture

No less significant a part of our planet is covered with grass, lands tropical, temperate and Arctic meadowland. Each has its own significance as it constitutes grazing lands for livestocks which contribute to varied resources for the subsistence of mankind.

Pasturelands, in general, are shrinking due to the following reasons:

(i) Spread of croplands to feed multiplying mouths,
(ii) Close and unsupervised grazing by animals,
(iii) Droughts,
(iv) Improper land tenure system,
(v) Nomadic grazing,
(vi) Indiscriminate burning, and
(vii) Invasion of impalatable species.

Pasture has, therefore, been preserved for feeding livestock, the source of varied food and other needs. It is essential to conserve pasture to safeguard livestock for their varied products.

## Landuse and Forest

Lands unfit for cultivation or for growing grasses should be put under forests. Forests are doubly blessed as preservers of varied resources and givers of rain. Forests control soil erosion when left to grow, but accelerate soil erosion when depleted. Forests have been cut for cropland but their complete elimination adds problems like soil erosion which initiates replantation of trees. If the present rate of deforestation is allowed to continue, it will lead to reduction in the supply of timber, firewood as also erosion of land due to misuse.

Lewis and Harison state that of the two types of errors in landuse, namely, trying to use land for a purpose to which it is not suited and using improper methods of farming, grazing and forestry, the first type of error is the greatest. Therefore, the first essential step for preventing improper use of land or for correcting misuses that are already established is to classify the land according to its best use.

## Landuse and Water Resources

The use of water resources is most significant of all landuses in India. Water in India is needed for the survival of crops, therefore, the numerous resources of water require proper planning for their use in times of need. Rain, besides being variable, uncertain and erratic, is periodical in India. This is a great alarm for an Indian farmer to be cautious in maintaining and preserving rainwater in all safety valves like ahar, pyne, tal and ponds for meeting the vagaries of the monsoons.

Floods are frequent and underground water, *patal ganga*, is perennial. They need control and planning for their use. Multi-projects seem to scatter on the face of India, but still many hyphens appear in this sphere. Every safety-valve like ahar, pyne, karha, etc., not only needs revival in India but requires countless blue dots to meet droughts like the one in 1979. According to Scoffield's article, "Soil Water Supply and Soil Solution in Irrigation Agriculture", the plans for an irrigation enterprise should include –

(i) Survey of soils and classification of land;
(ii) Investigation of the available irrigation water; and
(iii) Re-examination of soils after completion of irrigation projects to determine the trends or changes that may occur from time to

time in the root zone in consequence of inadequate sub-soil drainage on the accumulation of soluble salts.

**Reclamation**

Reclamation is the need of the hour when millions of mouths are being added every year all over the globe. Reclamation of topographically usable land, marsh land, etc., is the need of the hour. Reforestation, planting of trees as a crop, irrigation, permanent mixed farming and removal of weeds through deep ploughing can be classified as reclamation measures. Reclamation of weed-infested, abandoned beds of the rivers Kosi, Kamla and the like is being attempted in India through the operation of the Central Tractor Organisation. Reclamation of bad land topography gullies, chasms and various associated features can be tamed for further utilisation.

**Landuse and Farm Management**

The study of landuse is confined to the problems faced by society as a whole in the process of putting various types of land to the optimum use. The farm management seldom deals with the problem faced by society. Farm management restricts itself to the problems faced by the farmer as a private entrepreneur. Land utilisation primarily relates to the public problems whereas farm management involves private problems as faced by an entrepreneur of a firm who in this case is a farmer.

**Urban Landuse**

The complexities of landuse in urban sectors took a different form than in the rural areas (Singh, 1967). Urban landuse, therefore, refers to the spatial distribution of city functions, its residential communities or living areas, its industrial, commercial and retail business districts or major work areas, and its institutional and leisure-time functions (Chapin, 1957). Many workers like Burgess, Hoyt, Mckenzie, Harris, Ullman and Firy have presented their concept of urban landuse. Hoyt submits an explanation of city structures and its landuse configuration in terms of universal economic forces, which tend to govern the make-up, and change this configuration. Burgess' concentric zone concept and Hoyt's sector theory and to a certain extent Harris, Ullman extension of Mckenzie's multiple nuclei concept tend to balance towards "economic determinism of landuse, with the implication that human value systems and group action are self-regulating and continued by dominant economic forces. In contrast, Firey sees these elements of human behaviour as key variables in the make-up and change of urban landuse patterns with spatial use locations determined to an important degree by social values and conditions social action in the city" (Chapin, 1957).

Limited land and many men in the urban ambit showed an excellence of men in making the best use of land. Each such urban sector not only created new towns, but added multi-functions. Landuse configuration, therefore, has undergone metamorphosis and re-orientation. The adding dots into multiple dots and giant size necessitated systematic planning after careful diagnosis researches and studies.

The concept of landuse, therefore, revolves round man's accomplishments in conversion of land's major use into another general use. Each stage of such a change may involve many and mighty problems to pave the path for attaining equilibrium in the use of land to bring harmony in the evolving structure of man's life and economy.

The study in the field of land economics primarily pivots along the axis of landuse and institutional conditioning of land. Landuse deals with the problems emerging in the process of selecting and translating into action the optimum utilisation of limited land between the alternative major layouts of landuse.

## Landuse Configuration

To portray the pattern and problem of landuse configuration of a country, it is essential to organise a field-to-field survey of the whole country, covering every acre and recording its use and, then, of preparing a report or description of the work (Stamp, 1962).

Landuse, therefore, is a factual record of each field on a plot of land. This stock-taking of land will present use and misuse of land. The task of preparing a country's land utilisation map is not easy for it involves a regiment of men and plenty of money. A large country like India may resort to an aerial survey of every inch of land and then a careful preparation of first-hand landuse map may be the outcome. These maps will act as guide maps in eradicating misuse of land by landuse planning. At present satellite maps of the whole of India are available. They should be used as preliminary general landuse maps of the country. These satellite diagrams can be enlarged for preparing detailed landuse maps after comparing them in the field.

## Tooling—An Essential Element

Tooling is an essential element in extending the concept of landuse. The Indian topographical maps do convey same information of arable, forested, fallow and settled lands. The primary land forms, rivers and secondary waterbodies are depicted on them. Despite the magnificence of Indian Topographical Maps published by the Survey of India, they add not much information in the sphere of landuse. The cadestral maps of every village (on a scale of 16″ = 1 mile) are wonderful maps from the point of boundary of lots bunds and settlements. They can easily be accepted for sample surveys or for a survey of a selected area for preparation of landuse map for using them a grammar for preparing landuse planning map. Such landuse maps will give a factual impression landuse category. These maps may be photographed by drawing outlines of land utilisations to the same proportion as they are needed so that the necessary details may not be lost. Many complicated landuse categories may be eliminated for preparing general landuse maps with some specific details. A cartographical refinement may ultimately indicate the conventional choice and technique (Singh, 1967).

Landuse maps on the patterns of Great Britain have been prepared under the direction of Dr. S.P. Chatterjee but they hardly reach the accuracy, finish and quality of landuse maps of Great Britain prepared under late Sir L.D. Stamp. Recent landuse maps though prepared with care by the National Atlas Organisation seldom reach the quality of landuse maps of Great Britain as Great Britain has a better organisation, better personnel and better finance. Landuse maps of the respective areas under study have been prepared by Messrs Shafi, Bharadwaja, Saran, Sinha and Mrs. Indu Priya to illustrate their Ph.D thesis in London, Patna and Magadh universities. Landuse tools applied by researchers in different Indian universities exhibit some cartographic appreciation, but the Map Volume of Mrs. Priya's Ph.D. thesis (unpublished ) is of a high order.

Using air photographs R.R. Rawson and Dr. K.R. Sealy of the London School of Economics and Political Science prepared and published a landuse map of the whole of Cyprus (Stamp, 1960).

Dr. P.P. Karan (1960) made a study of Nepal and prepared a land utilisation map of Nepal using the air photographs. An aerial photographic survey has also been carried on by the National Atlas Organisation for preparing landuse maps. Satellite diagrams of whole of India should be used by the National Atlas Organisation or any other equivalent organisation to prepare a land utilisation map of India.

Landuse survey is being carried out by different countries. Pakistan commissioned an aerial survey of the vast areas of the Indus valley for increasing the production for domestic use and for export. Canada, the land marked with countless hyphens and barren grounds, has also carried out an aerial survey of the extensive forest tract and landuse maps have been prepared for further use.

The primary aim of tooling is not only to prepare cartographically sound maps, but such maps should be used for better production and field to feed the teaming millions. If Punjab has become the ace State in India in production, its map has to be engraved in every State of India to return higher production. Another example of the use of landuse map of the Himalayan Region may be selected. The aerial map of the Himalayan Region will reflect shifting cultivation, depletion of forest, construction of roads, etc. What do they mean? They mean accelerated erosion, landslides, desilting, increased load of rivers and excess flooding of rivers in the plain. They are eating away the rich Himalayan ecology and erasing some tilled lands of the plain. Such maps will portray many more hidden facts for immediate planning. Afforestation of the Himalayan slope should be made universal tree cultivation. Essential roads with controlled gradient will modify landslides and soil erosion.

**Land Conservation**

The concept of landuse includes land conservation which means using land to the optimum extent so as not only to get the maximum benefit of the resources but also to preserve it as far as possible in an unhampered efficiency for the prosperity (*Readings in Land Utilisation*, 1957). The dual aspects of land conservation are: (1) maximum social net return over time; and (2) limiting present consumption for future. Land conservation envisages the upkeep of soil fertility.

The value of land from the utilisation point of view determines the weightage of land conservation. The protection of soil and the maintenance of fertility must be achieved by erosion controls, afforestation, regulation of running water to control flood, reclamation of new lands and so on.

Land is used and misused by multiplying men and hence the unwiedly web is created to present a swarm of problems. Landuse tends us to explore every aspect of resource appraisal. Many questions crop up in its studies, which cannot be investigated within the limited span of this paper (Singh, 1967).

An experiment carried with one crop in one region cannot govern the cutlivability of the earth's surface. Food and feed crops compete and likewise the tractor competes with other animal powers.

Land is limited and geographical factors governing the capability of land are ever oscillating to hazard the promised return. Monsoon is a great friend when it is certain and is a formidable foe when it is erratic, variable and uncertain. No better example can be given than to illustrate agriculture's gamble with monsoon. On the other hand months multiply. In most of the orient countries these swarms of human bees seldom have plenty of honey to feed them from the numbered hives. These are growing habitation colonies without extending land area (Singh, 1967). The increasing pressure on land is an index to poverty, death and destruction. Land's capacity is measured and its best use can yield a limited return which will not cope with the growing human heads.

Land resources vary from one configuration to another, from one climate to another, from one kingdom of vegetation to another. Surface resources are different from what the land contains within its vaults. People therefore, change their economy according to the resource pattern. A miner must exchange his product for food. A pastoralist also exchanges his produce. Exchange economy is sculptured in this variation of economy. Trade and exchange develop to attain inter-dependence in the penultimate phase (Singh, 1967).

The divide line between developing economies is slowly eliminated when the supply and market economy attains balance. Historical traditions, customs, etc., modify this divide line. Accessibility, facility and mobility are tri-elements to control the uses of land. Methods of farming, use of machine, change in technology, chemicalisation and growing labour force play their significant roles in determining the nature and character of landuse.

Manpower, machine and other powers together with

chemicalisation and mechanisation shall have to attain an equilibrium to use the land for an optimum return to feed the teeming millions. An imbalance in food and growing families has also to be directed by family planning and population checks.

### Land Classification and Landuse Planning

The story of the concept of landuse cannot be completed if a few words are not attributed to land classification of landuse planning. With the multiplication and intensification of landuse problems, land classification has become increasingly varied in technique and application. The uses of land, therefore, need proper handling and careful planning. Each country has its own characteristics of land classification. The rural lands of Plain Bihar are classified into "Bhit", "Dit", "Tans", "Tikul", "Dhanhar", "Rabi land", "Bunds", varying water bodies as "Jhil", Chaure", "Charan", "Ahar", "Pokhar", "Tal", "Talai", etc., whereas Highland Bihar denotes two primary plains, i.e., upland (Tans), lowland (Don). Each one of them has further been classified into Don I, II and III and Tans I, II and III, according to configuration, slope, soil and productivity. It is difficult to review classification but it has been mentioned to show its effectiveness in the study of landuse. Classification should be based upon best use of land.

The Indian agriculture has recently been injected with green revolution but its true benefit has been earned by Punjab and Haryana. Other States have to accelerate their cropping with chemicalisation and mechanisation to equal Punjab, the ace agricultural State of India.

Our land needs planning on every front. The diagnostic features of various uses of land verge towards many misuses, underuses and overuses of land than towards optimum use.

Land with its varied topography, slope, field patterns, soils, temperature, precipitation, natural cover and countless creatures have to be planned to engrave an economy where man can maintain a standard of existence. He has to build his home in a land which can give him a balanced diet, dress to protect him and shelter for rest.

#### REFERENCES

*The Creation*, Editor's Note, Special Double issue. The Bible Life International, Vol. 38, No. 7, April 19, 1965, p. 20.

Chapin, F. Stuart, 1957 : *Urban Land Use Planning*, Harper and Brothers Publishers, New York, pp. 3-4.

*Reading in Land Utilisation*, 1957 : The Indian Society of Agricultural Economics, Bombay, p. 43.

Stamp, L. D., 1962 : "The Land of Britain, its use and misuse", Third Edition, Longmans Green and Ltd. London, p. 4.

———, 1960 : *Applied Geography*, p. 48.

Singh, R. P., 1967 : "Concept of Landuse", *Patna University Journal*, Vol. 23, No 1, Jan. 1967, pp. 52-62.

Zimmermann, K. W., 1933 : *World Resources and Industries*, Harper and Brothers, New York and London, pp. 82-100.

# 5

# APPROACHES TO LANDUSE SURVEY

ABDUL MAJEED

LAND RESOURCES play a strategic role in the socio-economic development of the nations. Among the land resources agricultural land resource have always played a vital role since time immemorial engaging the largest percentage of the inhabitants of the world. Therefore, in any plan which aims at utilising the resources in a scientific, rational and economic manner, the landuse surveys are of fundamental importance. Such being the importance of it, different countries of the world have conducted landuse surveys on different lines. A critical review of these landuse surveys indicates that there are mainly three approaches to it. They are:

1. Census Approach,
2. Unit Area Approach,
3. Sampling Approach.

## Census Approach

In the census approach the survey of all units, i.e. field-to-field survey in the region or country is carried out through personal investigation on the spot according to the schedule or a general questionnaire. It resembles the method by which the 10 yearly population statistics and the five-yearly agricultural, cattle or livestock statistics are collected.

This approach of landuse survey, perhaps, was firstly adopted by Dudley Stamp in Britain in the 1930s (Stamp, 1962). The objective of the survey was to undertake a national count of land resources, using the methods familiar to all geographers relying essentially on fieldwork and direct observation. Its immediate aim was to record the then existing use of every acre of land in England, Wales and Scotland.

It was not carried out with any ulterior aim : its purpose was to find out the facts, irrespective of what use might afterwards be made of its findings.

The British Ordnance Survey maps, drawn on the scale of six inches to one mile (i.e. 1 : 10,560) were used as base maps for this survey. These maps, extending three miles from west to east and two miles from north to south, cover only six square miles each and hence 22,000 such maps cover whole of England, Wales and Scotland. These basic six-inch sheets show field boundaries together with other details and, therefore, the apparent simple task of the investigators was to identify each field and record its use according to the given instructions.

The survey was carried out with full cooperation of the Directors and Secretaries of Education in each country and the most amazing part was the voluntary help of 22,000 workers recruited from different organisations.

Stamp categorised landuse into seven types and in order to maintain uniformity and comparability he evolved letter symbols and a colour scheme for various types of landuse and all the field investigators followed the same scheme for recording the landuse of the area allocated to them.

The field work began in 1931 and the substantial area of England was surveyed in the same year while the major parts of Wales and Scotland were completed by the end of 1932. Thus it may be said that, in general, the survey refers to the years 1931 – 33. However, the survey of the whole country had been completed before the outbreak of the Second World War.

When the field sheets were received they were carefully checked for agreement along the margins. The results were then reduced by a staff of specially trained cartographers to the scale of one inch to one mile in colour, and the work of publishing the one-inch maps was undertaken. The whole of England, Wales and small portions of Scotland had been covered in sheets to the scale of one inch to one mile (i.e. 1 : 63,360) while the major portion of Scotland (occupying simply by moorland or rough grazing and virtually uninhabited) was contained on the general map on the scale of 1 : 625,000, or approximately 10 miles to one inch.

In the final analysis of these landuse maps, Stamp has classified the land into three major categories which have further been classified into 10 types. Whilst the first four of the types are regarded as constituting the Good Quality Agricultural Lands, and Types 5 and 6 the Medium Quality Agricultural Lands, and Types 7 to 10 the Poor Quality Agricultural Lands, it is not true to say that there is a steady change in quality from No. 1 the best to No. 10 the poorest. For certain purposes No. 1 is the best, for others No. 3 or even No. 4.

As a national survey it was unique and of incomparable value.both for the war effort and as a historical record of a time when agriculture was at a low ebb. Since then the survey has been the source of inspiration for national surveys all over the world and for countless pieces of individual or group research (Stamp, 1960, 1967).

Realising the importance of such national survey Great Britain was again put in the forefront of the nations for fine landuse mapping in 1960 and the subsequent years. Miss Alice Coleman adopted the Census approach in the *Second Land Utilisation Survey of Britain.* Miss Coleman gave considerable thought before starting the Second Land Use Survey of Britain and she came up with a more rational scheme of classifying the use of land, recording 13 main groups with litters/symbols and colours on the six-inch field sheets.

Except, the third, fourth, fifth and thirteenth types all the other types were subdivided. However, the main groups are not only similar in most respects to the classes of the original survey, but differ only in minor details—the adoption of grey for settlement as used on the base maps, and addition of classes (3), (4), (5) and (13)—from the scheme recommended by the World Land Use Survey.

The field work completed within two years of the start of the survey. The cost of the survey, like the earlier one, was largely saved by the contribution of the surveyors themselves who had given their labour and provided their own maps. Generosity thus made the survey possible which otherwise would have been financially untenable.

After completing the field investigations the base maps were thoroughly checked and the final maps prepared. In these maps 64 categories of information were represented including a record of individual crops, of factory class by 14 types of products and noting of the distribution of a dozen kinds of natural vegetation. Looking to the wealth of detail to be shown, special measures were called for in order to maintain clarity and legibility. The printed maps are on the large scale of 1 : 25,000 (approx. 2.5″ to a mile) in contrast to 1 : 63,360 (one inch to a mile) sheets of the first survey. These details were incorporated partly by the large-scale of maps and partly by grouping the categories at two different levels of intensities, namely, the first level with 13 broad categories each represented by a distinctive colour, adhering closely to the scheme of World Land Use Survey, and the second level consisted of sub-divisions represented by variations of tone within the main colour or superimposed symbols or by other subdued cartographic devices, without hampering or obscuring the unity of the main groups and yet remaining distinct to permit the tracing of more accurate and sensitive correlations.

The value of this survey lies in its analysis of the landuse in a Britain greatly changed from that of the 1930s, a time of depression in agriculture of piecemeal and even in some places unplanned suburban growth (Coleman and Maggs, 1960; Coleman, 1961).

Mention must be made of the Land Use Survey of Poland conducted under the direction of Kostrowicki and Dziewonski on the county basis in 1953. This survey has followed almost the same principles as those of the Land Use Survey in Great Britain thus adopting the same approach, with the only difference that the former is more detailed as regards the utilisation of arable land. The crops are grouped on the basis of agronomic foundation and the crops are classified into three sections based on the physical requirements of the crops, their methods of cultivation and crop rotation and the amount of input applied. The three sections are:

(1) Intensive crops (potatoes, maize, sugarcane, vegetables and fibres); (2) Structure building crops (clover, lucerne, peas and beans); (3) Extractive crops (wheat, rye, barley and oats). This grouping constitutes the basis for defining the orientations of arable land classification map. A further distinction is made on the basis of an analysis of the share that particular groups of crops have in the total sown area and also the relative domination of particular crops in each of the groups.

In this survey six main types of landuse were evolved, which are: (1) Agricultural Land, (2) Forest, (3) Waters, (4) Settlement and Associated Non-Agricultural Land, (5) Unproductive Land, and (6) Special Areas. Most of these types were divided and subdivided. Moreover, the Polish scheme categorises its information into six main areas of interest such as the physical basis of landuse or the inherent characteristics of the land; the cultural basis of landuse or the social background to landuse, and so on, which are obviously closely interrelated.

In mapping landuse, Kostrowicki and his colleagues found it difficult to apply directly the categories laid down by Stamp (adopted by the Commission on World Land Use Survey), not only because of the huge task in mapping, but because of the tiny units and intimate mixtures that characterise the systems of farming in Poland. Hence they evolved a complicated system of cartographical representation, based mainly on statistics. The result is a series of very complex maps. These maps show the form of landuse by groups of colour, the property boundaries and the technical and organisational aspects by block symbols, and the predominant uses by suitable colour gradations within the main form.

Maps are being published on three scales : (1) a general landuse map on the scale of 1:1,000,000 showing the main types of use and based on survey maps of 1:300,000; (2) a generalised map at 1:300,000 showing a more detailed map at 1:50,000 giving a more exact representation of landuse in selected areas only and based on surveys of 1:50,000 and 1:10,000. (3) The small-scale maps present the general picture of agricultural landuse for the entire country while the more exact pattern in the local regions is portrayed on the larger-scale maps.

The Polish system is the most comprehensive and detailed survey and delves most deeply into the form of landuse and concomitant technical and economic aspects (Dziewonski, 1957; Reed, 1964; Stamp, 1964; Shafi, 1969; Romanowski, 1971).

**Unit Area Approach**

This is another major approach to landuse survey which has extensively been used in the United States of America and hence is also called American Approach. Under this approach the present as well as the potential use of the land is recorded. The landuse survey is combined with the "Land Capabilities Survey". Their purpose is to record not only the use of the selected units of land at the time of survey but also to indicate the use for which the land is best suited.

Under this approach the entire area of survey is divided into several blocks which are further divided into small units. They are known as "Unit Areas". The size of these unit areas may be constant or vary from region to region. They may also be political units such as counties, or they may be arbitrary units laid out in the form of grid. In either case the desired data are plotted in the centre of each unit.

Although this approach has been adopted in many areas either at State level or at regional level, here the author is concentrating on three major surveys—Michigan Land Economic Survey; Survey of the

Tenessee Valley Authority; and the Rural Land Classification Programme of Puerto Rico.

*Michigan Land Economic Survey*

The project came into existence because of a need for adequate information on which to base a State-sponsored programme to determine which of the lands of northern Michigan might best be devoted to agricultural, recreational or other uses. Consequently, in 1919 the Michigan Academy of Science appointed a Special Committee to investigate means by which the unfortunate economic situation in northern Michigan growing out of the great areas of non-productive land could be remedied. In 1920 this committee, after a series of discussions on landuse problems of northern Michigan, recommended that an inventory of the lands of this region and their resources was necessary. An attempt was made to represent in the group all agencies capable of contributing to the objective of a complete inventory. Thus agricultural specialists, foresters, biologists, geologists, soil scientists and economists as well as geographers were engaged.

Michigan Land Economic Survey was started in 1922. The county was taken as the unit of operation and one or two counties depending on their size were inventoried in each season. At a time an area of one-quarter township, or nine square miles was assigned to each surveyor.

Recognising three major uses of the land in northern Michigan, viz. forest production, agriculture and recreation, surveys were carried out, with a view to ultimately using the land for one of the three, as a guide to the facts to be recorded. Every fact having a possible bearing upon the use of the land for any known purpose was obtained.

Since no accurately detailed base maps were available for most of the northern Michigan, complete maps of the cultural base data-roads, rail roads, town and villages, stores, houses, and the like as well as the natural base data — lakes and streams were carefully prepared by the Land Economic Survey. Base mapping was done on the scale of 2″ to a mile.

To provide a detailed knowledge of the distribution and character of different soils, the mapping of soil types was carried on by soil surveyors. Classifications were based upon the same criteria used by the United States Bureau of Chemistry and Soils, and wherever a soil was found that had been previously mapped and named by this Bureau elsewhere the same series name was applied. Soils were mapped on a scale of two inches to the mile—the same as for base maps—and all the areas of distinct type 10 acres or more in size were shown.

Cover was mapped simultaneously with the soil, which included not only the forest growth, but every other form of native vegetation. The forest mapper also showed the location of pastures, cropped land, orchards and abandoned agricultural land. The different types, density and sizes of forest cover were mapped in considerable detail. There were 36 different cover symbols. These symbols permitted of combination to express almost any conceivable condition of density of land, average diameter, class or association of species.

A hydraulic engineer completed a water power survey of the area, and the cost of construction of plants and the market available to each site were included in the survey.

Because recreation was recognised as one of the three major uses of the land, on every lake of any size a special survey of the lake for recreational development was made and the information was assembled in blueprint form as a special lake map for each of the more important lakes.

The geological inventory included a delineation of the surface geology as well as the location and examination of deposits of limestone, shale, sand gravel, marl, peat, artesian waters, and any metallic ores that might have been found.

The land economist prepared an ownership map of the county, made a survey of market and transport facilities and conditions, and obtained the records of shipments of commodities in and out of the county, as indicative of its business.

The complete field sheets were joined with those of the adjacent areas, inked by the field mapper and turned over to the checkers for examining the errors and omissions.

The survey thus collected all the information relating to soil, slope, vegetation and similar natural features (Barnes, 1929, DeVries, 1928; James and Jones, 1954; McMurry, 1936).

### *Survey of the Tennessee Valley Authority*

This is the second major American survey where unit area approach has been adopted. The survey was a lineal descendant of the Michigan project. Like the Michigan survey, it was concerned essentially with an appraisal of resources.

In 1936, under the direction of G.D. Hudson, the Tennessee Valley Authority developed the "unit area" method of land classification, which recognised distinct areas of limited extent each differing from its neighbours due to a combination of particular physical features. This unit area method represented an attempt to achieve the greatest possible accuracy in field investigation within certain practical consideration of time and cost. Its unique qualities were the application of fractional code notations to land units of not less than 200 acres, the use of aerial photographic mosaics as base maps, and a consideration of those items best calculated to serve land planning.

In this survey the factors relating to physical and human environment were tabulated and numbered and the numbers for each unit area of land were expressed in a fractional code. The digits in the denominator represented physical conditions, while digits in the numerator stood for human environment. The complete symbol for each land unit is in three parts: (i) a Roman numeral, (ii) a short fraction, and (iii) a long fraction.

## Roman Numerals

Roman numerals from I to V have been used to show the severity or absence of problems in the area. These numerals show the conditions noted against each: I—problems insignificant; II—problems not critical; III—moderately critical; IV—very critical; V—so critical that a radical change in landuse is desirable.

In part, these Roman numerals represent a summary of their accompanying short and long fraction.

## The Short Fraction

Actually the short fraction is the summary of the conditions the long fraction reveals. There is only one Arabic numeral digit ranging from 1 to 5 in the numerator and denominator each. The numerator shows the effectiveness of the present agricultural use of the land which has been classified into five categories as noted against each numeral: 1.—Excellent; 2.—Good; 3.—Medium; 4.—Poor; 5.—Very Poor. It means if the Arabic numeral 3 is given in the numerator, the use of the land in the area is medium.

Similarly the denominator which also ranges from 1 to 5 shows the physical conditions of the land for arable farming as noted against each numeral: 1.—Excellent; 2.—Good; 3.—Medium; 4.—Poor; 5.—Very Poor. It implies that if 4 is given in the denominator of the short fraction, the physical condition of the land is poor.

**The Long Fraction**

The long fraction refers to features relating to the types of landuse and natural physical conditions of the area.

The numerator refers to the method of landuse and its general characteristics. It goes up to five digits and their subsequent orders related to items which are grouped under the heads given below:

*First Digit*

It represents the major landuse which has been classified into 10 types, and has been shown by Arabic numerals which are as follows: 1.—General Farming; 2.—Animal Industry; 3.—Cash Crop Farming; 4.—Part-time Farming; 5.—Subsistence Farming; 6.—Forest Land; 7.—Recreational Area; 8.—Rural Village Area; 9.—Urban Area; 10.—Manufacturing and Mining Areas. (The tenth type is denoted by 0 instead of 10). It means if the Arabic numeral 5 is given in the first digit of the numerator the major landuse is subsistence farming.

*Second Digit*

It represents the agricultural emphasis (denoted by letters) which are as follows: A.—Corn; G.—Grain (small); B.—Beef Cattle; D.—Dairying; S.—Sheep; H.—Hogs; M.—Mules and/or Horses; P.—Poultry; T.—Tobacco; C.—Cotton; W.—Truck; O.—Orchard; N.—No Emphasis; F.—Forage. It implies that if letter G is given in the second digit of the numerator, the agricultural land gives the major emphasis to grain crops.

*Third Digit*

It represents the field size ranging from 1 to 4 (denoted by Arabic numerals) which are as follows: 1.—Large; 2.—Medium; 3.—Small; 4.—Very Small.

*Fourth Digit*

It represents the amount of idle land classified into four categories and denoted by Arabic numerals. These categories are: 1.—Little Idle Land; 2.—Limited Idle Land; 3.—Considerable Idle; 4.—Excessive Idle.

*Fifth Digit*

This digit represents the quality of farm steads and equipments which have been classified into 5 categories noted against each Arabic numeral: 1.—Excellent; 2.—Good; 3.—Medium; 4.—Poor; 5.—Very Poor.

The denominator of the long fraction goes up to seven digits (all in Arabic numerals) and their subsequent orders relate to items which are grouped under the heads given below:

*First Digit*

It represents the slope ranging from 1 to 5. These five categories are noted against each numeral: 1.—Relatively Level; 2.—Relatively Level to Undulating; 3.—Undulating to Moderately Hilly; 4.—Hilly; 5.—Steep.

*Second Digit*

It represents the drainage of the area which has been classified into five categories: 1.—Thorough; 2.—Adequate; 3.—Poor; 4.—Very Poor; 5.—Excessive.

*Third Digit*

It represents the rate of erosion. They have been classified into five categories and are denoted by Arabic numerals ranging from 1 to 5. These five categories are noted against each numeral: 1.—Little or No Observable Erosion; 2.—Little Denudation by Erosion; 3.—Sheet Erosion and Ephemeral Gullies; 4.—Excessive Sheet Erosion and Gullying; 5.—Excessive Gully Erosion.

*Fourth Digit*

It represents the amount of stoniness in the unit area and has been categorised into four types which are denoted by Arabic numerals ranging from 1 to 4. These types are: 1.—Free from Stones; 2.—Moderately Stony; 3.—Stony; 4.—Very Stony.

*Fifth Digit*

It represents the extent of rock exposure which has been classified into five types. These five types have been shown by Arabic numerals ranging from 1 to 5 which have been noted against each numeral: 1.—Little or No Rock Exposure; 2.—Limited Rock Exposure; 3.—Considerable Rock Exposure; 4.—Excessive Rock Exposure;

5.—Rock Exposure Barren.

*Sixth Digit*

This digit in the denominator of the Long Fraction represents the depth of the soil found in the unit area which has been categorised into four types and denoted by Arabic numerals ranging from 1 to 4. These types are: 1.—Deep (6 feet or more); 2.—Moderately Deep (3 to 6 feet); 3.—Shallow ( 1 to 3 feet); 4.—Very Shallow (less than 1 foot). It implies that if numeral 2 is given in the sixth digit of the denominator, the soils are moderatory deep (3 to 6 feet).

*Seventh Digit*

It represents the soil fertility which has been classified into five types. These five types have been denoted by Arabic numerals ranging from 1 to 5 and have been noted against each numeral: 1.—Exceptionally Fertile; 2.—Fertile; 3.—Moderately Fertile; 4.—Low in Fertility; 5.—Very Low in Fertility. It implies that if numeral 3 is given in the seventh digit of the denominator, the area is moderately fertile.

Thus the "fractional code" system of mapping is a shorthand method of classifying aerial association. The following is an example of the complete symbol expressed in a fractional notation recorded on a unit area land:

$$\text{II} \ \frac{3}{3} \ \frac{1\text{G}123}{2221123}$$

In this symbol the Roman numeral at a glance reveals that in the area the problems are not critical. The numerator of the short fraction denotes that the area is of medium class, i.e. in the area there are medium to larger fields, little or no idle land and efficient agricultural use under a regime well suited to the land. The denominator of the short fraction reveals that the area has the medium quality of land, i.e. there are lands suited to less intensive forms of arable farming or to general farming under proper land management.

The numerator of the long fraction denotes an area of general farming with major emphasis on grain crops. The farm fields are of large size, there is limited idle land and the quality of the farms is medium. The denominator shows that the land is relatively level to undulating with adequate drainage and little denudation by soil erosion; it is free from stones and there is little or no rock exposure; soils are moderately deep (three to six feet) and moderately fertile.

Adopting the above discussed method detailed field studies were made over an aggregate of approximately 550 square miles in three widely separated and distinctly different areas in the Tennessee River Basin. The aerial mosaics were used as base maps. They were on the scale of 1:12,000 and were printed on heavy duty paper with a matt finish.

After the completion of the survey the tentative boundaries of homogeneous units were determined by examining large-scale base maps and aerial photographs. The size of the unit varied with the type of the region and the scale of the base map.

Finished mosaics were reproduced photographically to a scale of one inch to approximately two miles (1:125,000). In some cases the reproductions were used as they were, i.e. 1:12,000. In other cases they were coloured with transparent dyes to bring out the distributions of individual items or combinations of items. The following colour scheme was adopted in the TVA to emphasise the five major classes of agricultural land:

*Agricultural Land*

| | | | |
|---|---|---|---|
| Class I | – | General Farming | Blue |
| Class II | – | Animal Industry | Green |
| Class III | – | Cash Crop Farming | Brown |
| Class IV | – | Part-time Farming | Yellow |
| Class V | – | Subsistence Farming | Red |

*Non-agricultural Land*

Forest Land; Recreational Area; Urban area; and Manufacturing and Mining Areas — Uncoloured

Thus the unit area method of land classification as evolved by TVA contributes both to the field of geography and to the field of planning. In geography it affords another tool of investigation that partially fills, at least, the gap between methods of detailed field analysis and methods of reconnaisance survey. In planning, its use yields results that are vital to the development of an effective land planning (Hudson, 1936a, 1936b; McMurry, 1936).

**Rural Land Classification Programme of Puerto Rico**

This survey also has adopted the unit area approach to landuse. In this survey field observers indicate both landuse characteristics and major resources. The survey also classified lands according to their agricultural usefulness.

Recognising the need for current information on the landuse and their distribution throughout the island of Puerto Rico, the Department of Geography of North-western University (Illinois, USA) was invited in 1949 to consider the feasibility of making a landuse map of Puerto Rico on the scale of 1:10,000. The Department of Geography came to the conclusion that the best technique of field mapping to be used was the unit-area method, with notations in the form of a fractional code. This technique of geographic mapping had been widely used in the United States, but had not been used in tropical areas. The methods of delimiting unit areas and of recording data in fractional notation to be used in the Rural Land Classification Programme of Puerto Rico were the same as those used by Hudson in the survey of the Tennessee Valley Authority in 1936. However, the scale of mapping and classification of features to be included in the unit areas were adapted to the conditions found in Puerto Rico.

The survey was started in September, 1949, and completed in August, 1951, that is, it took only two years to cover the whole of the island of Puerto Rico (which is 3,421 square miles in area). Unquestionably this programme provided the most complete inventory of landuse and probably the most detailed survey of its kind in a tropical area of this size. The unit area method of mapping and the use of aerial photographs for plotting data in the field contributed greatly to the speed and accuracy of the operation.

The island was divided into 18 mapping units. These mapping units varied in size from a minimum of 125 square miles to 289 square miles. The field work in each mapping unit was done by a field team of three men.

Aerial photographs were used for plotting data in the field and the landuse and the physical characteristics of the land were recorded on the aerial mosaics in a fractional notation. In the numerator of the fraction the landuse was recorded while the physical characteristics of the land were recorded in the denominator of the fraction.

In the Rural Land Classification Programme of Puerto Rico eight categories of landuse were recognised and mapped. They were as follows: (1) Cropped land, (2) Pasture and harvested forage land, (3) Forest and brush land, (4) Non-productive land, (5) Rural public and community service land, (6) Quarrying and mining land, (7) Urban and manufacturing land, (8) Land of miscellaneous uses such as canals, water tanks, roads, houses, etc. Each of the first seven categories was divided into classes on the basis of type of use. These categories were indicated on the field photographs by letter-number symbols in the numerator of the fraction.

The physical characteristics of the land included: (1) Soil types, (2) Degree of slope, (3) Conditions of drainage, (4) Rate of erosion and (5) Amount of stoniness and rock exposure. With the exception of the soil-mapping units each physical characteristic of the land was divided into classes numbered in the key in the order of its value for agricultural use. The classes of soils, called the soil-mapping units were an adaptation of the 53 soil-mapping units developed by the United States Soil Conservation Service of Puerto Rico.

The six classes of slope were: (1) Level to undulating, (2) Undulating to rolling, (3) Rolling to hilly, (4) Hilly to steep, (5) Steep to very steep, and (6) Very steep.

In the field mapping the apparent drainage condition was mapped, instead of the inherent natural drainage condition. The five classes of drainage were: (1) Well drained, (2) Adequately drained, (3) Poorly drained, (4) Very poorly drained, and (5) Excessively drained.

The following seven classes of deposition and erosion were (+) alluvium and collovial deposits: (1) No observable erosion, (2) Little sheet erosion, (3) Moderate sheet erosion, (4) Relatively severe sheet erosion, (5) Severe sheet erosion, and (6) Very severe sheet erosion.

The eight classes of stoniness and rock exposure recognised and mapped were: (1) No stones or rock outcrops, (2) Moderately stony, (3) Stony, (4) Very Stony, (5) Limited rock exposure, (6) Considerable rock exposure, (7) Excessive rock exposure, and (8) Rock exposure barren.

As referred earlier the landuse and physical characteristics of land are indicated by letter-number symbols in the form of a fractional notation. The appropriate numbers used to designate the items in the unit area are selected from the keys of landuse and land characteristics. The entire notation, for example within a unit area, appears as:

$$\frac{252}{52-1312}$$

The numerator of the abovementioned fractional notation (referring to the categories and classes of landuse) appears as 252, signifying pasture land (2), tall marsh grasses used as pasture (5), average quality

(2) or major category of landuse (first digit), major class of landuse (second digit) and quality of landuse (third digit) respectively. The denominator of the fractional notation (referring to the categories and classes of land characteristics) appears as 52–1312, signifying organic soil of the sea coast (52), level slope (1), poor drainage (3), no observable erosion (1), moderately stony (2), or soil-mapping unit (first digit), class of slope (second digit), class of drainage (third digit), class of erosion (fourth digit), and class of stoniness and rock exposure (fifth digit) in that order.

After completing the survey a preliminary set of the landuse maps at the scale of 1:10,000 (at the same scale as that of the aerial photographs) was prepared with the data on landuse which was recorded on the aerial photographs in the numerator of the fractional notation. For the purpose of preparing this set certain classes of cropped land, pasture land and forest land were combined and the quality of the crop or pasture was excluded so that the landuse pattern might be simplified. Thus instead of 59 classes only 22 classes were formed and they were shown on the landuse maps by letter symbols. These letter symbols were used as an economy measure and in order to have the maps transparent for overlay purposes.

Two generalised landuse maps, one at the scale of 1:200,000, were completed for the purpose of giving at a glance the broad patterns of landuse for the island as a whole. An annual rainfall map with graphs of mean monthly rainfall for more than 400 stations was constructed on an overlay to the 1:50,000 landuse map.

Thus in the "Rural Land Classification Programme of Puerto Rico" the unit area approach to the landuse survey was adopted and with this approach the Puerto Rican Survey made a successful attempt to indicate both landuse and land characteristics according to its agricultural usefulness, (Northwestern University, 1952).

**Sampling Approach**

In sampling approach a limited selection of items or cases is made out of a total population (here population refers to the total number of items). The limited selection (sample) is generally sufficient for making a generalisation about the whole population. Moreover in several cases the number of individuals in the population, as, for example, the average yield of all the plots of a geographic region, is so numerous that measuring of all of them becomes rather impossible or, so to say, impracticable, laborious, cumbersome, time-consuming and costly, and for this very reason a systematic and accurate sampling approach is adopted to the landuse surveys. Although such an approach does not lead us to a total landuse picture relating to actual distribution in an area, it is undoubtedly helpful in certain respects. It reveals clearly the landuse problems which are similar in areas having homogeneity in regard to physical and social conditions. In fact this is in many cases preferable to a full survey of a particular problem for reasons of available time, manpower and finance.

Although there are many sampling methods only two of them have been adopted in landuse studies, viz. Random Sampling and Systematic Purposive Sampling and hence we call it as: (i) Random Sampling Approach, and (ii) Systematic Purposive Sampling Approach.

*Random Sampling Approach*

The statistical method of random sampling, which consists in the sample being selected at random, in a mechanical manner, by different processes of selection, so that every unit has an equal chance of being selected. In a random sampling the villages are chosen from a variety of villages, or the agricultural families are taken from all types of cultivators, according to the principle of proportionate representation. This random method of sample selection is considered as the nearest approach to exactness. Randomisation implies the elimination of the unreliability of the human factor—the elimination of the investigator's bias, prejudice or personal consideration. The method is to show the percentage with reference to the total number. Obviously the bigger the size of the sample, the smaller will the range of probable error tend to be. It is said that random selection results in the emergence of the representative sample, if the sample is fairly large.

However, J.L. Buck adopted this approach for the Land Use Survey of China in 1929. The selection of this technique is of great importance for such a big country as it is not possible to conduct a survey of all the villages due to lack of technical skill and resources.

The basic purpose of this monumental survey (which was carried out during 1929 to 1933) was to make available knowledge about China's agriculture for its improvement and as a basis of national agricultural polices. In this survey the entire field was not covered but *hsien* or the district was regarded as the unit and an intensive study was made of certain selected localities or samples with a number of farms. Data were collected by the sampling method from 16,786 farms scattered in 168 localities which were part of 154 *hsien* lying in 22 provinces and a generalised picture of landuse was obtained. The sampling was as representative of major types of farming in China as possible. The collection of data relating to population, food, standard of living and marketing were also included in the survey. The agricultural problems were predominantly emphasised and a number of geographical problems were studied simultaneously. These problems were connected with farm study, agricultural population, food, locality and *hsien* survey. Along with the assimilation of agricultural information, regional investigation regarding floods, droughts, climatic conditions, forested areas, classification of land, cattle, manure and labour conditions was also carried out.

The areas studied were determined by types of farming areas, which were differentiated by crops using 20 per cent or more of the farmers' labour and extending over several *hsien*. Preliminary approximate boundaries to such areas were determined by obtaining estimates from persons familiar with conditions in the *hsien* and in adjacent *hsien*. After having determined the representativeness of an area, the particular villages accepted for study were selected according to the availability of a suitable local investigator and the personal contact which could be made by the regional investigator.

For each type of farming area, a representative village (or group of small villages or hamlets) was selected and 100 farms were studied in detail by use of "Farm Schedule".

Buck's work may be described as monumental as there is an extensive array of facts on all aspects of landuse (Buck, 1937; Cressey, 1939).

*Systematic Purposive Sampling Approach*

This is the other sampling approach to landuse surveys in which, first, the area for survey is thoroughly studied in the geographical perspective and then it is divided as far as possible into homogeneous strata. From each stratum samples of villages, which include clusters of fields, are selected with a definite purpose in view and finally a complete enumeration of these clusters is taken.

In India M. Shafi conducted *Land Use Survey in Eastern Uttar Pradesh* during 1952-53. He adopted the Systematic Purposive Sampling Approach to this landuse survey, which was a pioneer work of its kind undertaken in India. After a detailed study of the structure and relief, drainage, climate and soils of the entire area (comprising five districts of the Ganga-Ghaghra interflue) Shafi divided the whole area into four homogeneous strata and selected the representative villages from each stratum. In all, 12 villages were chosen. The setting of the selected villages was such that four villages lay on the well-drained land, five on ill-drained land, two on the Ganga and the Ghaghara *Khadar*, and one in the black soil region. The selection of these representative villages was based on the Systematic Purposive Sampling Method.

After the selection of the villages, Shafi carried out personally a complete field survey of each village tract in summer and winter (i.e. in the *Kharif* and *Rabi* seasons) and recorded precisely the existing landuse on the base maps. Base mapping was done on a scale of 16″ to a mile, i.e. 1:3,960 on which generally the cadastral maps exist in India.

Shafi recognised various types of landuse and in order to show them on the maps he evolved symbols. These types were recorded by his Land Use Survey and were shown by symbols on the general landuse map on the scale of 1:3,960 for every selected village.

Under cultivated land the area occupied by each crop in summer as well as in winter was recorded and shown on two different maps (one showing landuse in the *Kharif* season and the other in the *Rabi* season). The crops were shown by letter symbols such as S for Sugarcane, M for Maize, W for Wheat and so on, on the maps of Kharif and Rabi seasons.

Along with these two maps the land cropped twice in the year of survey was also shown on the third map on the same scale (i.e. 1:3,960) using the same symbol scheme.

In addition collection of data relating to land classification and irrigation were also included in the survey and they were shown on the maps on the scale of 16″ = 1 mile. As regards land classification Shafi collected information on soil characteristics, on the availability of irrigation water and on the supply of manures to and the crops grown in each field of the villages and on the basis of this inquiry he classified the village lands into four categories which are:

| | |
|---|---|
| A | Good Quality Land |
| B I, B II | Medium Quality Land |
| C | Poor Quality Land |

Shafi's Land Use Survey is based on painstaking intensive field study and puts forth detailed accounts of land utilisation in a dozen villages chosen from four representative but contrasted regions through the Systematic Purposive Sampling Approach.

Landuse survey on similar lines have been carried out since 1956 pertaining to various parts of India, (Shafi, 1960, 1966; Mohammad, 1978).

## REFERENCES


Barnes, C. P., 1929 : "Land Resource Inventory in Michigan" *Economic Geography*, Vol. 5, No. 1, pp. 22-35.

Buck, J. L., 1937 : *Land Utilisation in China*, University of Nanking, China, pp. viii-xi.

Coleman, A., 1961 : "The Second Land Use Survey : Progress and Prospects", *Geographical Journal*, Vol. CXXVII(2), pp. 168-86.

Coleman, A. and Maggs, KRA, 1960 : *Land Use Survey Handbook*, (London), pp. 1-4.

Cressey, G. B., 1939 : "Foundations of Chinese Life", *Economic Geography*, Vol. XV(1), p. 25.

DeVries, W., 1928 : "The Michigan Land Economic Survey", *Journal of Farm Economics*, Vol. 10, pp. 516-24.

Dziewonski, K., 1957 : "Detailed Survey of Land Utilisation in Poland", *Proceedings of International Geography Seminar*, Aligarh, pp. 562-66.

Hudson, G. D., 1936a :"The Unit Area Method of Land Classification", *Annals of the Association of American Geographers*, Vol. 26, (2), pp. 99-112.

———, 1936b : "Methods Employed by Geographers in Regional Surveys", *Economic Geography*, Vol. 12 (1), pp. 98-104.

James, P. E. and Jones, C. F., 1954 : *American Geography Inventory and Prospect*, Syracuse University Press (New York), pp. 258-77.

McMurry, K. C., 1936 : "Geographic Contributions to Land Use Planning", *Annals of the Association of American Geographers*, Vol. 26, No. 2, pp. 91-98.

Mohammad, N., 1978 : *Agricultural Land Use in India*, New Delhi.

Northwestern University, Department of Geography, 1952 : *The Rural Land Classification Programme of Puerto Rico*, Northwestern University Studies in Geography, No. 1 (Evanston, Illinois), pp. 1-20.

*Proceedings of the International Geography Seminar* (1957) : Department of Geography, Aligarh Muslim University, Aligarh, (Dziewonski, K., "Detailed Survey of Land Utilisation in Poland)", pp. 562-66.

Reeds, L. G., 1964 : "Agricultural Geography : Progress and Prospects", *Canadian Geographer*, Vol. VIII, No. 2, p. 54.

Romanowski, J. I., 1971 : "Geographical Research and Methodology on East, Central and Southeast European Agriculture", in *Eastern Europe : Essays in Geographical Problems* (ed. Hoffman, G.W.), Methuen & Co. Ltd., London, pp. 137-69.

Shafi, M., 1969 : "Landuse Planning, Land Classification and Land Capability—Methods and Techniques", *Geographer*, Vol, XVI, p. 2.

———, 1960 : *Land Utilisation in Eastern Uttar Pradesh*, Department of Geography, Muslim University, Aligarh, pp. 48-226.

———, 1966 : "Techniques of Rural Landuse Planning with special reference to India", *Geographer*, Vol. XIII, pp. 15-24.

Stamp, L. D., 1964 : Landuse in Eastern Europe, *Geographical Review*, Vol. 54, No. 1, pp. 116-17.

———, : *The Land of Britain—Its Use and Misuse* (3rd edn.), Longmans, London, pp. 20-41 and p. 353.

———, 1960 : *Our Developing World*, London, pp. 152-53.

———, 1967 : *Applied Geography*, England, pp. 38-49, 107-111.

# 6

# LAND CLASSIFICATION FOR AGRICULTURAL PLANNING

N. A. SIDDIQI

DEVELOPMENT OF land resources has very often been left in the past entirely to trial-and-error methods. This naturally has led to improper use of land in many parts of the world, with the result that an important basic resource has gradually deteriorated. In countries where efforts are being made to modernise agriculture through the adoption of scientific methods, there is an excellent opportunity for correcting past errors in landuse and to get rid of further errors through detailed mapping and classification of land. In this paper an attempt has been made to study the various methods of land classification which have been proposed in different countries with a view to deriving maximum benefits from every type of land whether agricultural or non-agricultural.

Land classification envisaged here aims at grouping of lands according to their suitability for producing plants of economic importance. Such a classification must be the starting point of agricultural planning. Land classification attempts at integrating lands according to their native characteristics, pre-existing use and yield capacity. These classes are then assessed in relation to their suitability to different uses envisaged in the plan.

Land classification for agricultural planning should always be related to the objective of the plan. According to Kellog (1940) the first requirement for the successful development and application of any technical system of land classification is a clear understanding of the problems for which the classification is needed. The United Nations Conference on Food and Agriculture held at Hot Springs, Virginia, in 1943 made several recommendations for the long-term organisation of agriculture that may be taken to reflect the common aspiration of landuse planners everywhere. The main theme of all the recommendations was the maintenance of soil fertility. In fact, land classification for landuse planning and agricultural development should be based on soil fertility characteristics rather than those characteristics of the land that may be significant to the specific objective of the plan. For example, the specific objective in any region might be to cultivate perishable crops which need to be carried to the consumer quickly. In this case, roads and railways would be important characteristics in land classification but the basal classification would still be a fertility classification showing which areas would successfully and safely produce the crop in question and which area should be developed for alternative uses.

In land classification two sets of factors are important, viz. physical, chemical and biological, on the one hand, and economic and social, on the other. Soil results on account of the combination of various physical factors which cannot be profoundly altered by man. A physical classification of land is thus permanent and, once made, can serve as the basis on which social and economic classification can be superimposed. Alteration in the social and economic conditions may necessitate changes in the classifications based on the second set of factors.

Almost every soil can be transformed into optimal state of fertility but in practice the application of the required measure is often limited by economic factors, e.g., the Thar Desert could be made fertile by irrigation and manuring, but it may not be an economic enterprise. Chemically and physically poor sandy soil may be classed as suitable for intensive market gardening if favourably situated with regard to the market and supply of manure, but quite unsuitable if market facilities are poor or the cost of manure is high. The significance of these factors will always vary with the social and economic development of the State or region.

Gray and Regan (1940) are of the opinion that the land classification is aimed at the determination of the best use from both the entrepreneurial and social standpoints.

Astor and Rowntree have explained the differences between the optimum and maximum fertility of an agricultural soil. Optimum fertility levels are determined by the ratio between input, i.e. total cost of production, and output, i.e. total return. They vary not only between one crop and another, and one soil type and another, but also between the same crops and soils in different localities. Maximum fertility is thus determined simply by yield per acre and the weight of the crops is not always as important as the quality.

In recent years attempts have been made to classify land from different points of view and for many purposes and a variety of methods have been developed. These various methods of land classification may ultimately help in landuse planning.

## Land Classification in the USA

The United State National Resources Planning Board in its report on "Land Classification in the United State" (1941) distinguishes five types of land classification as per details given below:

I. Land classification on the basis of inherent characteristics.
II. Land classification on the basis of present use.
III. Land classification on the basis of use capabilities.
IV. Land classification on the basis of recommended use.
V. Land classification on the basis of programme effectuation.

The first type of land classification deals with problems related to contour map of slopes in the field, the kind of soil with which the field is endowed, texture, structure, depth, other physical characteristics and chemical properties of the various layers of soils. Besides these characteristics, temperature, precipitation, susceptibility to frost and geologic origin may also be taken into consideration. The second basis of land classification is considered in terms of the present use to which the land is put. A map based on the consideration of present landuse, e.g. crop land, pasture land, lands for industrial establishments and residential uses, etc., may be prepared.

Type II classification read against the background of Type I classification provides more information than either type alone. A detailed study of lands classified on these bases would reveal that the differences in landuse were not accidental but were related to variations in soil characteristics.

A comparison between the land class map based on the degree of roughness or smoothness (topography) of the land surface and a present landuse map would bring out a strong relationship between surface configuration and landuse.

A use-capability map deals with the type of use that is needed in order to avoid loss in yield per acre. Farmers as users of land must take care of fields so that they do not suffer from erosion, rather the soils are improved and the fertility is conserved. This objective can be achieved only when farmers get sufficient profit from the land. Profits or earnings depend on several factors besides use-capability of land as might be indicated by type III land classification map. Prices and costs may not be such as to justify the use of land for inter-tilled crops, for example, even though the use-capability map would indicate this use to be best adopted biologically.

Sometimes it would not be advisable to classify as suitable for farming a tract of land of high-use capability that is large enough to support a farm, for the tract is the only one for miles around and the cost of public services would be too high to justify its occupation as a farm.

In the classification of agricultural land for recommended use, the size of farms and the standard of living of the peasantry are some of the important factors that must be considered besides those influencing a Type III classification. In such areas where living standards are low, less capital needed for farming and less income is required for capital maintenance and for living than in use whenever living levels are high. Poor and less productive lands may, therefore, be recommended as more suitable for continuous agricultural use in regions with a low level of living than in regions where living standards are high.

Land classification of Type IV divides the land into classes according to the use that is recommended which offers the best opportunity for people to earn income and accumulate capital. A map showing land that is adapted to agriculture of different degrees of intensity and land that is adapted to less intensive uses, such as extensive grazing or forestry, is a typical example of Type IV land classification map.

In such regions where soils, landuse, landuse capabilities and living levels are well understood through previous studies, Type IV classification of rural land can sometimes be accomplished without the aid of actual maps of these more basic factors; but these factors should always be considered, even if only by field inspection, in making a Type IV map. In making a Type IV map of a settled rural area, a map of variation in capital already accumulated by farmers on the land, in itself a Type II classification, is a useful aid. Type IV classification is, thus, the result of measuring a cause-and-effect chain starting with inherent factors (soil, topography, climate, etc.) and passing through existing landuse, apparent landuse capabilities and result factors such as accumulated farm capital (buildings, livestock, groves, etc.). Judgment based on these observations and on previous studies provides the basis for the recommendations that are made.

Type V land classification is more than a recommendation; it is, in fact, a plan. Land classification of Types I, II, III and IV must be made by experts in the field with a thorough understanding of the factors in a particular type and kind of classification. Type V classification, however, is made by agricultural planners, that is, by people who know how to carry out various programmes. A farmer's plan for the use of land on his farm for the coming year is Type V land classification. He has to rely, to a large extent, on a Type III map of landuse capabilities in making such a plan.

Type IV land classification for recommended use is needed in connection with land reclamation and settlement programmes as a means of avoiding the gross errors in land utilisation which have been so common in the past. Whenever land is to be newly developed and settled, it should be classified on the basis of recommended use. This method will lead to correcting the misuse and developing of each land for the purpose of which it is best suited.

Rural development plans are generally related to the use of land and should be guided by land classification for recommended use, i.e. Type IV. The effectiveness of agricultural extension work can be much improved by adjusting the programme to the widely varying needs and financial means of rural people on different land classes in the same community. Rural road building plans, electrification schemes, health centres and schools planning will better succeed and better serve the rural community if they are adjusted to varying land resources as indicted in Type IV land classification. Zoning of agricultural land against encroachment of urban bases, zoning forest land against agricultural uses, disposal of sub-marginal agricultural land through re-forestation or establishment of grazing reserves in activities should never be undertaken without prior classification of land according to recommended use.

**Land and Classification in Great Britain**

During the Second World War, Great Britain, under the guidance of the late Prof. L. D. Stamp, conducted the "Land Utilisation Survey" and a simple classification of land was arrived at. This classification can be used as a basis for de-limitation of the tracts of good agricultural land which could be avoided in building developments. According to this classification all the lands were divided into three major categories I, II and III. Category I related to good quality land, Category II to medium and Category III to poor quality land. Land in the different categories has the following characteristics:

Category I: Good Quality Land (Highly productive under good management):

- Site
  - (i) Not too elevated,
  - (ii) Level, gently sloping or undulating,
  - (iii) Favourable aspect.
- Soil
  - (i) Deep,
  - (ii) Favourable water condition,
  - (iii) Texture, mostly loams.

Category II: Medium Quality Land (Medium productivity even when under good management):

- Site
  - (i) High elevation,
  - (ii) Steepness,
  - (iii) Unfavourable aspects.
- Soil
  - (i) Shallowness,
  - (ii) Defective water conditions.

Category III: Poor Quality Land:

Land of low productivity by the extreme operations of one or more factors of site and soil.

Regional classifications of land in Great Britain have generally been based on the abovenoted three categories. In Northern Ireland and Scottish lowlands a slight deviation in the system of land classification was allowed. Here, instead of adopting the major categories system, letters have been used. In the Northern Ireland system of land classification, Category A, Category B ($B_1$, $B_2$, $B_3$, $B_4$ Classes), Category C ($C_1$ and $C_2$ Classes) and Category D ($D_1$ and $D_2$ Classes) have been used, while in Scottish lowlands, symbols A, B, B +, B, B +, B–/B/B –/C, C, C, C, D, D, and D for the corresponding categories in N.I. system have been adopted.

**Land Classification in the Soviet Union**

K.V. Zvorykin (1963) has tried to assess the worth of individual land in terms of its value with a view to recommending measures for increasing yields from different classes of land. He suggested that mapping of the land on the basis of its physical attributes and quality of existing cultivation is the first stage in the land classification work. The suggested bases are:

(i) Relief and slope,
(ii) Soil variety,
(iii) Landuse type,
(iv) Vegetation,
(v) Tillage condition of the land,
(vi) Depth of water table.

Taking into consideration the abovenoted points landuse is mapped according to the practices of the past few years in the second stage. In the third stage, data regarding crop yields and labour and costs required to grow crops on different lands with unlike natural properties is collected and processed. This helps towards selecting the most economical landuse methods for different lands. In the fourth stage, properties of land are assessed with a view to explaining differences in fertility and workability of the soil. Climatic, water, chemical and soil properties are the important items of this stage of the work.

In the fifth stage, the actual land classification work is taken up and all the land units or "modifications" found in the first stage are combined on the basis of information from the later stages of the work into higher-rank classification units. Modifications with similar observed properties and fertility are grouped into a "land class". Land classes suitable for identical uses but differing in fertility are grouped together as "land types". Land classes that favour different combinations of crops would then be assigned to different land types. Besides the different categories of lands, agro-production land groups are also recognised by Zvorykin. He believes that a solution of the problems of classification and land appraisal is hindered by undue emphasis on soil types. Soils map, however, may be of great assistance in land classification provided that reliable values can be associated with each unit of the soil classification.

**Land Classification in the Tropical Areas**

The problem of land classification in the tropical regions can be considered on a different footing, for conditions in these areas are quite different from those prevalent in the temperate latitudes. In view of the conditions prevailing in the tropical regions it is rather difficult to arrive at any definite land classification scheme in order to step up production. However, a multiple land classification system may be recommended. Multiple classification envisages a long-term optimum and a more immediate aim for a given area. This view is supported by the report of the landuse survey team in British Honduras.

## Techniques

### *Unit-Area Land Classification*

This method of land classification was developed for the Tennessee Valley Authority. In this method factors relating to physical and human environment are tabulated and numbered and the numbers for each unit area of land are expressed in a fractional code. The digits in the denominator represent physical conditions, while digits in the numerator stand for human environment. Whenever change occurs, a new unit area is delimited. A complete symbol for each land unit is in three parts, i.e. a Roman numeral, a short fraction and a long fraction. The three parts can be indicated here as:

$$\mathrm{V}\ \frac{2}{3}\quad \frac{1\mathrm{G}^{*}\,234}{2331123}$$

The Roman numerals from I to V show the severity or absence of problems in the area, e.g. I—problems insignificant; II—problems not critical; III—moderately critical; IV—very critical; V—so critical that a radical change in landuse is desirable. The short fraction is a summary of long fraction. In the numerator, areas are classified into five classes (1 to 5), i.e. excellent, good, medium, poor and very poor, while in the denominator, seven digits have been used and each digit relates to a condition, i.e. 1—slope; 2—drainage; 3—erosion; 4—stoniness; 5—rock exposure; 6—soil depth; and 7—soil fertility. Each of these physical conditions expressed by seven digits is further classified into five categories in the case of slope, drainage, erosion, rock exposure and soil fertility while stoniness and soil depth have four subdivisions

* G stands for grain (small).

each. Owing to lack of space, the details of the subdivision have not been given here.

The suitability of this method depends upon availability of facilities for detailed field-work, aerial photographs, etc. This method has, after necessary modifications of the digit, been used with success in New Zealand by K.B. Cumberland (1944).

### *Quantitative Land Classification*

In the present discussion so far no attempt has been made to relate one class to another quantitatively but many such attempts have been made entirely to devise an objective and quantitative form of land classification. The methods used may be classified as "inductive" or "deductive" or a combination of both types.

### *Inductive Methods of Land Classification*

This is also known as "Storie Index" (Storie, 1933) method. It has been used in different countries including the USA. It is based on soil profile. Regardless of actual cropping experience with the soil, Storie compiles the rating from percentage values accorded to certain characteristics of the soil profile. Three "factors" are used : one refers to the general characters (excluding texture) of the profile; second refers to the surface texture; and third refers to "soil modifying conditions", such as drainage, acidity and erosion. The values of these three factors expressed as percentages of the optimal conditions for plant growth are multiplied together to obtain the rating and the product is expressed as the percentage of the maximum. The advantage of multiplying instead of adding the points awarded to each soil characteristic is that thereby any one abnormal factor can dominate the final rating. A soil, thus, may have excellent profile conditions if it is rated at 100 for the first factor, excellent texture if rated at 100 for second factor, but because of acidity or alkalinity it is rated at 40 for the third factor. The product of three factors gives a percentage rating of the soil of 40 despite the other conditions being favourable, the alkalinity renders the soil unproductive. On addition basis the rating will be 80 per cent. These results can be illustrated mathematically as follows :

$$\text{Case I} \quad : \quad \frac{100 \times 100 \times 40}{100 \times 100 \times 100} = 40 \text{ per cent}$$

$$\text{Case II} : \frac{100 + 100 + 40}{100 + 100 + 100} = \frac{240}{300} = \frac{8}{10} = 80 \text{ per cent.}$$

It will be seen that although quantitative, such methods have been unable to avoid a high degree of subjectivity in the award of the points or in deciding optimum conditions. In spite of this drawback, the greater degree of comparability can be achieved through this method.

*Deductive Method of Land Classification*

The U.S. Soil Survey "Productivity Ratings" illustrate this type of classification as follows :

| Soil Type | Corn | | Wheat | | Rye | | Alfalfa | | Sugar beet | | Productivity grade | |
|---|---|---|---|---|---|---|---|---|---|---|---|---|
| | A | B | A | B | A | B | A | B | A | B | A | B |
| Miami Loam | 70 | 90 | 70 | 100 | 80 | 100 | 70 | 90 | 60 | 70 | 3 | 1 |

*Source*: Jacks, G. V., *Land Classification for Landuse Planning,* Imperial Bureau of Soil Science, Technical Communication No. 43, 1964, p. 69.

The figures below A and B in each column refer, respectively, to the percentage of the standard yield of crop obtainable (A) with the common practices of management (average farming) in the area and (B) the best current practices. The standard yields refer to average yields obtained without the use of amendments.

Suppose the standard for corn is 50 bushels per acre, average yield under average management is 35 bushels and, under best management, 45 bushels. The "Productivity Grade" can be calculated by a simple percentage weighting of the crop ratings according to each crop's local importance. The values 70 and 90 under A and B respectively can be calculated as follows:

$$A = \frac{35 \times 100}{50} = 70$$

$$B = \frac{45 \times 100}{50} = 90$$

Similarly, values of A and B in the other columns can be calculated. The difference between A and B ratings is a measure of the soil's response to good management.

"Productivity Grade" indicates the soil's general productivity for the common crops of the region. Soils with a weighted average between 100 and 90 are graded 1; between 90 and 80, 2; and so on. Those with low "A" and high "B" productivity grade would be classed as low in fertility but productive. Productivity grading is some arbitrary procedure, depending on the validity of the system of assessment. From the example mentioned earlier it is clear that the Miami Loam is used predominantly for cereal crops, for its productivity grade corresponds to its productivity ratings with respect to the cereals.

**Land Classification in India**

In India, land classification has not so far been attempted on a scientific basis. Different States during the period of settlement operations or consolidation of holdings have adopted arbitrary principles to classify land for the purpose of fixing revenues. While determining rental values of different categories of lands, several considerations influence the basis of land classification. However, soil fertility is the usual basis of land classification. Lands in the vicinity of settlement area are classed as good quality lands; adjacent to this zone, we have the second zone of medium quality lands; and a third zone situated at the periphery of the settlement area consists of poor quality land. These three categories of lands varying in soil fertility are placed in concentric zones.

The problem of land classification in Uttar Pradesh has, however, been tackled by a good many researchers during the last one-and-a-half decades. Prof. M. Shafi has done a classic work on "Land-Use in Eastern Uttar Pradesh". The author has also attempted land classification in the "Central Ganga-Ghaghara Doab" in connection with his doctoral thesis. In almost all the works on landuse, land classification has been attempted on the basis of sample village surveys. A detailed study of the village was made with the help of the soil map and lands were grouped in different categories on the basis of soil fertility.

The regional soil laboratories located at different places in Uttar Pradesh are charged with the task of preparing soil maps of the entire State. An appreciable amount of work has already been done and soil maps of some districts have been published. These maps can help in the work of land classification and ultimately in the agricultural development of the State.

In the light of the above discussion it seems desirable to suggest a tentative plan for the land classification in the developing countries. It is possible only after a detailed survey has been undertaken on a local and regional basis. However, the following bases of land classification could be examined and adopted after some modifications if necessary:

I. Land classification on the basis of soil fertility.
II. Land classification on the basis of land capability.
III. Land classification on the basis of economic landuse.

Soil fertility classification can be based on the following characteristics:

(i) Depth of the entire soil,
(ii) Moisture-holding capacity of the soil,
(iii) Permeability of the sub-soil to water,
(iv) Natural soil drainage,
(v) Organic matter content,
(vi) Degree of erosion,
(vii) Alkalinity, and
(viii) Natural plant nutrient content.

It is important to note that not only the individual characteristics but also their combination in a given soil affect the growth of crops. Detailed soil maps help farmers to adopt the rotation of crops in relation to the characteristics of the soil. In order to know the different soils with different degrees of fertility and land capability, a soil map on a large scale must be prepared so that it can show the significant soil variations within field and farms.

Land capability classification related to the use and treatment of the land, the effects of various physical characteristics must be considered in making the classification. Hence, this type of classification must be based on inquiry into the relation of land characteristics to plant growth. This classification is intended to help the farmer in planning his agriculture and treatment of his land.

In land capability classification, land is to be divided into land capability units, each unit consists of one kind of land from the viewpoint of the use and treatment the land should receive if it is to be held in maximum continuous production. This type of classification is useful in projects for irrigation, drainage, forest clearing or some other form of amelioration. In such cases the classification shall take into consideration those particular characteristics of the soil and other features of the land that will determine it suitability after improvement has been effected.

Economic landuse classification determines and maps the local variations in the capacity of land to produce income against productive expenditure. It is a fact that soil and topographic features vary to a great extent within short distances in most parts of the world. In rural areas there will be land on which the income per farm would not be sufficient to maintain the capital required for the type of farming that is adapted to the region, nor to maintain a farm family at a level of comfort commensurate with the regional standard of living. Such land may be classified as sub-marginal for plough-farming. This category of land may be earmarked for forest or grazing of livestock depending upon the degree of moisture supply. This type of land is usually placed in a low economic land which is of a high productive capacity and which provides farmers with a level of living. Many encourage its permanent occupation by farmers for plough-farming.

The economic classification of land is a sort of prediction of what the comparative economic productive capacity of the land will be during the foreseeable future. The type of classification has a special significance in the case of developing countries like India.

A detailed and complete classification of land based on the criteria mentioned earlier not only indicates the distributional patterns of crops, farms, rainfall, soil, etc., but also explains the relationship between various factors in such a way that it is possible to trace major influences which have shaped the pattern of present land utilisation and which will indicate the possible path of future agricultural planning.

## REFERENCES

Astor, V. and Rowntree, B. S. : *Mixed Farming and Muddled Thinking,* Mac-Donalds, London.

Cumberland, K. B., 1944 : "The Survey and Classification of Land in New Zealand; a basis for planning." *Trans. Royal Soc. New Zealand,* 74.

Gray, L. C. and Regan M., 1940 : "Needed points of development and reorientation in land economic theory," *J. Farm Eco.,* 22, pp. 34-46.

Hudson, G. D., 1936 : "The Unit Area Method of Land Classification", *Ann. Assoc, Amer. Geogrs., 26.*

Jacks, G. V., 1946 : *Land Classification for Landuse Planning,* Imperial Bureau of Soil Science, Technical Communication, 43.

Kellog, C. E., 1940 : *The Theory of Land Classification, Missouri Agri. Expt. Station Bulletin,* 421, pp. 164-73.

National Resources Planning Board, 1941 : *Report on Land Classification in the United States,* Washington.

Shafi, M., 1964 : "Land Use Planning, Land Classification and Land Capability—Methods and Techniques", *The Geographer,* XVI.

Stamp, L. D., 1948 : *The Land of Britain, Its Use and Misuse.*

Storie, T. E., 1933 : An Index for Rating the Agricultural value of Soils, *California Expt Station Bulletin,* 556.

Symons, L., 1968 : *Agricultural Geography.*

Zvorykin, K. V., 1963 : "Scientific Principles for an Agro-production Classification of Lands", *Soviet Geography, Review and Translation,* 4.

7

# LANDUSE SURVEY AND PLANNING

R. M. POFALI

A NUMBER of Land Development Projects are initiated in India. It is essential that information regarding the present landuse should be collected and landuse potentials computed under a wide variety of environments, i.e. physical, economical, and social, to have broad ranges in landuse planning. Landuse potentials based on experimental data are ideal but may not be possible at all times. However, interpretation of land and environmental features may provide some indices of potentials.

The concept of landuse survey is old. In the past, the decision regarding landuse was made by the individual so far as farm planning was concerned. The farmer is the best judge of his land, as he is well acquainted with it. Accordingly, he puts it to various uses as per his needs. But due to increased need of food, fuel and fibre and due to tremendous pressure on the limited land resources, the need for rational planning has become imperative. The main aim of landuse survey is to know the distribution of the land under various uses in specific socio-economic and environmental conditions and to put the resources to improved uses based on scientific principles and without any degradation. The landuse planning is thus needed with changing needs and pressures and involves competing uses for the same land (FAO, *Soils Bulletin No. 32*).

The function of landuse planning is to guide decisions on landuse in such a way that all the available resources of the environment are put to the most beneficial use, simultaneously conserving them for future. It is the primary duty of the landuse planner to study the various land characteristics in respect of various natural aspects and suggest the best use under defined management practices for sustained utilisation. Any alternative use of land suggested should be without disturbing the ecological as well as socio-economic balance of the area.

### Aim of Landuse Survey and Planning

The landuse survey deals with inventorying, classifying and mapping of lands according to their present utilisation and is needed for their proper assessment. Generally, the landuse surveys are carried out by traversing the area as well as collecting data from the administrative records. It is time-consuming and many a time erroneous. By the time the data are compiled, they become outdated. But in recent times the data can be collected from multidate as well as sequential aerial photographs and LANDSAT imageries.

Collection of landuse data is one aspect. The more important aspect is the analysis and assessment in respect of landuse trends and changes, effect on land quality, environment and socio-economic structure. Such studies are the forerunner to introducing any radical changes in the landuse pattern. Eventually, the ultimate objective ought to be to meet the increasing needs of the local population and yet maintain ecological balance and satisfy the conservation needs.

Landuse survey should provide data on the present landuse and also the land available for future. It should help in predicting the suitability of the lands for various uses, the management practices necessary and the possible likely changes in landuse consistent with the environmental features and changing socio-economic conditions.

### Levels of Landuse Survey and Mapping

The level at which different landuses are classified and mapped is based on the type of survey and the intensity, the time available and scale of the map. Before deciding the scale of mapping the landuse, the objectives are defined. At the national or regional level when the data are required quickly the scale of mapping may be small and units of landuse broad to have a quick-look analysis of the land utilisation. In small and specific areas requiring detailed information, large scale with smaller units are desirable. Table. 7.1 gives the various scales on which the types and details of landuse can be mapped.

### Importance of Landuse Survey and Planning

Land is an important basic natural resource for human survival. It comprises the physical environment including climate, relief, soils, water and vegetation all of which have influence on landuse potential. Initial knowledge of landuse pattern may be obtained through a panoramic view from small-scale imagery in respect of cropped land, forest land, variations in topography, soil, plant cover and climate.

A detailed inventory and proiection of natural resources, their analysis should help in the rational planning of resources consistent with human needs and socio-economic policies.

TABLE 7.1: **Landuse mapping: scales**

| S.No. | Purposes | Scale | |
|---|---|---|---|
| 1. | Detection of crop and vegetation diseases, crop identification, crop yield forecasting, sample studies, settlement pattern, and water resources | 1 : 5,000 | Large |
| 2. | Crop identification, land analysis, soil erosion and conservation survey and planning, study of salinity problems, identification of various species of trees, detailed soil survey | 1 : 10,000 | Large |
| 3. | Landuse comparison, area analysis, agricultural landuse, land development programme, urban impact on agriculture, forestry, pasture, landuse trends, soil conservation, surveys, semi-detailed soil surveys | 1 : 25,000 | Medium |
| 4. | Broad landuse study for agriculture, forestry, pasture, reconnaissance soil mapping, landscape analysis, drainage study | 1 : 50,000 | Small scale |
| 5. | Broad classification of landuse system in relation to landscale, soils, hydrology, geology, etc. | 1 : 10,00,000 | Very small |

Land in a wider concept includes lithosphere, biosphere and atmosphere. Taking these components into consideration, during a landuse survey, each unit with specific characteristics should be analysed and mapped separately. For classification of units, help of various natural resource surveys should be taken to assess the suitability of each unit under defined management practices.

## Major Kinds of Landuse and Land Utilisation Types

Mapping of landuse and other resources require development of systematic legend which can be useful for the planners. The legend may vary according to the purpose of survey and scale of map. The dichotomous systems with three to four levels of classification will serve the purpose. The example of levels I and II as adopted by the U.S. Geological Survey is given in Table 7.2.

TABLE 7.2: **Landuse classification system**

| Level I | Level II |
|---|---|
| **Urban and Built-up** | **1. Residential** |
| | **2. Commercial and Services** |
| | **3. Industrial** |
| | **4. Transportation and Communication** |
| | **5. Institutional** |
| | **6. Mixed** |
| | **7. Others** |
| **Agricultural Land** | **1. Crop Land and Pastures** |
| | **2. Orchards and Ornamental Areas** |
| | **3. Feeding Operations** |
| | **4. Others** |
| **Range Land** | **1. Herbaceous Range** |
| | **2. Shrub Bush Land Range** |
| | **3. Mixed** |
| **Forest Land** | **1. Deciduous** |
| | **2. Evergreen** |
| | **3. Mixed** |
| **Water** | **1. Streams and Canals** |
| | **2. Lakes** |
| | **3. Reservoirs** |
| | **4. Bays and Estuaries** |
| | **5. Others** |
| **Wet Land** | **1. Forested** |
| | **2. Non forested** |
| **Barren Land** | **1. Salt Flats** |
| | **2. Beaches and Mud Flats** |
| | **3. Sandy Areas other than Beaches** |
| | **4. Bare Exposed Rocks** |
| | **5. Strip Mines and Gravel Pits** |
| | **6. Transitional Areas** |
| | **7. Mixed** |
| **Tundra** | **1. Shrub and Bush Tundra** |
| | **2. Herbaceous Tundra** |
| | **3. Bare Ground Tundra** |
| | **4. Wet Tundra** |
| | **5. Mixed** |
| **Permanent Snow and Ice** | **2. Glaciers** |

*Source:* Inter-agency Report USGS — 253.

The lower level of classification should be prepared by the users according to needs and as per the details needed for the purpose. For landuse survey the legend reflects the land forms and physiographic system at the first level of classification so that it is possible to correlate the existing landuses with various aspects of land resources. The second level of classification should include the utility aspect followed by subsequent levels as per needs and mapable details (Table 7.3, Fig. 7.1).

TABLE 7.3: Landuse classification

| Level | Unit | Classification |
|---|---|---|
| 1. | Physiography | Hilly area<br>Plain land |
| 2. | Utility | Forest<br>Arable land |
| 3. | Identification | Sal<br>Teak<br>Sugarcane<br>Wheat<br>Jowar<br>Cotton |
| 4. | Management | Deciduous<br>Irrigated arable<br>Dry farming land |

This system of mapping land units is convenient as it allows generalisation to the scale of maps. For detailed mapping, four level classification may be necessary which provides quantitative land evaluation under various uses. The detailed mapping consists of a set of technical specification in a given physical, economic and social setting.

Attributes of land utilisation types to be considered are: (i) Produce, i.e. crops, livestock, timber or other services like recreation, wildlife preservation, (ii) Market orientation whether subsistence or commercial production, (iii) Capital, labour, power intensities, (iv) Technology employed, (v) Land holding, (vi) Land tenure and income level. Demarcating the land utilisation types in this manner will help in future planning of the land utilisation on a scientific basis. An important study in the land utilisation type relates to multiple landuse requiring different management practices. For sound planning, the data on such aspects of landuse should be collected and decisions regarding the best use of land arrived at after research and experience.

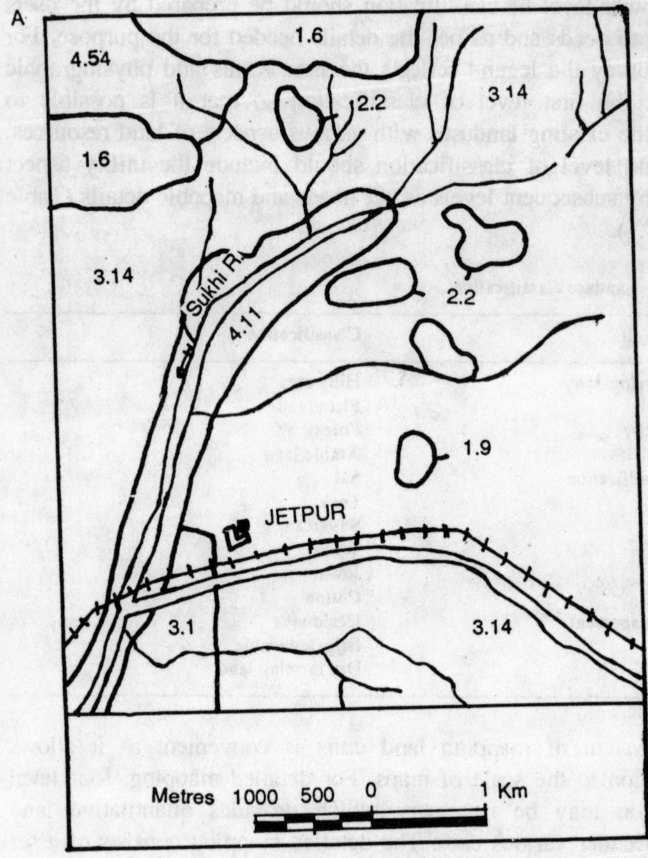


**FIG. 7.1: Physiography and landuse.**

*Legend*

| *No.* | *Geomorphic Unit* | *Landuse* |
|---|---|---|
| **1.6** | **Hilly area** | **Dense vegetation cover** |
| **1.9** | **Isolated hillocks** | **Sparse vegetation cover** |
| **2.2** | **Disintegrated hillocks** | **Thin vegetation cover** |
| **3.1** | **Pediment area** | **Cultivation on the lower slopes** |
| **3.14** | **Pediment area** | **Patchy cultivation** |
| **4.54** | **Fluvioaelion plain land** | **Cultivated** |
| **4.11** | **Bad-land topography along the river course** | **Sparse vegetation** |

**Prerequisites of Landuse Survey and Planning**

Information collected by various organisations engaged in natural resource surveys is very useful in predicting the use of land and relating the various land characteristics with suitability, capability, potentiality and productivity for various uses like farming under rain-fed and irrigation, pasture, forestry, engineering works, settlements under constraints like erosion or salinity hazards. Therefore, landuse requires a multi-disciplinary approach to arrive at sustained landuse planning. Landuse survey information may include details on location and accessibility, climatic zone, relief, soils, hydrology, vegetation and geology under environmental approach and information regarding population and its structure, economic infrastructure, government policies and administrative machinery under socio-economic and political approach. In the present paper, only the environmental aspect in landuse survey and planning is dealt with in detail to find out the suitability, productivity, capability and potentiality of land for suitable landuse planning especially in the field of agriculture using the Vidarbha region of Maharashtra as a casc study.

**Environmental Approach in Landuse Survey and Planning**

Physiographic or environmental approach provides the basic material for preparation of landuse maps for sustained landuse planning. As such, inventory of natural resources is fundamental in the landuse planning. For this purpose data on the following important basic items should be collected.

*Geographical Location of the Area of Study*

Location of the unit area may be viewed from the point of probable productivity as assessed by climatic parameters and marketability as assessed by accessibility and profitability. Analysis of topographical maps provide information in respect of these items.

*Climate*

Rainfall, humidity and evapotranspiration data obtainable from the Indian meteorological records are useful for a variety of purposes, viz.

rainfall pattern, available moisture in the particular period, drought-prone areas, etc. This data can be superimposed on thematic maps.

### *Hydrology*

Water plays a vital role in determining the landuse pattern of an area. There is a direct link between water and plant relationship. The data on drainage network, types of drainage channels, reservoirs, ponds, lakes, tanks, ground water potential of the areas, storage capacity of soils, etc., should be collected from the organisations carrying out surveys on hydrology and the same be presented in the form of maps for interpretation and correlation.

### *Land Characteristics*

The landscape analysis maps prepared from existing topographical maps, aerial photographs and LANDSAT imageries provide useful information regarding the land characteristics. It is generally observed that the pattern of landuse is governed by the relief and morphometrics, and morphological characteristics of the area. It is generally found that most of the hilly terrain, ridges, domes, rocky table lands are either under forest or pasture according to the climatic condition of the area. Due to the shallow nature of soils, low moisture-holding capacity, coarse texture, excessive run-off and steep gradient these areas are prone to accelerated erosion and unfit for agriculture. Landscapes recognised as valleys, alluvials, fans, deltas, flood plains, outwash plains, loess plains are under intensive agriculture under conditions of low climatic constraints. A detailed landscape analysis map coupled with climatic data serves as a good guide to carry out landuse survey quickly and information so collected can be utilised for sound landuse planning of the area. It also provides information regarding the erosion hazards, soils cover, salinity problems, permeability and ground water.

### *Geology*

Before proceeding for landuse survey, geological information about the area is necessary as it has a bearing on the formation of soils, surface drainage and underground water resources. Geological structure plays a vital role in the morphology of land form, vegetation pattern and other properties of land. It reflects the land utilisation pattern in the form of mineral resources, industrialisation, etc. All the information can be obtained from the geological departments at the Central and State levels.

### *Soil*

Soil is the important medium for plant growth and it plays a vital role in the evaluation of land for various purposes including agriculture, forestry, pasture, industry and settlement, directly or indirectly. In landuse survey and planning, more attention is given to the type of soil, its physical and chemical characteristics. Also the problems related to soil erosion, drainage and salinity need to be thoroughly studied and interpreted for sustained landuse planning. Hence data on soil resources is very essential as well as fundamental in landuse planning. As it is not possible to collect data on all aspects of natural resources, the same can be acquired from the various State and Central agencies and interpreted for suitability of crops, productivity and potentiality of each type of soil under various land utilisation types.

### *Role of Soil Survey in Landuse Planning*

The process of soil survey characterises various types of soils associated with the landscape units and helps in production of different scale maps. The units delineated on the maps depend on the scale and provide information on the soil characteristics. Thus the soil map shows the location and extent of each kind of soil and a good soils survey interpretation helps in making policy decisions in different disciplines where land is one of the factors. Soil survey interpretations provide information on soil potential, productivity and the limitations like erodibility, etc., in their sustained use.

The soil surveys are conducted at various levels of intensity depending upon the purpose for which the information is required. Low-intensity surveys are made to assess the broad characteristics of soils for regional planning in the scale of 1 : 50,000 using the Survey of India toposheets or aerial photographs. The mapping unit for this survey is generally series association (Fig. 7.2). But high-intensity surveys are carried out to workout the local problems or where the land is to be used intensively. In farm planning, evaluation of problems like salinity, erosion hazards and for watershed development,

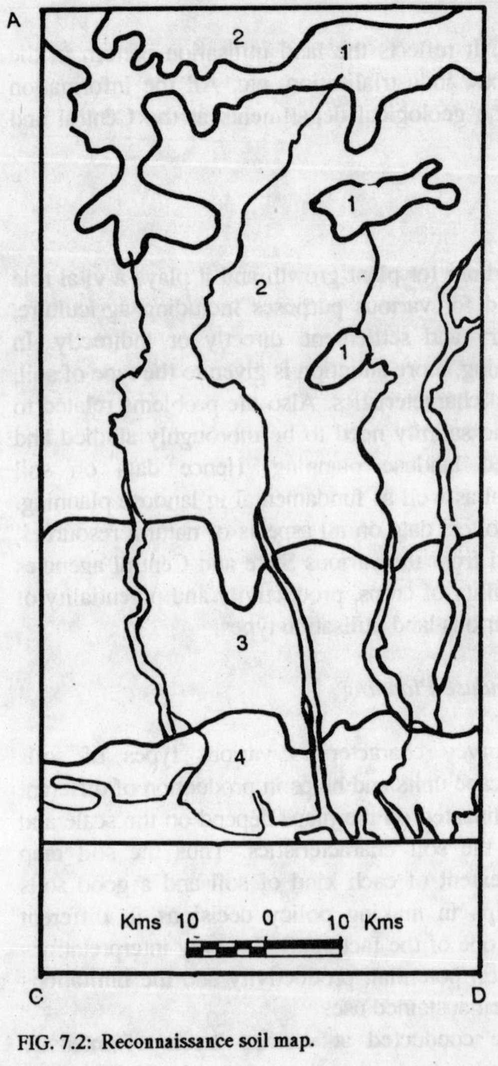


FIG. 7.2: Reconnaissance soil map.

*Legend*

| *Map Symbol* | *Mapping Unit Description* |
|---|---|
| 1. | Shallow to very shallow, fine to loamy soils and shallow fine loamy soils on gently to moderately sloping residual hillocks and ridges. |
| 2. | Moderately deep to deep fine loamy soils and shallow to moderately deep coarse loamy soils on very gently to gently sloping pediment and nearly level to very gently sloping pediment plain. |
| 3. | Deep coarse loamy soils, deep to very deep fine loamy soils and very deep, fine loamy soils on very gently sloping to nearly level deltaic plain. |
| 4. | Moderately deep, fine loamy soils on a nearly level tidal inlet. |

high-intensity surveys on base maps in the scale of 1 : 5,000 using village maps or 1 : 15,000 aerial photographs are used. In this process a mapping unit consists of phases of series showing soil texture, depth, slope, erosion, stoniness, rockiness, etc.

In landuse planning, the data on soils plays a vital role as characteristics and qualities of soils affect the performance of landuse. The important soil characteristics such as depth of soil, texture, structure, amount of organic matter, amount of clay and its type, plant nutrient status, permeability and moisture-holding capacity determine the type of cropping pattern, irrigation needs and level of management for economic returns. After knowing all the details, optimum use of land can be suggested which may envisage change in the present use pattern. In addition to socio-economic characteristics, data on landscape characteristics, geology and climate including temperature, precipitation and its distribution, wind, humidity, length of day, cloud cover, etc., are also important (Steele, 1967, *Soils Bulletin 8*).

Crop performance under different conditions helps in formulating yield models. The soil survey data can be utilised for rating the soil for individual or group of crops. This is of great help in the land development programme. Various methods on rating and prediction of productivity have been evolved. The principal classification systems currently used are given below:

### *Storie Index*

The Storie index expresses numerically the relative degree of suitability or value of soils for general intensive agricultural landuse. The rating is based on the soil characteristic, and is obtained by evaluating only four factors, viz. depth, texture, slope and drainage (Storie, 1933).

The index rating is obtained by multiplying the above four factors. Thus any one factor may dominate or control the final rating. Soils are placed in grades according to their suitability for general intensive agricultural landuse and the grading decreases as the number increases as given in Table 7.4.

The other factors like availability of irrigation water, climate and distance from market are not taken into consideration which may determine the value of a particular land under certain landuse and its intensity in a given locality. Therefore, the Storie index has many limitations.

TABLE 7.4: Storie index system

| Grades | Index Rating |
|---|---|
| 1 | 80 to 100 |
| 2 | 60 to 80 |
| 3 | 40 to 60 |
| 4 | 20 to 40 |
| 5 | 10 to 20 |
| 6 | Less than 10 |

## *Land Capability Classification*

It is commonly carried out for agricultural use. Classification of land in terms of use capability is based upon the inherent characteristics of the soils, as well as land setting.

The land capability classification system developed by the U.S. Department of Agriculture is widely used. Broad grouping of soils is useful for developing and interpreting soil map for making generalisations about the types of landuse.

Soils are grouped at three levels of generalisation. Classes at the more general level are based on the general suitability of soils for long-time production of common crops and are separated according to the degree of permanent limitations in their sustained use. The broad capability classes are designated by Roman numerals, beginning with Class III for the soils that have lowest limitations to Class VIII having maximum limitations for the use related to agriculture, pasture or forestry. The Class VIII lands are mostly hilly, rugged, rocky and suitable for wildlife preservation.

The capability sub-classes are based on the kind and intensity of limitations in the use of the soils or in management requirement. The sub-classes are designated by small letters added to the class number. The limitations taken into consideration are mainly hazards of erosion (e), wetness (w), soil limitation depth, texture or other features (s), and the climate (c).

The system also provides further grouping, called capability unit within a sub-class. It is management group and may contain several similar kinds of soils. Soils in any one capability unit are adopted to the same kind of common cultivated crops and pasture and require similar alternative systems of management.

This system of capability groupings of soils is based in relation to an assumed set of technology and economic conditions. The individual soil mapping unit is the building stone in this system worked out during the processes of soil survey of high intensity. This system is based mainly on the effects of combination of climate, permanent soil characteristics on risking soil damage, productive capacity and soil management. It does not suggest the most profitable use of soils. With reclamation of limiting characteristics, capability groups can be changed. In this system also, distance to market, availability of means of communications, skill and resources of individual farmer and other characteristics of land ownership or land management are not taken into consideration. (*Agri. Hand Book No. 210*, 1961).

Riquier (1970) has developed a system to evaluate productivity and potentiality which is an improvement over the earlier methods and takes into consideration some of the important factors briefly described below (Requier, J. et al., 1970).

*Nine-factor Method*

The concept of productivity means the initial land capability to produce a certain amount of crop per hectare per annum. It is natural fertility of virgin land under simple farming practices while the potentiality is the future productivity of land taking into account physical and chemical characteristics of soil as modified by conservation practices or by improvements.

Accordingly nine factors considered for determining soil productivity are moisture (H), drainage (D), effective depth (F), texture/structure (T), base saturation (N), organic matter content (O), mineral exchange capacity/nature of clay (A), mineral resources (M), and Slope (S), Productivity is measured as a product of these factors after their interpreted contribution for production.

Productivity is calculated on the existing soil condition and potentiality evaluated taking into consideration the improvements that are possible by the management practices and the impediments that could be removed. Each factor is assigned marks from a maximum of 100 and the fractions multiplied cumulatively and expressed as per cent, etc., to obtain ratings. The ratings thus obtained are categorised into productivity and potentiality classes as given in Table 7.5.

TABLE 7.5: Productivity and potentiality rating

| Productivity | Classes | Rating | Potentiality |
|---|---|---|---|
| 1 | Excellent | 65 — 100 | I |
| 2 | Good | 35 — 64 | II |
| 3 | Average | 20 — 34 | III |
| 4 | Poor | 8 — 19 | IV |
| 5 | Extremely Poor | 0 — 7 | V |

This method of evaluation was tested in the Vidarbha region of Maharashtra to know the productivity and potentiality on two geological formations in respect of agriculture and pasture development. The ratings are presented in Table 7.6.

The study reveals that the basaltic uplands of undulating character in the low-rainfall zone have poor productivity and average potentiality so far as agriculture is concerned. However, for pasture development though the productivity is average the potentiality is good. The study also reveals similar results in the case of undulating uplands having a complex geology. The coefficient of improvement is comparatively high in the case of the basaltic, undulating uplands for the development of pasture.

Thus it can be summarised that the soil survey data give an excellent information for the interpretation of various soilscape units for judicious landuse planning.

## Soil and Land Irrigability Classes

The soil survey data helps in evaluating soils for their suitability for irrigation. First, the soils are classified according to increasing limitations from A to D. The soil classes are then interpreted along with site and necessary characteristics into land classes from 1 to 5. Land Class 1 is most potential land for irrigation and land Class 5 unsuitable for irrigation.

The soil and land irrigability class may be assigned to each unit mapped and a separate map for command areas may be prepared. The soil and land irrigability classes help in evolving a rational water management technology.

The advantage of evaluating soils in terms of various interpretative groupings may be illustrated with an example. Figure 7.3 exhibits the various types of soils associated with tableland, pediment, subdued

TABLE 7.6: Productivity and potentiality rating for two of geomorphic units in part of the Vidarbha region, Maharashtra

| Geomorphic Unit | Landuse | | H Moisture | D Drainage | P Effective Depth | T Texture/ structure | N Base Saturation | O Organic matter | A C E. C. | M Mineral | S Slope | Value | * 10% | Value | Rating | Coeff. of Improvement |
|---|---|---|---|---|---|---|---|---|---|---|---|---|---|---|---|---|
| Undulating uplands (Basaltic) | Cultivation | P | 0.90 | 1.00 | 0.50 | 0.80 | 0.80 | 0.70 | 1.00 | 1.00 | 0.90 | 18.12 | — | 18.12 | 4 | 1.29 |
| | | P' | 0.90 | 1.00 | 0.50 | 0.80 | 0.80 | 0.80 | 1.00 | 1.00 | 0.95 | 21.19 | 2.20 | 23.39 | III | |
| | Pasture | P | 0.80 | 1.00 | 0.80 | 0.80 | 0.90 | 0.70 | 1.00 | 1.00 | 0.90 | 29.03 | — | 29.03 | 3 | 1.33 |
| | | P' | 0.80 | 1.00 | 0.80 | 0.80 | 0.90 | 0.80 | 1.00 | 1.00 | 0.95 | 35.02 | 3.5 | 38.50 | II | |
| Undulating Uplands (Gneiss complex) | Cultivation | P | 0.90 | 1.00 | 0.50 | 1.00 | 0.80 | 0.80 | 0.95 | 0.95 | 0.90 | 23.4 | — | 23.4 | 3 | 1.32 |
| | | P' | 0.90 | 1.00 | 0.50 | 1.00 | 0.80 | 0.90 | 0.95 | 0.95 | 0.95 | 27.8 | 2.8 | 30.6 | III | |
| | Pasture | P | 0.80 | 1.00 | 0.80 | 1.00 | 0.90 | 0.80 | 0.95 | 0.95 | 0.90 | 37.43 | — | 37.43 | 2 | 1.31 |
| | | P' | 0.80 | 1.00 | 0.80 | 1.00 | 0.90 | 0.90 | 0.95 | 0.95 | 0.95 | 44.44 | 4.4 | 48.84 | II | |

P : Productivity.
P' : Potentiality.
* : 10 per cent in value of the index for the management measures necessitated by conditions extraneous to the soil.

*Source: Soil Series Handbook* Vols. I and II, Regional Centre, Nagpur.

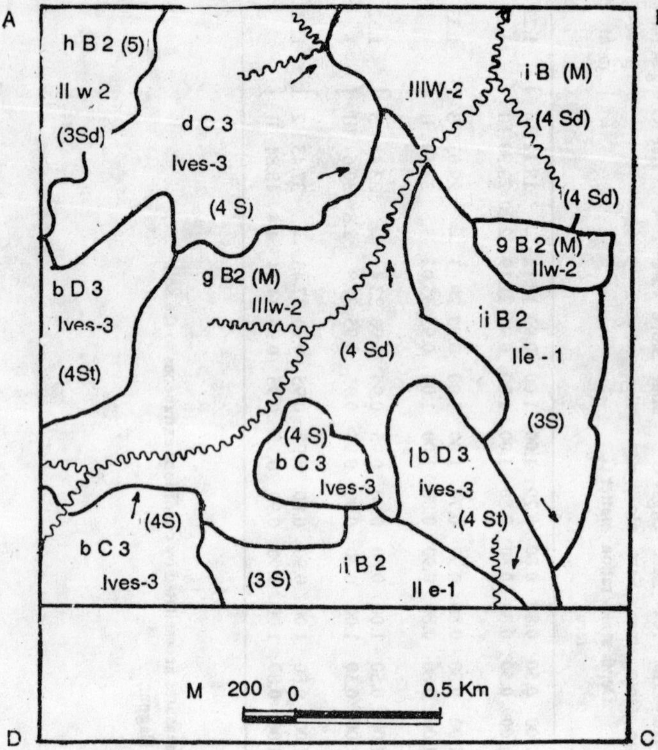


FIG. 7.3: Soil, land capability and irrigability map.

*Legend*

| | |
|---|---|
| hB2(s) | Silty clay soils along foothill nallas, deep on 1-3 per cent slope, moderately eroded, somewhat poorly drained and slightly saline. LC IIw-2, LI 3Sd. |
| iB(M) | Clayey soils on flood plain, deep to very deep on 1-3 per cent slope, slightly eroded, somewhat poorly drained and moderately saline, LC IIw-2, LI 4Sd. |
| bc3 | Gravelly loamy soils on tableland, shallow on 3-5 per cent slope, severely eroded, stony, LC IVes-3, LI 4S. |
| bD3 | Gravelly loamy soils on tableland, shallow on 5-10 per cent slope, severely eroded, stony, LC IVes-3, LI 4S. |
| iB2 | Clayey soils on flood plain, deep to very deep on 3-5 per cent slope, moderately eroded, LC IIe-1, LI 3S. |
| dC3 | Gravelly silty clay loam soils on tableland, shallow on 3-5 per cent slope, severely eroded, stony, LC IVes-3, LI 4S. |
| gB2(M) | Clay loam soils along foothills and tiny drainage lines, moderately deep on 1-3 per cent slope, moderately eroded, somewhat poorly drained and moderately saline. LC IIIw-2, LI 4Sd. |

LC-Land Capability, LI-Land Irrigability.

hills, along small nallas and on the flood plains. They differ from one another in their morphological characteristics, i.e. texture, colour, depth, erosion, etc., on the basis of which they are separated from each other by drawing phase boundaries.

In Table 7.7 are given the associated land capability units. The Roman numeral IV is the capability class, IVes, is the capability sub-class, while IVes-3 is the capability unit. The Unit IVes-3 indicates that the soils are shallow, well to excessively drained, medium to fine textured, severely eroded and occur on 3—10 per cent slopes. Shallow depth, severe sheet and gully erosion, low available moisture and gravelliness are the problems which render the soils cultivable only occasionally. Stony and rocky phases are encountered. After knowing all the details about the soils, suggestions for their improvement can be made.

TABLE 7.7: **Particulars of detailed soil maps**

| Sl. No. | Topography | Map Symbol | Land Capability Unit | Soil Irrigability Class | Land Irrigability Class |
|---|---|---|---|---|---|
| 1. | Gently undulating tableland | bD3 | IVes-3 | D | 4st |
| 2. | Near foothill | hB2(s) | IIw-2 | D | 3sd |
| 3. | Flood plain | iB(M) | IIw-2 | D | 4sd |
| 4. | Gently undulating tableland | bC-3 | IVes-3 | D | 4s |
| 5. | Flood plain | iB2 | IIe-1 | C | 3s |
| 6. | Gently undulating tableland | dC3 | IVes-3 | D | 4s |
| 7. | Midland (foothill) | gB2(M) | IIIw-2 | D | 4sd |

Regarding the above unit it may be inferred that this type of soil occurring on the tablelands requires intensive care. Growing of grass-legume mixture in contour furrows and planting agave hedge to check soil loss may be suggested. After regeneration of vegetal cover, occasional cultivation with erosion-resistant crops may be taken up. Gullied lands should be suitably reclaimed by check dams and gully plugging.

The soil and land irrigability class assigned to this unit is D4st. The letter D denotes very severe soil limitations for sustained use under irrigation, 4 denotes lands that are marginal for sustained use under irrigation while st is the land irrigability sub-class denoting soil and topographic limitations. (*Soil Survey Manual*, IARI, New Delhi, 1971). Therefore, the detailed soil map is more informative and gives a wealth of information regarding the better landuse planning (*NBSS & LUP, No. 365*, 1975).

The uses to which the soil survey data may be put are numerous to develop upon and beyond the scope of this paper. The data can be used for agricultural development, engineering works and structures, terrain analysis, assessment of suitability for industrial complexes, town planning, forestry, etc.

### *Assessment of Environmental Impact on Landuse Planning*

While suggesting changes in the landuse pattern the probable consequences of changes in the environment may be taken into consideration. The most important aspect is to assess the possibilities of environmental degradation. For example, in Bastar district, M.P., deforestation has posed a severe problem of erosion as the land has been cleared for agriculture. In Wardha district of Maharashtra it is noted that due to improper management of irrigation, salinity and alkali problems have arisen. Water table has risen in a short time. At places, the otherwise healthy orange orchards are adversely affected. In the Mahanadi catchment, the destruction of forest has led to heavy floods, sedimentation and excessive erosion. Such types of serious environmental degradations should be taken into consideration before initiating any changes in the landuse pattern.

### *Application of Recent Technology in Landuse Survey and Planning*

For landuse surveys, modern techniques are helpful for covering large areas within a reasonable time with sufficient accuracy. Aerial photographs and LANDSAT imageries offer special advantages.

Aerial photographs in the appropriate scale provide accurate information on farming activities, pasture development, forestry and other cultural developments and help to interpret regional landuse patterns. Spatial variations in landuse types in relation to natural and cultural environment are known by aerial photo interpretation. In any land development projects, ecological considerations are important to evolve integrated landuse planning. These considerations involve maintenance of sustained yields, optimum use of irrigation water, adoption of rational farming practices, etc., without detriment to resources. For this purpose survey data are required regarding various types of natural and cultural resources in the form of inventories obtainable through interpretation of aerial photographs with ground checks.

The units interpreted on aerial photographs can be transferred on the geodetic base and thus information regarding the landuse pattern and potential landuse can be collected within a short time as compared to conventional methods.

Trends in changes of landuse can be studied during the year with the help of multidate aerial photographs of the same area in different seasons. In addition sequential photographs of the same area over a period of years will be of help for monitoring changes in landuse and their effects. Because of flexibility in technique and substantial gains in accuracy, speed and economy over the conventional methods, the future of photo interpretation techniques in the landuse survey and mapping is assured (USDA *Agri. Hand Book, 294*, 1960).

During the past decade, considerable evidence has shown that remote-sensing from aircraft or space platforms can provide much needed qualitative and quantitative data regarding the natural resources, landuse, cropping pattern, forestry, pasture, crop condition, potential yield, water resources, mineral resources, etc., which are useful for the land utilisation pattern. The remotely sensed data provides an opportunity to the scientist, planner environmentalist, geographer, economist, and others to monitor the environmental impact caused by man's activities and the consequent changes in the land surface cover.

Multi-spectral scanners fitted in space-borne vehicles, are capable of collecting spectral information on narrower land in the various regions of the electro-magnetic spectrum (ultraviolet, visible, reflected and thermal infra-red) and the data obtained can be analysed rapidly with the help of a suitable computer programme. The digital as well as pictorial nature of remote-sensing data ideally provides a reference for mapping resources. Both sensor technology and data-analysis techniques are increasingly geared to quantify resources on macro- and micro-scales. Due to these techniques, the data on land resources and landuse can be made available to the planners and policy-makers within a short time.

A number of application studies in agriculture, forestry, soil-mapping, hydrology are carried out at some national institutes dealing with natural resources.

Satellite sensing is a viable technique for the developing countries to study the natural resources for generation of small-scale maps indicating broad landuse patterns for updating the information on broad landuse, pinpointing environmental degradation areas, conservation of potential agricultural land, etc.

In the field of agricultural planning, the remote-sensing technology has many direct applications, such as crop censusing and identifications, crop condition assessment, inventory of irrigated land, soil problems, floods and to assess natural catastrophies (Sahi, 1978).

TABLE 7.8: Explanation of map symbols

| | |
|---|---|
| *Soil Type* | *Texture (Surface)* |
| b | Gravelly loam |
| d | Gravelly silty clay loam |
| g | Clay loam |
| h | Silty clay |
| i | Clay |
| *Slope Class* | *Slope per cent* |
| B | 1—3 per cent very gently sloping |
| C | 3—5 per cent gently sloping |
| D | 5—10 per cent moderately sloping |
| *Erosion Class* | *Degree of Erosion* |
| 2 | Moderately eroded |
| 3 | Severely eroded |
| *Salinity Phase* | *Degree of Salinity/Alkali* |
| (M) | Moderate |
| (S) | Slight |
| *Soil Irrigability Classes* | |
| C | Severe soil limitations for sustained use under irrigation. |
| D | Very severe soil limitations for sustained use under irrigation. |
| *Land Irrigability Classes* | |
| Class 3 | Lands that have severe limitation for sustained use under irrigation. |
| Class 4 | Lands that are marginal for sustained use under irrigation because of very severe limitations. |
| *Sub-classes* | |
| S | Soil limitation |
| t | Topographic limitation |
| d | Drainage limitation |

TABLE 7.9: Interpretation of reconnaissance soil map for landuse planning

| Mapping Symbol | Physiographic Units | Slope Class | Soil | Present land | Suggested landuse | Remarks |
|---|---|---|---|---|---|---|
| 1. | Residual hummocky ridges pediments | B-D | Shallow to moderately deep coarse loamy soils on residual hummocks and ridges and gently sloping pediments respectively | Mostly gullied, rocky and gravelly wasteland under sporadic cover of xerophytes and sparse coarse grasses | Graded bunding along contours, plantation of suitable xerophytic species and agave along the bunds with the plantation of suitable grasses in the contour | |
| 2. | Pediment-pediment plain | A-B | Moderately deep to deep coarse to fine loamy soils on gently sloping pediment to nearby level pediment plain | Mainly cultivated under rainfed condition with patches of gullied land uncultivated | Gully-plugging, contour-farming with strip-cropping, crop-rotation practices should be introduced to maintain the soil fertility | |
| 3. | Deltaic plain | A-B | Deep to very deep coarse loamy to fine loamy soils on very gently sloping to nearly level deltaic plain | Good agricultural land suitable for various crops like, jowar, bajra, groundnut, cotton under rain-fed conditions | Strip-cropping, green manuring and scheduled crop rotation of the region under rain-fed conditions | |
| 4. | Coastal land (tidal inlet) | A | Moderately deep fine loamy soils occurring on nearly level tidal inlet situation | Salt wasteland with mangroves and halophytic species and grasses | Construction of levees across the inlets to check high tidal flow inside the land | |

## Conclusion

Landuse survey and planning is a very complex subject which requires multi-disciplinary study and integrated approach. The ultimate aim of landuse survey is not merely to show the distribution of landuses on the map but the interpretation of landuse pattern, study scientifically the important factors which have imposed a particular pattern in the arear, its impact on the economy of the region in the event of any change in the landuse patterns and the consequent effects on environment. The landuse survey should bring out information regarding the multiple pattern of land utilisation and utility of land to its maximum without degrading the environmental characteristics. All the data should be collected from various natural resource surveys and assessed for a sound and sustained landuse planning. For this the geographers should take the help of various organisations engaged in the survey of natural resources and adopt the new techniques which are developing fast.

### REFERENCES

FAO, 1978 : "A framework of land evaluation", *Soils Bulletin*, 32.

IARI, 1970 : *Soil Survey Manual.*

National Bureau of Soil Survey and Land Use Planning, 1975 : *Detailed Soils Survey Report of M.P.K.V. Research Station, Rahuri*, Ahmednagar, *No. 365.*

Riquier, J.D., Bramao and Cornet, J.P., 1970 : *A New System of Soil Appraisal in Terms of Actual and Potential Productivity*, FAO, United Nations, Rome.

Sahi, B., 1978 : "Bird's eyeview of Earth Resources", *Science To-day*, Vol. 13, No. 5, Bombay-1.

Steele, J.G., 1967 : "Soil survey interpretation and its use", *Soils Bulletin*, 8, FAO, United Nations.

Storie, R.E., 1933 : "An Index for Rating the Agriculture, Values of Soils", Calif Agri. Expt., *Soils Bulletin*, 566.

U.S. Deptt. of Agriculture, 1961 : "Land Capability Classification", *Agricultural Hand Book No. 210*, Soils, Cons. Service, U.S. Deptt. of Agriculture.

U.S. Deptt. of Agriculture, 1966 : "Aerial Photo Interpretation in Classification and Mapping Soils", *Agricultural Hand Book*, 294.

8

# LANDFORMS AND LANDUSE PLANNING IN WESTERN RAJASTHAN

SURENDRA SINGH AND P. C. VATS

GEOMORPHOLOGICAL STUDIES conducted in different parts of the western Rajasthan have revealed that identification of different landforms, their genetic classification, mapping and assessment is an essential pre-requisite for rational landuse planning and development of any region (Ghose and Singh, 1965; Sen and Singh, 1977; Singh and Ghose, 1982; Vats and Singh, 1983, and Singh 1985, 1987). In the present paper the morphological and landsat spectral characteristics of landforms and their impact on landuse planning in Barmer district of western Rajasthan has been discussed. In Barmer district like other parts of the western Rajasthan, two types of landforms viz., fluvial and aeolian have been identified and mapped by using aerial and space-remote sensing techniques and ground truth. These landforms had resulted due to climato-morphogenetic processes operated on various lithological formations under different climatic phases during late Quaternary period. Fluvial landforms were formed by fluvial processes during wet phase, whereas the aeolian landforms had been created by aeolian processes during prolonged dry phase. The variations in the genesis of the landforms cause the variations in their morphological properties and production potentials and limitations.

## Environment

Barmer district with an area of 28,387 km$^2$ is located between latitudes 24°58′ to 26°32′ N and longitudes 70°5′ to 72°52′ E. The climate of the district is characterised by extreme temperatures, high wind velocity and evapotranspiration and low and erratic rainfall. Average annual maximum and minimum temperatures are 34°C and 22°C, respectively. May is the hottest month when average maximum temperature goes up to 41.9°C and January is the coldest month and average minimum temperature comes down to 10.2°C. The average annual rainfall is 263.7 mm and it decreases towards west. More than 90 per cent rainfall occurs during the months of July to September. The average annual wind speed is 9.1 km/hr with a maximum of 13.5 km/hr in June and minimum of 5.3 km/hr in November. The Luni river and its major and minor tributaries form the drainage system of the district. Due to arid climate and inadequate rainfall, the drainage channels, by and large, are disorganised and buried under sand and alluvium.

## Results and Discussion

### *Landforms and Associated Characteristics*

Based on the morphological, slope, drainage and Landsat spectral characteristics, the district has been divided into ten landform units (Fig. 8.1) and their extent is given in Table 8.1. The paper print of Landsat band 5 image (Fig. 8.2) shows the spatial and spectral characteristics of different landforms.

1. *Hills*: This unit, comprising of rhyolite, granite and sandstone formations, is characterised by steep slope, well developed joints and fractures, domes and inselbergs. The height of these hills from ground level varies from 50 to 202 m. Due to constant weathering and erosion, the debris and detritus has been accumulated on the slopes of these hills in the form of 'talus'. The effect of diurnal change of temperature is distinctly visible in the form of 'spliting' and 'peeling' of rock masses. The hills and their circular and linear features are clearly visible in dark grey to dark tone on Landsat band 5 image (Fig. 8.2).

2. *Rocky/Gravelly Pediments*: These pediments have developed at the base of the hills of different formations due to the retreat of slope and corrosion. The surfaces of this unit are rocky and gravelly dotted with numerous rock outcrops. Along the joints and fractures, gravelly sediments have been encountered. The drainage patterns developed along the joints are parallel and dendritic. The pediments associated with sandstone formations have almost level slope varying from 0° to 1° and the slopes of granite and rhyolite pediments are variable ranging from 0° to 1° and 1° to 5°. This unit due to sharp break of slope between hills and pediments is distinctly visible in white grey to dark grey tone on Landsat band 5 image (Fig. 8.2).

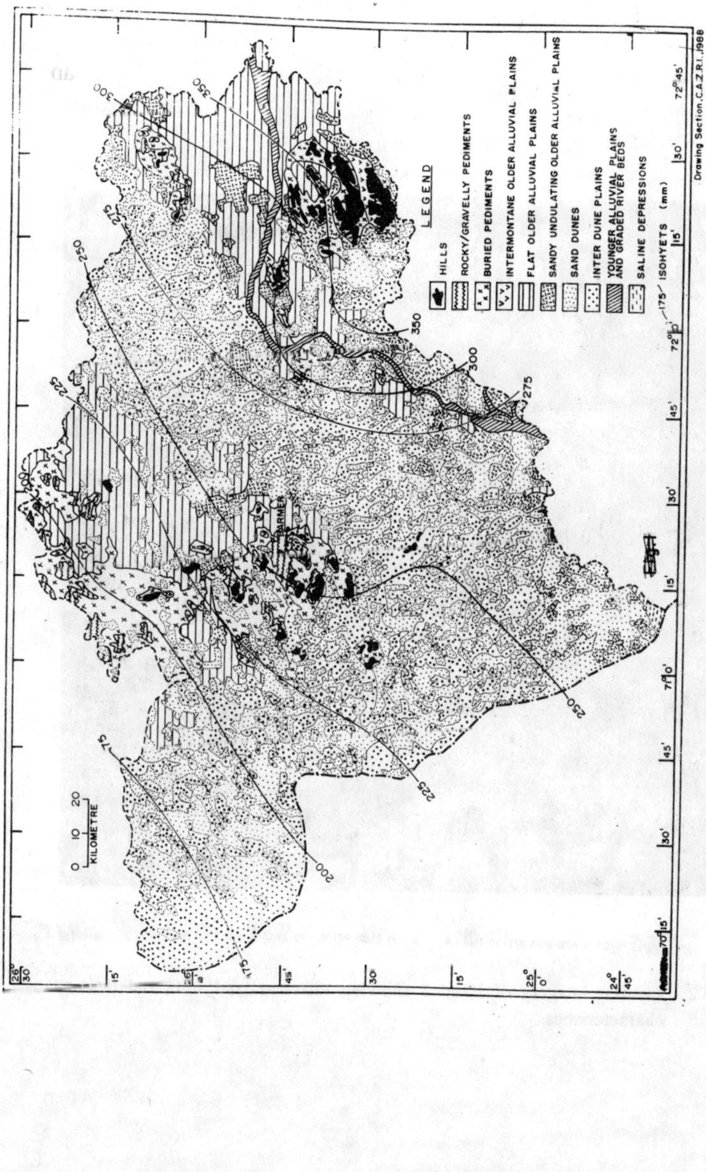


FIG. 8.1: Barmer district landforms.

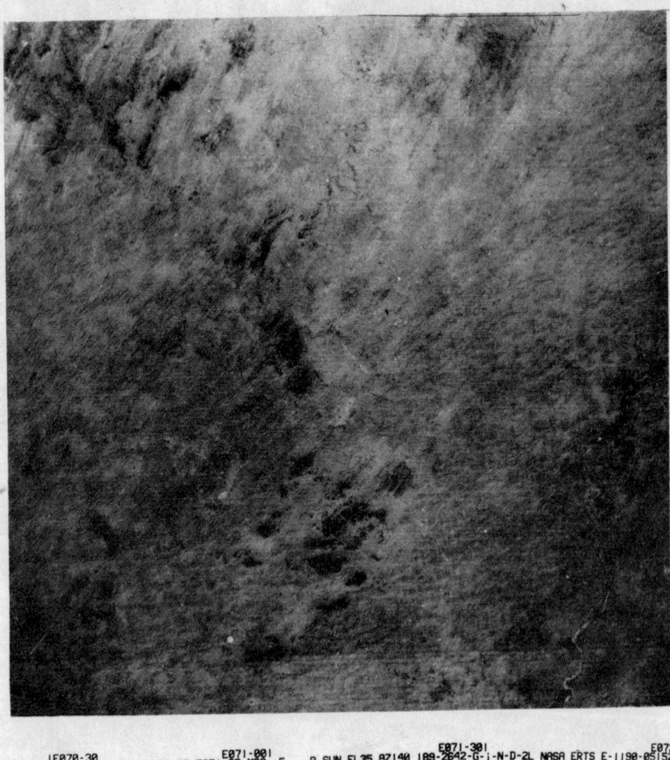


FIG. 8.2: Landsat imagery of Band 5 showing different landforms and their spatial characteristics.

3. *Buried Pediments*: The rocky surfaces covered with alluvial, colluvial and aeolian sediments due to fluvial and aeolian processes have been termed as, 'buried pediments'. Based on the nature and thickness of sediments, slope and drainage characteristics, these pediments have been further divided into flat and sandy undulating buried pediments.

Flat buried pediments are characterised by almost level slope 0° to 1°, shallow to moderately deep sediments underlain by rocky strata, absence of *Kankar* pan and presence of short drainage channels. It is free from hazards except slight to moderate sheet to rill erosion in pockets. The size of sediment particles varies from 0.04 to 2 mm with an average diameter of 0.58 mm. This unit due to sharp break of slope and dark grey to dark tone could be easily identified and separated from the adjoining rocky/gravelly pediments on Landsat band 5 image (Fig. 8.2).

Sandy undulating buried pediments due to the accumulation of aeolian sands against the hill slopes are characterised by undulations in the form of dunes and hummocks, deep to very deep light textured sediments, irregular slopes varying from 1° to 4° and wind erosion/deposition and water erosion hazards. The size of sediment particles varies from 0.04 to 0.50 mm with an average size of 0.14 mm. This unit due to higher reflectance from aeolian deposits appears in white to grey tone on Landsat band 5 image during dry season and could be easily identified and separated from rocky/gravelly and buried pediments (Fig. 8.2).

4. *Intermontane Alluvial Plains*: These plains occur between two rhyolitic hill ranges and extend from Kundal in one side to Ramania in the other side. Due to physiographic variations, these plains have been classified into flat and sandy undulating intermontane alluvial plains.

Flat intermontane alluvial plains constitute level slope of 0° to 1°, alluvial sediments underlain by *kankar* pan at 100 to 180 cm depth and parallel to sub-parallel drainage patterns. Slight to moderate sheet to rill erosion occurs in the pockets. The size of sediment particles ranges between 0.04 to 2 mm. In eastern parts of this unit, the aeolian activities have created the dunes and hummocks on the initial alluvial surfaces resulting into the formation of sandy undulating intermontane alluvial plains. The undulations, irregular slope, deep to very deep sandy deposits and severe to very severe wind erosion/deposition are the salient characteristics of these plains. The size of sediment particles varies from 0.04 to 0.50 mm with an average size of 0.14 mm.

The flat and sandy undulating intermontane alluvial plains could be easily separated from each other on Landsat band 5 image due to significant variations in their reflectance and tonal characteristics (Fig. 8.2).

5. *Flat Older Alluvial Plains*: These plains created by fluvial action of well integrated prior drainage system largely occur in the eastern and southeastern parts of the district. Based on the thickness of sediments and morphological and Landsat spectral characteristics, this unit has been further divided into four sub-units viz., shallow, moderately deep, deep and very deep. These four sub-units due to variations in nature of sediments and moisture show different spectral reflectance and appear in white to grey and dark grey to dark tone on Landsat band 5 image (Fig. 8.2).

Shallow flat older alluvial plains are characterised by almost level slope of 0° to 1°, coarse to medium textured sediments of 10 to 20 cm depth underlain by hard *kankar* pan and widely spaced drainage channels. At places, these plains have been degraded by sheet and rill erosion and *Kankar* pan is exposed. The diameter of sediment particles ranges between 0.04 to 1 mm and their average size is 0.43 mm.

Moderately deep flat older alluvial plains comprise light to medium textured sediments of 20 to 60 cm depth. The sediments are underlain by *kankar* pan and fullers earth and their texture varies from sand to loamy sand and sandy loam to loam. The slope is level ranging from 0° to 1° but in certain pockets due to accumulation of aeolian sand, the slope varies from 1° to 2°. The size of sediment particles is varying from 0.04 to 2 mm and 0.04 to 1 mm with an average diameter of 0.58 mm and 0.43 mm, respectively.

Deep flat older alluvial plains comprise loamy sand to sandy loam and loam to sandy clay loam sediments. Some parts of these plains have become saline due to impeded drainage conditions and floods of 1979. The salient characteristics of this unit are almost level slope of 0° to 1°, short drainage channels and saline and non-saline sediments, underlain *kankar* pan and weathered rock. The diameter of sediment particles ranges from 0.04 to 1 mm with an average size of 0.43 mm. Sheet to rill erosion occurs in certain pockets of such plains.

Very deep flat older alluvial plains are more extensive than other older alluvial plains. Light to medium and medium to fine textured sediments underlain by *kankar* pan and fuller's earth at 100 to 180 cm depth, almost level slope of 0° to 1°, sand sheeting and sheet erosion in pockets and widely spaced drainage channels are the major characteristics of these plains. The diameter of sediment particles ranges from 0.04 to 2 mm with an average size to 0.58 mm.

TABLE 8.1: Extent of different landform units and sub-units in Barmer district

| S.No. Landforms | Total area (sq. km.) | | Percentage | |
|---|---|---|---|---|
| 1. Hills | 367.85 | | 1.30 | |
| 2. Rocky/gravelly pediments | 607.89 | | 2.14 | |
| 3. Buried pediments | 2330.34 | | 8.21 | |
| (a) Flat buried pediments | — | 1510.50 | — | 5.32 |
| (b) Sandy undulating buried pediments | — | 819.84 | — | 2.89 |
| 4. Intermontance older alluvial plains | 65.05 | — | 0.23 | — |
| (a) Flat intermontane older alluvial plains | — | 8.55 | — | 0.03 |
| (b) Sandy intermontane older alluvial plains | — | 56.50 | — | 0.20 |
| 5. Flat older alluvial plains | 4924.18 | — | 17.34 | |
| (a) Shallow flat older alluvial plains | — | 302.88 | — | 1.07 |
| (b) Moderately deep flat older alluvial plains | — | 701.46 | — | 2.47 |
| (c) Deep flat older alluvial plains | — | 1585.31 | — | 5.58 |
| (d) Very deep flat older alluvial plains | — | 2334.53 | — | 8.22 |
| 6. Sandy undulating older alluvial plains | 1572.42 | — | 5.54 | — |
| 7. Sand dunes | 10336.91 | — | 36.42 | |
| 8. Interdune plains | 7652.19 | — | 26.95 | |
| (a) Flat interdune plains | — | 2859.60 | — | 10.07 |
| (b) Sandy undulating interdune plains | — | 4792.59 | — | 16.88 |
| 9. Younger alluvial plains and graded river beds | 388.15 | — | 1.37 | — |
| 10. Saline depressions | 142.02 | — | 0.50 | — |
| | 28,387.00 | | 100.00 | |

6. *Sandy Undulating Older Alluvial Plains*: Light to medium textured flat older alluvial plains degraded by aeolian action have resulted into the formation of sandy undulating older alluvial plains. The

undulations in the form of dunes and hummocks, deep to very deep aeolian and alluvial sediments, weakly developed concretionary layer, presence of hard *kankar* pan at deeper depth and irregular slope varying from 1° to 3° are the major characteristics of these plains. The size of sediment particles of undulating areas of such plains ranges between 0.04 to 0.50 mm and they appear in white to grey tone on Landsat band 5 image. The diameter of sediment particles of flat areas varies from 0.04 to 1 mm and they exhibit dark grey to dark tone on above Landsat image (Fig. 8.2).

7. *Sand Dunes*: This unit constitutes six types of sand dunes viz. obstacle, parabolic, longitudinal, transverse, barchan and shrub-coppice. The first four types of dunes are stabilised and they are 5 to 75 m in height, 200 to 6000 m in length and 80 to 500 m in width. The orientation of these dunes except transverse corresponds to the direction of the prevailing southwest winds from N 42°E to N 48°E, whereas the orientation of transverse dune is across to the prevailing wind direction from N 48°W to N 52°W. The density of these dunes is 0.3 to 2 D/km$^2$ and their intensity varies from 12 to 75 per cent. The size of sand grains within the dunes varies from 0.04 to 0.25 mm with an average size of 0.12 mm.

The stabilised dunes due to lower reflectance from the compact sand and vegetation appear in dark grey to dark tone on Landsat band 5 image, while the reactivated parts of these dunes due to higher reflectance from loose sand exhibit white to grey tone on above Landsat image (Fig. 8.2).

The unstabilised barchan and shrub-coppice dunes are 0.9 to 10 m in height, 40 to 100 m in width and 75 to 200 m in length. The barchan dunes with steep leeward side and gentle windward side are devoid of vegetation and the sands are loose through out the profile. (Fig. 8.2). Shrub-coppice dunes formed against shrubs and bushes do not have any definite orientation and occur in the sandy undulating older alluvial and interdune plains. The size of sand grains varied from 0.04 to 0.25 mm with an average size of 0.12 mm.

8. *Interdune Plains*: This unit is located among stabilised sand dunes and based on the nature and thickness of sediment, slope and drainage characteristics has been further divided into flat and sandy undulating interdune plains. Flat interdune plains are characterised by almost level slope of 0° to 1°, non-saline, saline and gravelly surfaces, loamy sand

to sandy and loam sediments underlain by *kankar* pan and gypsiferrous deposits and widely spaced short drainage channels. The size of sediment particles in these plains ranges from 0.04 to 2 mm with an average size of 0.58 mm. Sheet and rill erosion occurs in pockets.

Sandy undulating interdune plains due to deep to very deep aeolian sands and undulations in the form of sand sheets of 50 to 2000 cm thickness, dunes of 1 to 3 m height and hummocks of 0.5 to 1 m height. The slopes of these plains are irregular varying from 1° to 3°. In the flat areas of this unit, hard *kankar* pan has developed below the sediments while in undulating areas weakly developed concretionary layers occur at deeper depth. The size of sediment particles varies from 0.04 to 1 mm and 0.04 to 0.50 mm in flat and undulating areas, respectively. These areas due to variation in spectral reflectance appear in dark grey to dark and white to grey tone on Landsat band 5 image (Fig. 8.2).

9. *Younger Alluvial Plains and Graded River Beds*: Both saline and non-saline younger alluvial plains lie in narrow strips along the Luni river and its tributaries. The slope of this unit is almost level with 0° to 1° slope except those areas where riverine sediments in the form of ridges and hummocks have been deposited during the flood of 1979. The slope of these areas is irregular varying from 1° to 3°. Deep to very deep coarse textured sediments are associated with these plains and their width varies from 300 to 5000 m. At few places, salinity has developed due to impeded drainage conditions and sediments have been dissected into rills and gullies by flood waters. The size of sediments varies from 0.04 to 1 mm with an average size of 0.43 mm.

River beds of the Luni and its tributaries are graded with saline, non-saline and stony surfaces. The sand bars of different dimensions comprising of coarse textured sediments are present in the river beds. The banks of the river have been dissected by fluvial action and the amount of dissection varies from 4 to 7 m. The size of sediment particles ranges between 0.04 to 2 mm with an average size of 0.58 mm.

The younger alluvial plains appear in dark grey to dark tone on Landsat band 5 image due to good moisture regime and stand of crops and vegetation, while the graded river beds exhibit white to grey tone on the above image due to coarse sediments and higher spectral reflectance (Fig. 8.2).

10. *Saline Depressions*: The saline depressions of different shape and size largely occur along the confluences of the drainage channels. The wet and dry flat surfaces of these depressions constitute medium to fine textured sediments. The marginal areas have been degraded by aeolian action resulting into the formation of sand sheets and sand dunes. The size of sediment particles varies from 0.04 to 0.050 mm and 0.04 to 2 mm, respectively.

Due to wet and dry surfaces, these depressions appear in dark grey to dark and white to grey tone on Landsat band 5 image depending upon the variations in the spectral reflectance (Fig. 8.2).

**Assessment and Landuse Planning of Landforms**

Major landform units and sub-units identified in the district have different production potentials and limitations and could be developed into croplands, woodlands, pastures, silvipastures and desert catchments depending upon prevailing climatic conditions and landuse capability.

Fluvial landforms such as flat buried pediments, flat intermontane alluvial plains and younger alluvial plains, have good ground water and agricultural potentials. Geomorphic factors responsible for the development of these potentials are deep to very deep permeable sediments, sub-surface joints and fractures, surface drainage, almost level slope, fertile sediments and absence of hazards like water erosion, wind erosion/deposition and salinity. These landforms are used for growing irrigated and rainfed crops and also for 'Khadin' cultivation.

The yield of the irrigated crops like wheat, chillies, cumin and vegetables could be increased by using high yielding varieties, proper method of cultivation and suitable measures of soil conservation like field bunding, contour ploughing and trenching. Large acreage of these units could be brought under irrigated farming by digging more tubewells in the under-exploited and un-exploited ground water zones. Flat buried pediments which provide ideal geomorphic sites for run off farming (Khadin) could be properly utilised to upgrade the production. There is sufficient scope in this unit particularly in Shiva tehsil to construct new 'Khadins' by scientific methods and to increase the yield of crops like wheat and gram etc. by adopting proper soil conservation measures.

Some areas of these units particularly of the younger alluvial plains are degraded by wind erosion/deposition, salinity and flood hazards and are not fit for arable farming. Such areas could be utilised for the development of pastures and silvipastures by raising psammophytic and halophytic plant species like *Lasiurus sindicus, Cenchrus ciliaris, Zizyphus nummularia, Calligonum polygonoides, Sporobolus marginatus* and *Cressa cretica.*

Flat older alluvial and interdune plains located in the cast of 250 mm isohyet and having shallow, moderately deep, deep and very deep fertile alluvial sediments, almost level slope of 0° to 1°, less hazards and better rainfall form the backbone of the district in agricultural production. By and large, these plains are used for raising rainfed crops, and irrigated crops are only grown in the pockets where ground water is available. In certain parts of this unit, the salinity has developed due to impeded drainage conditions and the agricultural productivity has been depleted.

In the shallow older alluvial plains, large acreage of land, presently being used for cultivation, is fit for pasture development due to the limitation of soil depth, sheet erosion and exposed *Kankar*. The blocks of 1.0 × 1.0 m may be reseeded with improved grasses like *Cenchrus ciliaris* (358) and *Cenchrus setigerus* (175) with seed rate of 6 to 8 kg/ha and 5 to 6 kg/ha, respectively. The areas with sufficient soil cover and better moisture and fertility status could be put under rainfed and irrigated crops.

Moderately deep to deep and very deep flat older alluvial and interdune plains are largely suited for growing rainfed and irrigated crops like *bajra, mung, moth,* wheat and cumin etc. The yield of rainfed crops could be upgraded by using improved crop varieties like bajra (BJ 104), mung (S8, M8) and guar (FS 227), and dry farming techniques like field bunding to minimise runoff losses, and sowing in uniform row system with seed-cum-fertiliser drill (Singh, 1976). The large acreage of land could be brought under irrigated farming by developing the ground water potentials along the buried courses of the prior drainage channels. The high yielding varieties of irrigated crops should be used to boost up the the production. Such landforms lying in the west of 250 mm isohyet should be developed into pastures by using suitable grass species and cultivation should be avoided.

Aeolian landforms like sandy undulating older and interdune plains, sandy undulating buried pediments and sand dunes due to deep to very deep light textured sediments with poor fertility status, undulations and severe to very severe wind erosion/deposition and water erosion hazards have very limited agricultural potentials. The cultivation on higher parts of sandy alluvial plains and sand dunes has resulted into the sand movement and depletion of the biological productivity of these landforms.

Flat areas of aeolian landforms in the east of 250 mm isohyet and having level slope, less hazards and better fertility status of sediments and rainfall condition should only be put under rainfed crops by using proper soil conservation measures like wind breaks, strip cropping contour bunding, trenching and suitable dry farming techniques. The undulating areas and stabilised sand dunes due to high relief and vulnerability to wind action should not be put under cultivation. Flat fluvial and aeolian landforms located in the west of 250 mm isohyet should also not be used for cultivation due to poor rainfall and fertility of sediments and wind erosion/deposition hazard. Such landforms should be developed into silvipastures by using indigenous plant species like *Acacia senegal, Calligonum polygonoides, Zizyphus nummularia, Prosopis cineraria, Lasiurus sindicus* and *Cenchrus ciliaris* (Saxena and Singh, 1976). The shifting sands and dunes should be stabilised by using phyto-reclamation technique to check the depletion of the biological productivity of the adjoining land and to develop these landforms into silvipastures.

Hills and rocky/gravelly pediments due to severe to very severe water erosion in the form of sheet, rills and gullies, bare rocky surfaces and moderately steep to steep slope are not fit for cultivation. These landform units could be developed into desert catchments by using proper soil and water conservation measures and the surface water could be harvested at suitable geomorphic sites after detailed contour survey.

The Saline depressions at present are being used for extracting salts as they are not fit for cultivation. This unit after extracting salts should be levelled and developed into pastures by using halophytic plant species such as *Sprobolus marginatus, Cressa cretica* and *Salsola bryosma* and proper soil conservation measures.

## Conclusions

Fluvial and aeolian landforms identified in Barmer district have been created by various processes under wet and dry phases and have different morphological and Landsat spectral characteristics. These landforms have a direct bearing on the landuse patterns and planning due to variations in physical potentials and limitations. Fluvial landforms located in the east of 250 mm isohyet and characterised by almost level slope, fertile sediments, limited hazards, good ground water potentials and better rainfall conditions are fit for rainfed and irrigated crops. Aeolian landforms due to undulating topography, poor fertility of aeolian sediments and severe to very severe wind erosion/deposition hazards and fluvial landforms located in the west of 250 mm isohyet because of degradation and poor rainfall and moisture regime could be developed into croplands, pastures, silvipastures and hortipastures. Based on the suggested landuse planning, animal husbandry and agro-industries will be the main occupation of the people residing in the western part of the district, while in the eastern part, crop husbandry, animal husbandry and agro-industries will become the main source of livelihood for the people. The changing landuse pattern based on proper land management will play a significant role in the economic development of the district.

## REFERENCES

Ghose, B. and Singh, S., 1965 : Observations on geomorphology and landuse of Kitnod village (Aerial Photo-Interpretation and Analysis of Land). *Geographical Review of India*, XXVII (3), pp. 146-156.

Sen, A. K. and Singh, S., 1977 : Significance of geomorphic factors on landuse planning and development in Bikaner district, *Annals of Arid Zone* 16(1), pp. 13-24.

Saxena, S. K. and Singh, S., 1976 : Some observations on the sand-dunes and vegetation of Bikaner district in western Rajasthan, *Annals of Arid Zone* 15(4), pp. 315-322.

Singh, R. P. (Ed.), 1976 : Improved Dry Farming Agriculture for Western Rajasthan. *CAZRI Technical Bulletin No. 1.*

Singh, S. and Ghose, B., 1982 : Role of photogeomorphology in the reclamation and development of Rajasthan desert, Geological Survey of India, *Miscellaneous Publication*, No. 49, pp. 1-10.

Singh, S., 1985 : Geomorphology in the management of land and water resources in an arid environment — A case study of Jodhpur district, in Mohammad Shafi and Mehdi Raza (Ed.), *Dry Land Agriculture in India*, Rawat Publication Jaipur, pp. 173-190.

———, 1987 : Impact of geomorphology on landuse planning in the Rajasthan desert, *Annals of Rajasthan Geographers*.

Vats, P. C. and Singh, S., 1980 : Impact of geomorphology on land utilisation in Western Rajasthan — A case study of Nagaur district, *The Deccan Geographer*, 21(3), pp. 441-448.

# 9

# PRESENT AND SUGGESTED LANDUSE OF RAJASTHAN DESERT

AMAL KUMAR SEN

INFORMATION ON the district-wise areas of arid zone in Rajasthan has already been presented by Krishnan (1968). The boundary between arid and semi-arid zones in Rajasthan intersects, Jalore and Pali districts, goes roughly along the boundaries of Ajmer and Nagaur and then intersects Sikar and Jhunjhunu districts. This line passes through the towns of Bhinmal, Sojat, Beawar, Mali, Sri Madhopur and then between Chirawa and Khetri in Jhunjhunu district. The districts in Rajasthan lying wholly in arid zone are Ganganagar, Bikaner, Jaisalmer, Barmer, Jodhpur and Churu. The districts lying partly in arid zone and partly in semi-arid zone are Nagaur, Jalore, Jhunjhunu, Sikar, Pali and Ajmer with 96, 88, 69, 65, 48 and 9 per cent are falling under arid zone, respectively.

Assessment of use and misuse of land is the pre-requisite to plan the utilisation of resources. Land utilisation survey and mapping is the obvious requirement to make such estimates. The ultimate purpose of land utilisation survey and mapping is to arrive at systems of landuse and management best suitable to the kinds or resources and capabilities of the land composing it.

With the background of these principles, the specific problems of the arid zone were considered to standardise the mapping units for landuse survey. The problems of arid zone are many and were well discussed and summarised in the proceedings of the symposium on Problems of Indian Arid Zone (1972).

Considering these problems and on the basis of experiences drawn while conducting land utilisation mapping in the arid zone, the scheme for classification of landuse units for landuse maps as worked out (Table 9.1) has subsequently been modified (Table 9.2).

TABLE 9.1: Tentative scheme of land utilisation proposed for arid zone

1. *Settlement*
   (i) Rural settlement
      a. Villages with compact settlement (unplanned)
      b. Villages with scattered settlement (unplanned)
      c. Linear compact villages
      d. Temporary villages
      e. Deserted villages

   (ii) Urban areas
      a. Administrative
      b. Residential
      c. Mining
      d. Agricultural
      e. Religious centre

2. *Water Resources*
   a. Rivers
   b. Playas and depressions — salt lakes
   c. Wells
      i. Deep, ii. Moderately deep, iii. Shallow
   d. Tanks
      i. Permanent, ii. Seasonal

3. *Waste lands* (Also come under pasture land)
   a. Sand waste (including shifting sand dunes)
   b. Rocky waste
   c. Saline waste
   d. Gravelly waste

4. *Pasture land and grazing ground*
   a. Permanent pasture (Oran)
   b. Waste lands and current fallows
      (Suitable representation is to be made so that both are indicated)
   c. Hay, Thorn, Weeds and Shrubs — where the grasses are periodically out.

5. *Forest land*
   a. Natural forest
      i. Thin, ii. Dense
   b. Afforested land
   c. Plantation (roadside, railway side, etc.).

6. *Cultivated land*
   a. Single cropped area
   b. Double cropped area
   c. Current fallow (also fall under pasture land and grazing ground).

Cultivated lands, as maped, are not necessarily divided. In order to avoid annual survey and mapping, it has been proposed to map these three categories into one unit. The range of cultivation intensity as tentatively standardised for mapping is presented in Table 9.2 (Sen, A.K. 1978).

TABLE 9.2: **Intensity of cultivation**

| Mapping symbol | Percentage cultivable | Remarks |
|---|---|---|
| A | 100 | Very good intensity, double cropped land. |
| $A_1$ | 60 to 80 | Good intensity, mostly permanent cultivated lands, also include short fallow. |
| $A_2$ | 40 to 60 | Cultivated and current fallows, fair intensity. |
| $A_3$ | 30 to 50 | Cultivated and current fallow, moderate intensity. |
| $A_4$ | 20 to 30 | Moderate to poor intensity, cultivated, current and long fallows, marginal lands. |
| $A_5$ | 10 to 20 | Poor intensity, long fallows in marginal land. |
| $A_6$ | Below 10 | Very poor intensity, long fallows. |

## Major landuses

### *Landuse Classification through Secondary Data*

The landuse in different areas of the hot arid zone is being initially studied on the basis of secondary data available with the Directorate of Economics and Statistics of Government of Rajasthan. On the basis of such data, choropleth maps are often prepared and areas under different categories of landuse and sometimes their changes are shown in graphical forms. The extent and pattern of landuse classes cannot be depicted in these maps, but these maps are often helpful to visualise the different types of present landuse, their comparative analysis and thus provide good indications to formulate the mapping units to be drawn for landuse surveys and their consequent mapping.

The present landuse and its relation to agricultural aspects including crop yields and cropping pattern can be best studied or geographically analysed by classifying the area into agro-ecologically regions. The arid zone of Rajasthan has earlier been classified into two macro and four micro agro-ecological regions. This is shown in Table 9.3. The landuse characteristics of these regions are shown in Tables 9.4 and 9.5.

TABLE 9.3: Agro-ecological regions

| Macro region | Micro region | | Districts included | Area in sq.km. |
|---|---|---|---|---|
| Arid | (a) | Kharif monocropping | Barmer, Jodhpur, Churu Jaisalmer, Bikaner | 133,722 (64.20%) |
| | (b) | Irrigated kharif and rabi cropping | Ganganagar | 20,674 (20%) |
| Transitional belt between arid & semi-arid | (a) | Kharif monocropping | Jhunjhunu, Sikar Nagaur | 31,244 (15.00%) |
| | (b) | Irrigated kharif and rabi cropping. | Pali and Jalore | 82,780 (10.80%) |

TABLE 9.4: Landuse in agro-ecological regions of Rajasthan desert (% to total geographical area)

| Agro-ecological regions | Forest | Net irri-gated | Gross irri-gated | Tree orchards | Fallow land | Culti-vable waste | Net for culti-vation |
|---|---|---|---|---|---|---|---|
| Arid-kharif Monocropping | 0.26 | 0.27 | 26.32 | 0.003 | 21.21 | 33.76 | 18.24 |
| Arid-irrigated kharif and rabi cropping | 0.13 | 24.61 | 27.90 | 0.13 | 20.94 | 21.67 | 4.40 |
| Transitional belt between arid and semi-arid | | | | | | | |
| Kharif monocropping | 91.09 | 2.72 | 65.22 | 0.02 | 15.55 | 6.66 | 8.72 |
| Irrigated kharif and rabi cropping | 2.85 | 6.84 | 39.89 | 0.05 | 28.05 | 7.58 | 14.10 |

TABLE 9.5: Cropping pattern in agro-ecological regions of Rajasthan desert

| Agro-ecological regions | Crop | Area | Production in m. tonnes | Yield in kg/ha | % of area to total cropped area of region crops | % of production to total production of region (different crops) |
|---|---|---|---|---|---|---|
| Arid-kharif monocropping | Wheat | 374.05 | 44264 | 1183 | 1.54 | 9.09 |
| | Barley | 2710 | 2296 | 847 | 0.11 | 0.47 |
| | Gram | 56466 | 19215 | 340 | 2.34 | 3.94 |
| | Bajra | 2306677 | 415715 | 180 | 95.13 | 85.47 |
| | Jowar | 21293 | 4947 | 232 | 0.87 | 1.01 |
| | Maize | 173 | 119 | 687 | 0.01 | 0.02 |
| Irrigated kharif rabi cropping | Wheat | 175592 | 192094 | 1091 | 21.72 | 36.58 |
| | Barley | 32533 | 28578 | 878 | 4.01 | 5.44 |
| | Gram | 441385 | 25554 | 578 | 54.50 | 48.68 |
| | Bajra | 153126 | 45996 | 300 | 18.91 | 8.76 |
| | Jowar | 5573 | 1724 | 309 | 0.68 | 0.32 |
| | Maize | 1168 | 1070 | 916 | 0.14 | 0.20 |
| Transitional belt between semi-arid kharif mono-cropping | Wheat | 42002 | 43784 | 1042 | 3.22 | 10.34 |
| | Barley | 44217 | 46682 | 1055 | 3.38 | 11.00 |
| | Bajra | 1053998 | 293760 | 278 | 80.63 | 2.90 |
| | Jowar | 78587 | 12386 | 156 | 6.02 | 2.90 |
| | Maize | 1125 | 751 | 667 | 0.08 | 0.17 |
| Irrigated kharif rabi cropping | Wheat | 76734 | 103093 | 1343 | 9.91 | 28.57 |
| | Barley | 259127 | 25127 | 969 | 3.34 | 9.96 |
| | Gram | 12679 | 6470 | 510 | 1.63 | 1.79 |
| | Bajra | 555520 | 194717 | 350 | 71.74 | 53.96 |
| | Jowar | 73770 | 15651 | 212 | 9.52 | 4.33 |
| | Maize | 29636 | 15770 | 532 | 3.82 | 4.37 |

## *Arid Region*

The arid region coincides with the area receiving mostly less than 350 mm of annual rainfall. The sandy and hot desert conditions prevail mostly throughout the region. Rainfall decreases towards the west and the region is mostly dominated by the cultivation of bajra crops which covers more than 90 per cent of the total cultivated lands. The agro-ecological conditions of the micro regions are described next. In this zone 47.78 per cent of the total exploitable potential is trapped bearing 52.22 per cent as surplus. About 78 per cent of the tapped water is utilised for irrigation and 22 per cent for livestock and domestic purposes.

*Kharif Monocropping*

The region includes 64.20 per cent of the total area of the western Rajasthan. The density of cultivated lands varies from 3.4 per cent to total area in Jaisalmer district to 63 per cent in Churu district (Table 9.6). The details are shown in the map. The landuse details as worked out from the secondary data (revenue data) are shown in Table 9.6.

TABLE 9.6: **Landuse and density of cultivated lands (% to total geographical area)**

| Districts | Density culti- vation | Intensity of crop- ping | Forest | landuse | | | | |
|---|---|---|---|---|---|---|---|---|
| | | | | Settle- ment non- agri. land | Net avail- able for culti- vation | Fallow land | Single crop- ped | Double cropped |
| Bikaner | 16.23 | 100.00 | 0.36 | 3.27 | 1.28 | 77.82 | 16.23 | — |
| Barmer | 43.45 | 100.00 | 0.31 | 2.41 | 12.24 | 40.94 | 43.35 | 0.10 |
| Churu | 63.36 | 100.10 | 0.05 | 5.51 | 1.17 | 28.86 | 63.30 | 0.06 |
| Jodhpur | 43.35 | 100.50 | 0.04 | 3.37 | 12.00 | 41.18 | 43.18 | 0.17 |
| Jaisalmer | 3.40 | 100.00 | 0.48 | 1.21 | 44.60 | 50.25 | 3.49 | 0.17 |

*Cropping Pattern and Cropping Yields*

Bajra, jowar, wheat, barley, maize and gram are the most important agricultural products of the region. The main product is, however, bajra. About 99.50, 99.00, 98.00, 96.50 and 93.50 per cent of the total cropped area are put under bajra in Bikaner, Churu, Barmer, Jaisalmer and Jodhpur districts respectively. The total cropped area under wheat and barley is very insignificant (only 0.20 per cent of the total cropped area in Jodhpur). The yield of the main crops produced in this sub-zone is shown in Table 9.7.

TABLE 9.7: **Kharif and monocropping region production, crop yields in kg/ha**

| Crop | Districts | | | | |
|---|---|---|---|---|---|
| | Jodhpur | Barmer | Jaisalmer | Bikaner | Churu |
| Wheat | 1225 | 1090 | 1491 | 976 | 964 |
| Bajra | 210 | 211 | 58 | 66 | 158 |
| Jowar | 153 | 206 | 383 | 1756 | 233 |
| Maize | 687 | 708 | — | — | 500 |
| Barley | 902 | 975 | 1500 | 639 | 586 |
| Gram | 524 | 526 | 405 | 516 | 931 |

### *Arid Irrigated Kharif and Rabi Cropping*

This includes the district of Ganganagar in the north. The entire north-west portion is served by Ganga canal system. Except a small part in the south the rest are being irrigated by the Rajasthan canal system. The area is, however, well within the desertic tract receiving rainfall less than 150 mm in the west to 350 mm in the east. Climatically, it is a true desert.

### *Cropping Pattern and Landuse*

About 76 per cent of the total area is cultivated, the intensity of cropping being 1077. Land utilisation of the district indicates (in per cent to total area) forest 0.14, settlements and non-agricultural lands 3.97, land not available for cultivation 0.27, fallow land 290.2 and cultivated land 76.36 (single cropped 54.46 and double cropped 21.80). Net irrigated and unirrigated lands comprise 24.61 and 27.90 per cent respectively of the total rural area. Only 0.58 per cent of the total area constitutes waste lands (Fig. 9.1).

### *Crops and Crop Yields*

Rice covers 198 sq. km. (irrigated), Jowar, 21 sq. km., bajra 1053 sq. km., wheat 2625 sq. km. (irrigated), barley 213 sq. km. (irrigated). The yield of the crops are rice — 2000 kg/ha, jowar — 500 kg/ha, barley — 900 kg/ha, bajra 1000 — kg/ha, wheat — 1400 kg/ha, sugarcane — 40000 kg/ha, gram — 1000 kg/ha, cotton American — 1675 sq. km, deshi — 1000 kg/ha, and jowar — 1000 kg/ha.

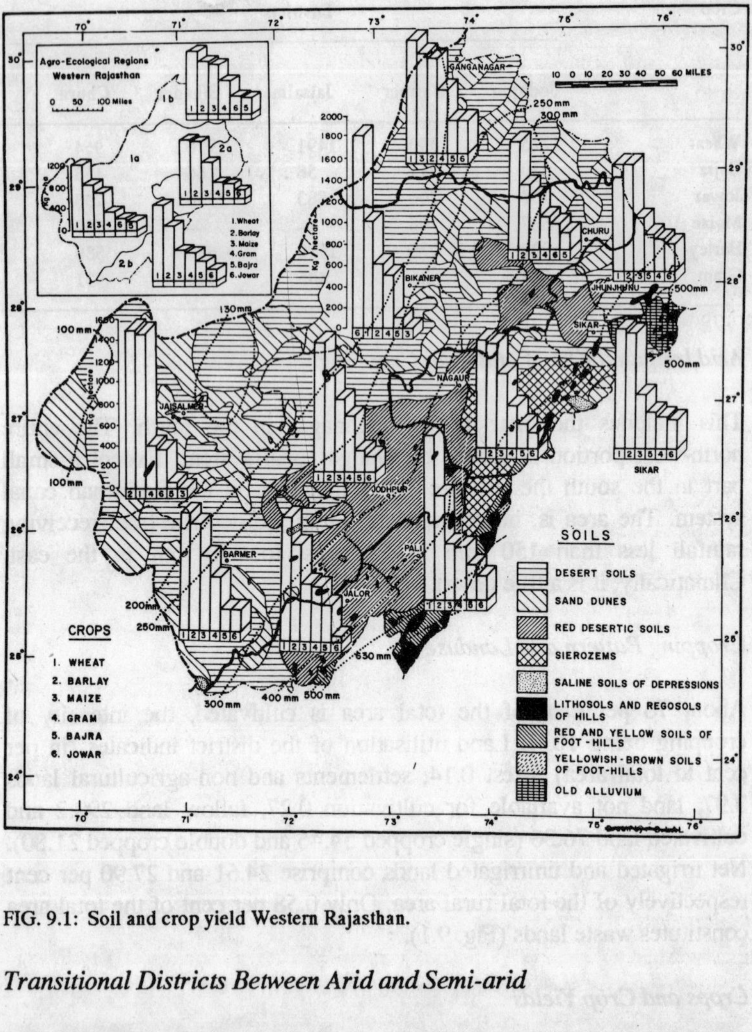


FIG. 9.1: Soil and crop yield Western Rajasthan.

## *Transitional Districts Between Arid and Semi-arid*

This agro-ecological zone lies between 350 mm to 500 mm rainfall belt and comprises about 25 per cent of the total area of western Rajasthan. This arid tract practically constitutes the desert (Raheja and Sen, 1964).

Eastern part belongs entirely to steppe rocky desert. This region has a good agricultural potentiality but the cultivation of the sand dunes and the marginal lands of the true arid lands has already become a menace of soil erosion and fresh sand deposition in this terrain. In a number of areas the condition has further deteriorated and requires special attention.

*Kharif Monocropping*

This sub-agro-ecological group includes Jhunjhunu, Sikar and Nagaur districts in the north, Cropping pattern and landuse. Crop yields are shown in Table 9.8. Density of cultivation, landuse and intensity of cropping are presented in Table 9.9.

*Irrigated Kharif and Rabi Cropping*

This micro unit includes Pali and Jalore districts. The topoclimatology is more or less similar with the exception that the eastern part receives more irrigation and drainage lines are more integrated and developed. The soils are same. Actually more than 45 per cent of these districts comes under semi-arid zone. Three to nine per cent of the total area is irrigated: 7 per cent by tank and 2 per cent by wells and tubewells. Red desertic and sierozen soils are deep here. About 64 per cent and 73 per cent of the estimated groundwater is exploited in Jalore and Pali districts respectively. Per cent of gross irrigated area by sources accounts food crops 85 per cent, cotton 5 to 8 per cent and others 10 per cent. Landuse, density of cultivated lands and intensity of cropping is given in Table 9.10 and crop yields in Table 9.11.

A further more detailed analysis are obtained when the data are cartographed districtwise only on a large scale map. The maps of Bikaner, Jodhpur and Nagaur are presented for comparative analysis.

The maps indicate that there are no significant changes in areas in landuse practices. In Jodhpur and Nagaur districts, cultivated lands along with short and long fallows account more than 75 per cent, whereas in Bikaner the unit accounts below 50 per cent (48.15%) except only in Nokha tehsil (77.11%) which is adjacent to Nagaur, Wastelands in Bikaner exceeds other units. Sandy wastes predominant in Bikaner (50.05%).

TABLE 9.8: Landuse

| Districts | Density of cult. (% to total area) | Intensity of cropping | Forest | Settlement non-agri. land | Net available for cult. | Fallow land | Single cropped | Double cropped |
|---|---|---|---|---|---|---|---|---|
| Sikar | 68.38 | 103.8 | 1.03 | 2.83 | 14.83 | 12.85 | 66.32 | 2.06 |
| Jhunjhunu | 75.05 | 103.7 | 2.86 | 1.51 | 14.83 | 5.72 | 71.68 | 3.37 |
| Nagaur | 60.71 | 100.6 | – | 6.19 | 8.11 | 26.32 | 60.49 | 0.22 |

TABLE 9.9: **Transitional belt between arid and semi-arid kharif monocropping production, kg/ha**

| Crop | Districts | | |
|---|---|---|---|
| | Sikar | Jhunjhunu | Nagaur |
| Wheat | 1115 | 1005 | 1012 |
| Bajra | 356 | 328 | 230 |
| Jowar | 287 | 266 | 155 |
| Maize | 553 | 560 | 715 |
| Barley | 1098 | 1005 | 995 |
| Gram | 303 | 297 | 416 |

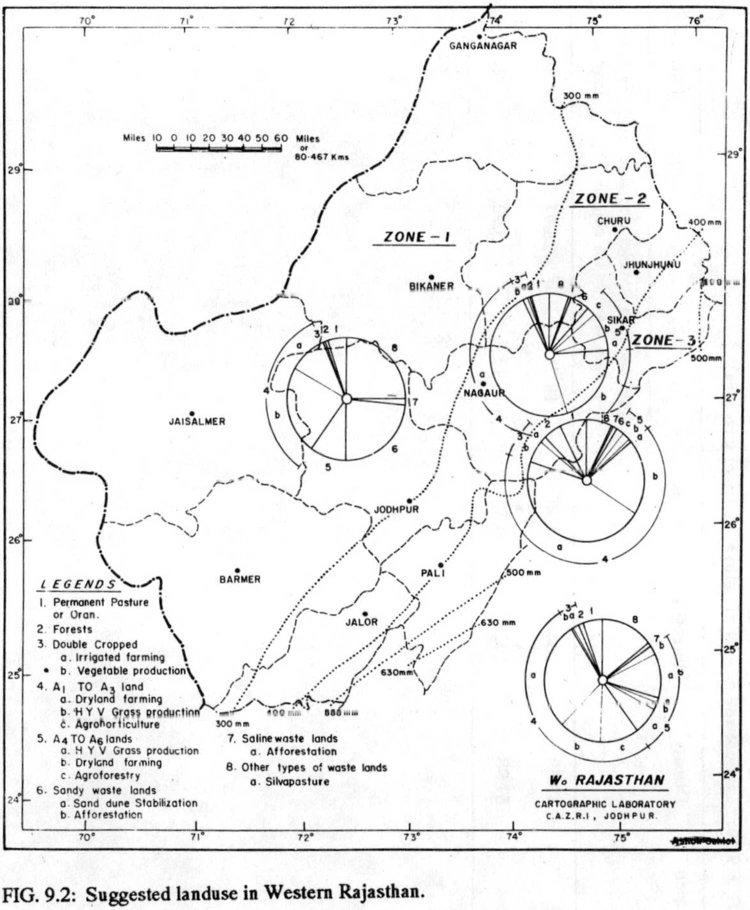


FIG. 9.2: **Suggested landuse in Western Rajasthan.**

TABLE 9.10: Landuse and density of cultivation (% of total area)

| Districts | Density | Intensity | Forest | Landuse | | | | |
|---|---|---|---|---|---|---|---|---|
| | | | | Settlement as non-agri. land | Net available for cult. | Fallow land | Single cropped | Double cropped |
| Pali | 42.39 | 104.8 | 4.90 | 19.71 | 28.07 | 28.07 | 40.84 | 1.55 |
| Jalore | 57.98 | 107.6 | 0.37 | 3.12 | 13.92 | 24.91 | 56.06 | 1.98 |

TABLE 9.11: **Transitional belt between arid and semi-arid/irrigated kharif and rabi cropping production, kg/ha**

| Crops | Districts | |
|---|---|---|
| | Pali | Jalore |
| Wheat | 1552 | 1170 |
| Bajra | 337 | 335 |
| Jowar | 234 | 52 |
| Maize | 532 | 536 |
| Barley | 975 | 909 |
| Gram | 502 | 532 |

## *Landuse Map of Western Rajasthan based on the Interpretation of Landsat Imageries*

Based primarily on the interpretation of five landsat imageries (1:1,000,000) supplied by ISRO, a tentative landuse map of Rajasthan has been prepared on 1:1,000,000. This is cartographed on the basis of total variation of the images aided by the study "convergence of evidence". Ground truths of the areas, already surveyed by Basic Resources Studies Division of Central Arid Zone Research Institute are also considered. No relationship between the images and its corresponding ground features is established to map the *oran* and the *abadi* areas. These, however, cannot be mapped on 1:1,000,000 scale. The map, however, has been found very useful to make the initial landuse classification for surveying further details and to pinpoint the problem areas. It is possible to find out and map only the areas having more than 100 km.

Present landuse of Western Rajasthan (based on landsat imageries)

| | *Areas to sq. km.* | |
|---|---|---|
| | 4268.88 | (1.90) |
| | 50776.74 | (18.62) |
| (40.80%) | 49874.92 | (19.17) |
| (30-50%) | 31562.71 | (11.56) |
| (20-30%) | 24095.45 | (8.83) |
| (10-20%) | 29404.55 | (10.56) |
| Below 10% | 8798.63 | (3.91) |

| | | |
|---|---|---|
| Sandy waste (duny complex and grazing land) | 26440.05 | (11.76) |
| Saline | 1802.41 | (0.81) |
| Gravelly and rocky waste | 11786.85 | (5.09) |
| Stoney, sandywaste and cultivated | 9459.40 | (4.56) |
| Hills | 6637.19 | (3.29) |

**Data Generated**

The 11 arid districts of western Rajasthan have been grouped into 3 rainfall zones. The zones are : (1) the arid districts of Jaisalmer, Ganganagar, Barmer and Bikaner experiencing less than 300 mm annual rainfall, (2) the arid districts of Nagaur, Jodhpur, Churu and Jalore experiencing 300-400 mm rainfall, and (3) the arid districts of Jhunjhunu, Pali and Sikar experiencing an annual rainfall of more than 400 mm. The degree and extent of landuse units can be best cartographed by means of field survey and remote-sensing devices. Based on the intensity of survey topographical maps, aerial photographs and landsat imaging interpretations, collection of field data and available informations, etc., three types of the landuse maps can be prepared from the restricting details viz. reconvinance, detail reconvinance and detail landuse maps. These are large scale maps being 1:25,000, 1:1,00,000 to 1:50,000 and 1:15,000 or more in general in case of reconnaissance, detail reconnaissance or detail maps respectively. These maps show the extent and degree of landuse classes and thus depicit the actual pattern of landuse. These are helpful to delineate or find out problem areas, planning and execution works.

Based on such survey and remote-sensing devices the area under present landuse units are presented in Tables 9.12 and 9.13. The present landuse and its relation to agricultural and livestock farming aspects can be best studied or geographically analysed by classifying the area into agro-climatic and agro-ecological regions. The arid zone of Rajasthan has earlier been classified into two macro and four micro-agro-ecological regions (Table 9.3). The landuse characteristics of these zones are also shown in Tables 9.12 and 9.13.

Tables 9.12 and 9.13 indicate that permanent pasture or oran land (II), wastelands including ten sub-classes (V) primarily constitute the grazing ground. Besides the marginal lands where intensity of cultivated lands are less than 20% (h.i.j. under IV) also comes under grazing ground during the drought years.

TABLE 9.12: Present landuse in arid district of Western Rajasthan

| Sl. No. | Arid irrigated Kharif and Rabi Crops | Transitional between Arid soil Semi and Transitional Kharif monocropping | | | | | | Arid & Semi-irrigated arid rabi cropping |
|---|---|---|---|---|---|---|---|---|
| | Ganganagar | Jhunjhunu | Sikar | Nagaur | Total | Eld | Jalore | Total |
| | 1 | 2 | 3 | 4 | 5 | 6 | 7 | 8 |
| I | 100517 | 16621 | 24729 | 72400 | 113750 | 40043 | 35050 | 83093 |
| | (4.92) | (2.00) | (3.19) | (4.09) | (3.63) | (3.91) | (3.38) | (3.64) |
| II | 25900 | 44509 | 40200 | 00400 | 173185 | 92463 | 51139 | 143602 |
| | (1.27) | (7.53) | (6.22) | (4.55) | (5.53) | (7.52) | (4.84) | (6.23) |
| III | — | 34243 | 13206 | 2600 | 50049 | 74199 | 16414 | 90613 |
| | | (5.70) | (1.70) | (0.15) | (1.59) | (6.03) | (1.55) | (3.96) |
| IV | 1613569 | 47225 | 617748 | 1520811 | 2618112 | 836913 | 857045 | 1693955 |
| | (79.04) | (79.69) | (79.72) | (86.43) | (03.49) | (81.83) | (81.18) | (74.10) |
| Va) | 157481 | 81327 | 53448 | 14900 | 149755 | 54705 | 78622 | 133327 |
| | (9.75) | (17.22) | (8.66) | (0.90) | (6.54) | (6.54) | (9.18) | (7.07) |
| b) | 717452 | 265430 | 196704 | 00100 | 1263134 | 321715 | 549886 | 871681 |
| | (27.46) | (56.19) | (31.0[illegible]) | (52.43) | (40.24) | (30.44) | (66.16) | (51.44) |
| c) | 440064 | 04250 | 101820 | 402133 | 500203 | 134399 | 140686 | 275005 |
| | (27.27) | (17.04) | (16.40) | (26.32) | (22.46) | (16.07) | (16.43) | (16.24) |
| d) | — | 11688 | 10030 | — | 111913 | 187421 | 70891 | 2258312 |
| | | (2.46) | (10.24) | | (4.27) | (22.39) | (8.29) | (15.25) |
| e) | 149987 | 8610 | 82190 | 223926 | 314726 | 47588 | 4024 | 51612 |
| | (0.30) | (1.02) | (13.30) | (11.65) | (12.83) | (5.69) | (0.46) | (3.05) |
| f) | — | 5300 | 37186 | — | 42480 | 22580 | 2200 | 24785 |
| | | (1.12) | (6.02) | | (1.62) | (2.69) | (0.35) | (1.45) |
| g) | 81672 | 8509 | 41880 | 66689 | 117078 | 28444 | 9357 | 37801 |
| | (5.06) | (1.80) | (6.78) | (4.36) | (4.47) | (3.40) | (1.09) | (2.23) |
| h) | — | 5498 | 2319 | — | 7817 | 20091 | 723 | 20814 |
| | | (1.12) | (0.37) | | (0.29) | (2.48) | (0.08) | (1.23) |
| i) | 44034 | 1745 | 1765 | 14366 | 17056 | 12319 | 598 | 12900 |
| | (2.73) | (0.87) | (0.20) | (0.94) | (0.68) | (.47) | (0.06) | (0.76) |
| j) | 14958· | 101 | 124 | 4917 | 5142 | 7643 | 67 | 7718 |
| | (0.93) | (0.02) | (0.02) | (0.32) | (0.19) | (0.91) | (0.00) | (0.46) |
| VI | 301451 | 24924 | 71041 | 84600 | 180565 | 170144 | 96709 | 274053 |
| | (14.77) | (4.20) | (9.17) | (4.78) | (5.76) | (14.49) | (9.15) | (12.02) |
| a) | 20737 | 7292 | 10499 | 0780 | 26491 | 18473 | 10937 | 29416 |
| | (95.33) | (29.26) | (14.79) | (10.89) | (14.67) | (10.37) | (11.30) | (10.70) |
| b) | 14073 | — | 2000 | 16509 | 18500 | 7546 | — | 7546 |
| | (14.69) | | (2.03) | (19.22) | (18.24) | (7.24) | | (2.75) |
| c) | — | — | 5205 | — | 5205 | 5250 | 8297 | 13547 |
| | | | (7.33) | | (2.08) | (2.95) | (0.57) | (4.90) |

(*Contd.*)

TABLE 9.12 (*Contd.*)

| 1 | 2 | 3 | 4 | 5 | 6 | 7 | 8 |
|---|---|---|---|---|---|---|---|
| d) — | 6330 | 26453 | — | 32788 | 46025 | 9421 | 554460 |
| | (25.40) | (37.23) | | (18.15) | (25.83) | (9.57) | (20.17) |
| e) — | 11231 | 6083 | — | 17814 | 100364 | 67234 | 167612 |
| | (45.06) | (8.56) | | (9.58) | (56.34) | (69.58) | (60.90) |
| f) — | — | 19627 | 20980 | 40527 | — | — | — |
| | | (27.62) | (24.70) | (22.44) | | | |
| g) — | — | — | 38500 | 38500 | — | — | — |
| | | | (45.52) | (21.32) | | | |
| h) — | — | — | — | — | — | — | — |
| i) — | 71 | 1101 | — | 1172 | 180 | — | 180 |
| | (0.29) | (1.54) | | (0.64) | (0.10) | | (0.07) |
| j) — | — | 73 | — | 73 | 306 | 006 | 112 |
| | | (0.19) | | (0.08) | (0.17) | (0.83) | (0.40) |
| Total Geo-graphical area (field survey) 2041446 | 592730 | 774925 | 1760011 | 3135666 | 1229762 | 1056301 | 2286119 |
| Total Geog-raphical area (Village) 2104531 | 564800 | 765803 | 1759200 | 3089000 | 1100800 | 1842000 | 2223600 |

Tables 9.14 and 9.15 which is the result of survey and remote-sensing clearly indicate the availability of grazing ground in western Rajasthan which is quite high in comparison to national figure. In western Rajasthan about 42 per cent is available for grazing. The figure is above 50 per cent in arid zone, while in arid Kharif monocropping belt more than 55 per cent of the total area is available for livestock farming. In Jaisalmer the available grazing ground exceeds 90 per cent, while in Bikaner it is above 60 per cent. The low figure in Barmer and Churu is, however, due to the increased use of marginal lands for cultivation which is a menance to desertification and soil erosion. Same is the case with the districts of Nagaur, Sikar and Jhunjhunu. In Pali and Jodhpur districts saline waste accounts for the availability of more grazing ground. The landuse data on the districts of Barmer, Jalore, Sikar and Jhunjhunu are based on compilation and projection of revenue data and interpretation of Survey of India, toposheets and landsat imageries.

TABLE 9.13: Landuse in arid districts of Western Rajasthan *(Area in hectare and per cent)*

| Sl. No. | Landuse categories | Arid Kharif monocropping | | | | | Total |
|---|---|---|---|---|---|---|---|
| | | Jaisalmer | Barmer | Bikaner | Jodhpur | Churu | |
| I | Settlements & Water features | 76152 (1.84) | 60112 (2.33) | 116111 (4.02) | 23690 (3.64) | 64659 (3.84) | 340784 (2.48) |
| II | Permanent pasture of oren | 145408 (3.52) | 214466 (7.61) | 26597 (0.92) | 133500 (6.02) | 49261 (2.92) | 569232 (4.14) |
| III | Forests | 52941 (1.28) | 14901 (0.53) | 8100 (0.28) | 830 (0.04) | 1940 (0.12) | 78712 (0.57) |
| IV | Cultivated land (including short and long fallow) also used as grazing land when not cultivated | 226476 (5.48) | 2025941 (71.91) | 1317184 (45.66) | 1376990 (62.09) | 1512254 (89.74) | 6458645 (46.91) |
| a) | Double cropped above 100% | 54 (0.02) | 23115 (1.14) | 404 (0.04) | 19890 (1.44) | 24772 (2.64) | 68235 (1.05) |
| b) | Monocropping with rotation intensity 80-100% | 143266 (63.29) | 918022 (45.31) | — | — | 82732 (5.42) | 1144100 (17.72) |
| c) | — do - 40-80% | — | 26001 (1.28) | 108230 (8.22) | 353410 (25.67) | — | 487641 (7.56) |
| d) | — do - 40-60% | 28894 (12.76) | 1465 (0.08) | — | — | 675008 (44.54) | 705367 (10.93) |
| e) | — do - 30-50% | — | 798 (0.04) | 123940 (9.41) | 398390 (28.93) | — | 523128 (8.09) |
| f) | — do - 60-80% | 51161 (22.59) | 734210 (36.25) | 612670 (46.51) | 312250 (22.68) | 326908 (21.61) | 20370999 (31.55) |
| g) | — do - 20-40% | — | 276541 (13.65) | — | — | 300691 (19.88) | 577232 (8.83) |
| h) | — do - 30-30% | — | 26404 (1.30) | 177560 (13.48) | 140170 (10.18) | — | 344134 (5.32) |
| i) | — do - 10-20% | 3021 (1.33) | 18235 (0.90) | 102140 (7.75) | 131480 (9.55) | 28675 (1.90) | 283551 (4.39) |
| j) | — do - below 10% | — | 1050 (0.05) | 192240 (14.59) | 21400 (1.55) | 73468 (4.06) | 288158 (4.40) |
| V | Waste land (also used as grazing) | 3634723 (87.88) | 501974 (17.82) | 1417020 (49.12) | 632790 (30.79) | 57100 (3.39) | 6290607 (4.40) |
| a) | Saline waste | 900315 (24.77) | 3623020 (72.17) | 1368559 (96.50) | 412830 (60.46) | 448440 (73.54) | 3088241 (49.08) |
| b) | Saline waste | 95097 (2.62) | 14326 (2.85) | 16570 (1.17) | 24470 (3.50) | 6591 (11.54) | 157054 (2.49) |
| c) | Stoney waste | — | 7629 (1.52) | — | 123100 (18.83) | — | 130725 (2.07) |
| d) | Gravelly waste | — | 78734 (15.68) | — | 122390 (17.93) | — | 201124 (3.19) |
| e) | Rocky waste | — | 4315 (0.86) | 31900 (2.25) | — | — | 36215 (0.57) |
| f) | Rocky and stoney waste | 58909 (1.62) | 12593 (2.51) | — | — | 2048 (3.65) | 73586 (1.16) |
| g) | Storney and gravelly waste | — | 21172 (4.23) | — | — | 853 (1.49) | 22025 (0.34) |
| h) | Rocky and gravelly waste | 664411 | — | — | — | — | 664411 (10.55) |
| i) | Sandy waste with open scrub land | 991960 (27.29) | 722 (0.14) | — | — | 2728 (4.78) | 995410 (15.86) |
| j) | Rocky and gravelly with open scrub lands | 924031 (25.42) | 181 (0.04) | | | | 924212 (14.69) |
| | Total geographical area (field survey) | 4135700 | 281729 | 2883872 | 2217608 | 1685114 | 13740989 |
| | Total geographical area (Village paners) | 3862260 | 20310 | 2737894 | 2262429 | 1686194 | 10837607 |

*Note:* The data on the districts of Barmer, Jalore, Sikar and Jhunjhunu 10 are based on compilation and projection of revenue data and interpretation on survey of India toposheets and landsat imageries. The revenue data are based on 4 years average (1977-78 — 1980-81). The data relating to others are based on field survey on the basis of aerial photo, land sat interpretation, and field investigations on sample basis to find out the relationship between the findings of laboratory interpretation analysis and their crops ponding ground featured.

*Sources:* (c) Supports of Land utilisation in Survey conducted by (Cartography Section). The Basic Resources Studies, *C.A.Z.R.I.*, ICAR, Jodhpur.

TABLE 9.14: Pasture grounds in West Rajasthan

| Districts | Total area in ha | Pasture lands in ha % Permanent pasture Oran | Waste lands | Marginal land | Total & per cent pasture lands to total area |
|---|---|---|---|---|---|
| 1A. Jodhpur | 2217800 | 133500 (6.02%) | 602790 (30.79) | 271650 (22.25%) | 1087940 (49.06) |
| Bikaner | 2005072 | 26597 (0.92%) | 1417020 (49.12%) | 294300 (10.20) | 1737997 (60.42%) |
| Barmer | 2817294 | 214466 (7.61%) | 501674 (17.02%) | 19285 (0.7%) | 735725 (26.13%) |
| Jaisalmer | 4135700 | 145400 (13.42%) | 3634723 (07.88) | 3021 (0.00%) | 3783152 (91.40%) |
| Churu | 1685114 | 49261 (2.92%) | 57100 (3.39%) | 102143 (6.1%) | 200504 (12.37%) |
| Total Arid Monocropping | 13740900 | 569232 (4.14%) | 6293608 (45.90%) | 690479 (5.02%) | 7553319 (55.06%) |
| 1B. Ganga-nagar | 2041446 | 25910 (1.27%) | 3011451 (14.77%) | 30792 (3.0%) | 386353 (19.04%) |
| Total Arid | 15782426 | 595152 (3.77%) | 6595059 (41.78%) | 6729171 (4.62%) | 7919672 (50.18%) |
| 2 A Sikar | 774925 | 40200 (6.22%) | 71041 (9.17%) | 1009 (0.24%) | 121130 (15.63%) |
| Jhunjhunu | 592730 | 44589 (7.53%) | 24924 (4.20%) | 1026 (0.30%) | 71339 (12.03%) |
| Nagaur | 1768011 | 80400 (4.55%) | 04600 (4.78%) | 19293 (1.1%) | 1842832 (10.43%) |
| Total Transitional Monocropping | 3135666 | 173109 (5.53%) | 180565 (5.76%) | 143602 (4.60%) | 497356 (15.06%) |
| 2a Jalore | 1056357 | 51139 (4.04%) | 96709 (9.15%) | 656 (0.06%) | 140504 (14.05%) |
| Pali | 1229762 | 90436 (7.52%) | 178144 (14.49%) | 19962 (1.62%) | 290569 (23.63%) |
| Total Transitional Kharif and Rabi Cropping | 2286119 | 143602 (6.20%) | 274853 (12.02%) | 20610 (0.90%) | 439073 (19.20%) |
| Total Transitional Arid & Semi-Arid | 5421785 | 316791 (5.84%) | 455418 (8.40%) | 164220 (3.03%) | 936429 (17.2%) |
| Total W. Rajas-than | 21204211 | 911933 (4.30%) | 7050477 (33.25%) | 093691 (4.21%) | 0056101 (41.76%) |

TABLE 9.15: Livestock population grazing area/livestock in ha pressure of livestock population on grazing land

| Districts | Total grazing ground in ha. | Livestock population | | | |
|---|---|---|---|---|---|
| | | 1972 (total) | Grazing land/live-stock | 1977 | Grazing land/ livestock |
| 1. Bikaner | 1737927 | 898760 | 1.93 | 1534903 (+70.70) | 1.13 |
| 2. Barmer | 735725 | 1974350 | 0.37 | 2534903 (+20.26) | 0.29 |
| 3. Churu | 208504 | 14263375 | 0.14 | 1574522 (+10.38) | 0.13 |
| 4. Jodhpur | 1087940 | 1893200 | 0.57 | 2442424 (+29.010) | 0.44 |
| 5. Jaisalmer | 3783152 | 616435 | 6.14 | 1300419 (110.85) | 2.91 |
| 6. Ganganagar | 306353 | 1533671 | 0.25 | 1645431 (+7.207) | 0.23 |
| 7. Sikar | 121130 | 1530176 | 0.00 | 1361092 (–1.05%) | 0.00 |
| 8. Jhunjhunu | 71339 | 978500 | 0.07 | 968588 (–1.01%) | 0.07 |
| 9. Nagaur | 104203 | 2304209 | 0.00 | 2371741 (+2.93%) | 0.07 |
| 10. Jalore | 140504 | 1234772 | 0.12 | 1322045 (+7.06%) | 0.11 |
| 11. Pali | 290569 | 1893650 | 6.15 | 1903093 (+4.72) | 0.14 |
| Total | 8856101 | 14584108 | 0.61 | 19836651 (+3.03) | 0.46 |

The revenue data are based on 4 years average (1977-78 to 1980-81). The data relating to other districts are based on field survey on the basis of aerial photo, landsat imageries, interpretations and field investigations on sample basis to find out the relationship on sample basis and also between the findings of the laboratory interpretation or analysis and their corresponding ground features. These are illustrated in Fig. 9.3. Areas shown under settlements and water features are not considered for planning for obvious reasons. Similarly, only 75 per cent of the wastelands are considered for silvipastoral use. $A_1$ to $A_3$ lands have an average cultivation intensity of around 65 per cent and are considered as agriculturally good lands. $A_4$ to $A_6$ lands have an average cultivation intensity of about 25 per cent and are considered as marginal lands. Wastelands are grouped under sandy wastes, saline wastes and others. 'Others' include rocky and stoney waste, stoney and gravelly waste, rocky and gravelly waste and sandy waste, with open scrub lands. In case of Ganganagar district, only 2,15,600 ha (forming about 13 per cent of the physical area) is considered as arid. Application of technologies for better use of these lands are suggested after a discussion with the senior experts at CAZRI and also referring to the related literature. Resource requirements and assessment of benefits are based on the estimates provided in "Economic Evaluation of Arid Zone Technologies".The monetary evaluation of resources for the suggested plan and the benefits from it is done based on 1981 year and prices. Quantities of only important resources are indicated. Expected production of some products are given under normal and subnormal (production rainfall) conditions.

## Suggested Landuse

The results are discussed under three heads, namely, 1. the suggested landuse plan, 2. resources for the suggested plan, and 3. benefits from the suggested plan.

### 1. *Suggested Landuse Plan*

Details of the suggested landuse plan for the three rainfall zones in western Rajasthan are presented in Table 9.9 and Fig. 9.3. The percentage under major landuses are abstracted in Table 9.16, as can be seen from Tables 9.16 and 9.17, the suggested landuse plan seems to be broadly in agreement with the basic tenets of landuse under desert ecosystem. In the suggested plan only one-third of the physical area in western Rajasthan will be under crop cultivation. The rest will be occupied by grassland and pasture (32%), forests (24%) and fruits and vegetables (11%). A comparison of the suggested plan with the existing landuse may be beneficial here. The existing landuse plan (Table 9.17) for western Rajasthan indicates that about 5 per cent of the area is under permanent pasture or oran, only 1 per cent under forests, 60 per cent under cultivation (with different intensities) and 34 per cent under waste lands. As per the FAO land capability classification for soil conservation, only 21 per cent of the arid lands is suitable for sustained cropping. Thus, in a region, where there should be more of grazing lands and forests, and less area under sustained cropping, the present landuse seems to have crossed the limits and is inimical to the growth and eco-balance in these areas. The study of trend in landuse also signifies the historical abuse of land in this region. It is observed that the net sown area in this region increased by 503 per cent between 1951 and 1961 and by an additional 8.3 per cent between 1961 and 1971 while the area under less intensive forms of

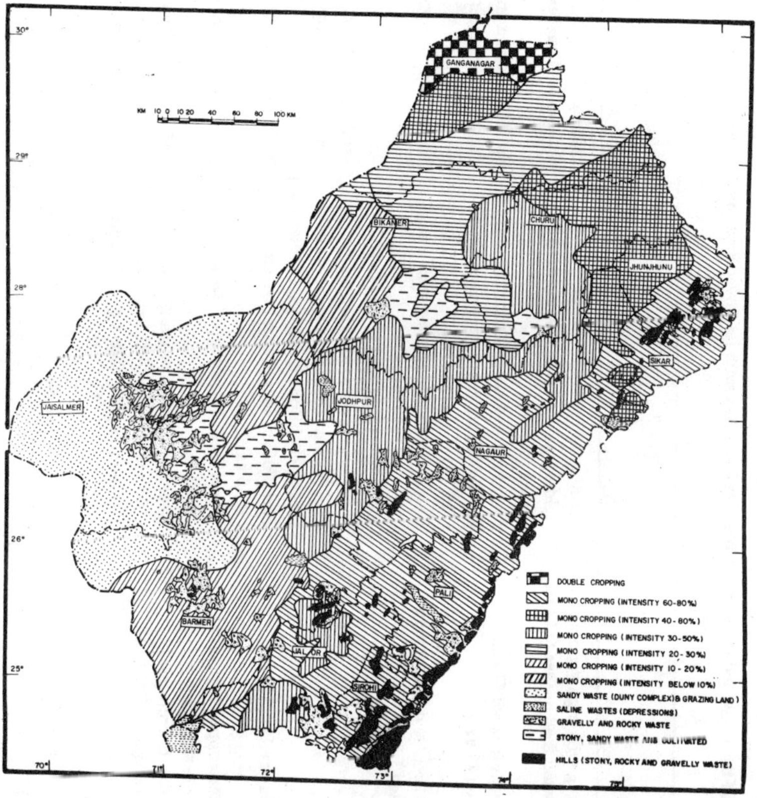


FIG. 9.3: **Present Landuse** of Western **Rajasthan** (Based on the interpretation of landsat imageries).

TABLE 9.16: Details of the suggested landuse for Western Rajasthan with improved arid zone technologies

(area in '000 ha)

| Sl. No. | Zone 1 | Zone 2 | Zone 3 |
|---|---|---|---|
| 1. | Fancing, reseeding with high yielding grasses with few too feed bushes (412 ha) | Fencing, reseeding with high yielding grasses with few top feed bushes (314 ha) | Fencing, reseeding with high yielding grasses with few top feed bushes (185 ha) |
| 2. | Protection and maintenance (76 ha) | Protection and maintenance (22 ha) | Protection and maintenance (122 ha) |
| 3. | Irrigated farming (43 ha) | Part (80%) under irrigated farming (111 ha) | Part (80%) under irrigated farming (152 ha) |
| | Kharif : Bajra (BL-104) intercropping with mung (5.5) | Kharif:Bajra intercropped with mung | Kharif:Bajra intercropping with mung |
| | Rabi : Mustard (T-59)/Cumin | Rabi : Mustard/Cumin Part (20%) under vegetable production under drip system (28 ha) (Tomato, Water-melon, cauliflower, papaya, Chillies etc.) | Rabi : Mustard/Cumin part (20%) under vegetable production under drip system (38 ha) Tomato, watermelon, Papaya,Chillies, etc) |
| 4. | Part (30%) under dryland farming (868 ha) | Part (70%) under dryland farming (3040 ha) | Part (70%) under dryland farming (1124 ha) |
| | Kharif:Bajra intercropped with mung part (70%) under high yield grasses intercropped with grain legume (2030 ha) | Kharif:Bajra intercropped with mung (2040 ha) | Kharif:Bajra intercropped with mung |
| | | Rabi: (Kharif fallow land): mustard (1000 ha) culture (482 ha) | Rabi:Mustard/Cumia Part (30%) under Agrshorti-culture (482 ha) |
| | | Bar intercropped with a gain legume | Bar intercropped with grain legumes. |

(Contd.)

TABLE 9.16 *(Contd.)*

| Sl. No. | Zone 1 | Zone 2 | Zone 3 |
|---|---|---|---|
| 5. | Pasture development with high yielding grasses intercropped with grain legume (833 ha) | Part (30%) under dryland farming (24 ha)<br>Kharif: Bajra intercropped with mung<br>Part (50%) pasture development with high yielding grasses intercropped with a grain<br>1. regime (29 ha)<br>Part (20%) under agro-Forestry (159 ha) | Part (40%) under dryland farming (52 ha)<br>Kharif : Bajra intercropped with mung.<br>Rabi : Mustard/Cumin<br>Part (40%) pasture development with high yielding grasses intercropped with grain legumes (52 ha)<br>Part (20%) under Agro-Forestry (26 ha) |
| 6. | Sand dunes stabilisation (2011 ha) | Afforestation (358 ha) | Afforestation (27 ha) |
| 7. | Afforestation (105 ha) | Afforestation (36 ha) | Afforestation (71 ha) |
| 8. | Silvipasture (2097 ha) | Silvipasture (297 ha) | Silvipasture (171 ha) |

* 1. Permanent pasture in an area (912 ha) culturally good lands; averages (intensity of around 25 per cent). lands (2566 ha).

2. Forests (219 ha) (cultivation intensity of around 65%).

3. Double Cropped (371 ha).

4. $A_1$ to $A_3$ lands (884 ha; Agri lands 1765 ha; average cultiation).

6. Sandy waste lands (2396 ha).

7. Saline waste lands (148 ha).

8. Other types of waste.

TABLE 9.17: **Details of existing landuse (average of 1978-81) in Western Rajasthan**
*(in per cent)*

| Details | Zone I | Zone II | Zone III | Western Rajasthan |
|---|---|---|---|---|
| Permanent pasture | 3.57 | 4.81 | 7.38 | 5.00 |
| Forests | 0.65 | 0.33 | 4.85 | 1.00 |
| Cultivated area | 44.99 | 81.00 | 76.85 | 60.00 |
| Waste lands | 50.79 | 13.86 | 10.92 | 34.00 |

use (barren, culturable and uncultivable waste lands, permanent pasture and fallow lands) declined by 18.9 per cent between 1951 and 1961 and by a further 6.2 per cent between 1961 and 1971. Hence, there is a need for a systematic landuse in these areas for higher and stable production and for maintenance of balance in the fragile ecosystem. It is to the credit of the suggested plan that it ensures not only higher supply of foodgrains but also fodder and fuel in the area. Besides these, more than enough of fruits and vegetables will be produced and supplied which will go a long way in enriching the diets of the desert people.

The enterprises suggested for individual zones are again, as can be seen, go well with the broad guidelines given by scientists. Zone I which receives the lowest of the annual rainfall constitutes nearly 50 per cent of the area in western Rajasthan. Annual rainfall received in this region ranges from 185 mm in Jaisalmer district to 294 mm in Bikaner district. Suggested animal husbandry is backed up by rangeland management and forestry in this region. Restriction of cultivation and extension of livestock rearing on protected grazing lands in this region is also suggested. As can be seen from column 3 of Table 9.2, nearly 50 per cent of the area is sown under grassland and pasture, 38 per cent under forests and only 11 per cent under crop farming. Most of the lands in this region will be occupied by silvipastoral cover largely favouring successful animal husbandry besides maintaining the eco-balance (Table 9.16). Similarly, for Zone 2, which is transitional integration of crops and animal husbandry is suggested. The suggested plan has about 14 per cent under grassland and pasture, 11 per cent under forests, 21 per cent under fruits and vegetables and 54 per cent under crop farming. It should be noted here that a large part of the area sown under fruits and vegetables is in fact envisaged under *ber* plantation which besides providing fruits will also add to the tree wealth of the region. Production of fruits and vegetables, it is hoped, will add to the quality of diets of the people in this region. For zone 3, which receives annual rainfall more than 400 mm, suitable technology for dryland production for improving and stabilising production is suggested. Emphasis on enterprises to different land units in this zone is more or less similar to one given in the suggested plan for zone 2. The only difference can be in respect of $A_4$ to $A_6$ lands where the intensity of cultivation has been assumed to be higher. Some of the merits of the suggested plan are that the area under grassland and pasture in the suggested plan is 10 per cent as compared to around 8 per cent as at present, percentage area under forests is double that of the existing landuse and about 21 per cent of the land is devoted to the production of protective feeds, namely, fruits and vegetables.

2. *Resources for the Suggested Plan*

An estimate of resources required to implement the suggested plan will be useful in procuring and making them available in time. The various resources required for the implementation of the suggested plan are shown in Table 9.18. One of the important inputs required is that of seeds and seedling. Shortage of quality fodder seeds, seeds of suitable variety of bajra and non-availability of seeds at right time are some of the important constraints in the adoption of technology in dry land areas of Jodhpur district. A systematic programme of seed production and distribution thus becomes very essential. Similarly, for afforestation programmes, adequate and well-developed seedling stock should be ensured. Establishment of tree nurseries at different places may be necessary. Tree seed collection and storage should also receive enough attention in this region.

Majority of the arid zone soils are reported to be low in organic carbon or available nitrogen. Adequate supply of organic manures including rural and urban compost may raise the fertility in soils. The total dung production for the 11 arid districts in western Rajasthan would be around 19 million tonnes (based on 1971 livestock census). But the requirements as per the suggested plan will exceed 40 million tonnes. The acute shortage of FYM on farmers fields is noted. In the absence of sufficient fuel in these areas, division of dung to fuel may further aggravate the problem.

As per the suggested plan, about 26 kg of NPK (16 N + $8P_2O_5$ +$2R_2O$) fertilisers are to be used per hectare of cropped land in these areas. But the existing use of NPK is only 13.5 kg (4.7 N + 6.8 $P_2O_5$ + $2K_2$) during 1976-77. Low or no use of fertilizers in these areas especially for kharif crops is found to be due to fear of failure and risk and uncertainty. With the availability of the improved technology, part of this fear may be shown to be unfounded. For higher production, there is an urgent need to stop up the rate of application of chemical fertilisers. Intensive extension efforts are needed to promote fertiliser use in dry land areas.

TABLE 9.18: **Production potentials of the suggested plan for Western Rajasthan**

| Details | | Unit | Zone 1 | Zone 2 | Zone 3 | Western Rajasthan |
|---|---|---|---|---|---|---|
| HYV grass | sit-1 | '000t | 1885 | 257 | 133 | 2275 |
| seed | sit-2 | '000t | 838 | 154 | 80 | 1072 |
| Acacia seed | | '000t | 3417 | 255 | 234 | 3906 |
| Eucalyptus seed | | '000t | — | 48 | 8 | 56 |
| Prosopis pods | | '000t | 6348 | 1181 | 103 | 7632 |
| Eucalyptus pods | | Cr. No. | — | 13 | 2 | 15 |
| HYV dry | sit-1 | '000t | 37702 | 5148 | 2651 | 45501 |
| grass | sit-2 | " | 15081 | 2059 | 1061 | 18201 |
| Bajra (grain) | sit-1 | " | 683 | 1792 | 997 | 3472 |
| | sit-2 | " | 228 | 597 | 332 | 1157 |
| Mung (Beans) | sit-1 | " | 638 | 1433 | 797 | 2868 |
| | sit-2 | " | 182 | 478 | 266 | 926 |
| Dry fodder | | | | | | |
| from | sit-1 | " | 456 | 1194 | 664 | 2314 |
| | sit-2 | " | 182 | 478 | 266 | 926 |
| Mustard | sit-1 | " | 43 | 111 | 1329 | 1438 |
| | sit-2 | " | 17 | 444 | 531 | 992 |
| Fuel | | " | 221210 | 26272 | 10798 | 253380 |
| Tomato fruits | | " | — | 1632 | 2251 | 3883 |
| Cauliflower | | " | — | 719 | 985 | 1704 |
| Ber fruits | | " | — | 9382 | 3471 | 12853 |

(t-tonnes, Cr. crores, No-number, sit- good rainfall situation, sit-2 worst rainfall situation)

Plant protection is yet another important input in modern technology. But its use now is restricted only to commercial crops like seera and raya. The increased income from HYV technology suggested in the plan may motivate these farmers to use them. However, like fertilisers, there is need for intensive extension efforts for this input also.

For better utilisation of limited wet periods and as means of transportation in arid areas, tractorisation has been suggested. Assuming 6 hours of working of tractor in a day and about 300 days of work in a year, the available tractor hours for the western Rajasthan would be of the order of 7410 thousand hours. But as per the requirements of operations in the suggested plan, there will be huge deficit of about 14071 thousand tractor hours. Acquisition of another 7800 tractors (against the existing 4118 tractors as per 1971 census) for western Rajasthan is needed to meet this deficit.

It is hoped that the estimates provided in this note may provide an idea of the extent of the problems of the desert in western Rajasthan. The initial capital required to implement the plan and thus initiate the process of agricultural growth will be of the order of about Rs. 2630 crores. A part of this investment will be made by private parties and hence, the entire amount need not be considered as exclusive burden on the government. But this part will be smaller. A state whose annual plan is limited to around Rs. 300-500 crores and whose current overdraft stands around Rs. 350 crores cannot think of touching even the fringe of this problem with its own resources. Hence, pooling of resources on a national or even international level becomes a must. Desert development may be taken up as a project of national importance. Permanent solution to recurring droughts and famines in these areas as said earlier, will only rest with how rational landuse plans are conceived, funded and implemented. The drain on the exchequer to provide relief will multiply if permanent solutions are not sought for. Very recently reported that during 1951-1981, about Rs. 3504 million were spent on famine relief in Rajasthan. Except saving life, it will not secure any enduring benefit.

### 3. *Benefits from the Suggested Plan*

Only tangible benefits are evaluated and presented in Table 9.19. The plan provides for increased availability of coarse feed grains in the area. The per capita annual availability of bajra (according to 1971 population census) increases from mere 75 kg new (1978) to 401 kg under favourable rainfall conditions and 135 kg under worst rainfall conditions. Similarly, the annual per capita availability of mung will be 334 kg and 108 kg under these situations, respectively, against the existing availability of only 47 kg of kharif pulses. The production of fruits and vegetables envisaged in the plan will not only meet their recommended daily rates (350 grams) but also result in surplus for export.

TABLE 9.19: **Net income and employment potentials of the suggested landuse plan for Western Rajasthan**

*(rupees in crores)*

| Details | Zone 1 | Zone 2 | Zone 3 | Western Rajasthan |
|---|---|---|---|---|
| Net income | | | | |
| sit-1 | 3040 | 2169 | 1820 | 7035 |
| sit-2 | 935 | 2066 | 1031 | 4032 |
| Employment (mandays in crores) | 231 | 108 | 50 | 389 |

The prospects of HYV fodder seed production even under worst rainfall conditions is around 1.07 million tonnes which leaves a margin for even export, estimated the deficit of fodder of about 7 million tonnes by 1985 for Western Rajasthan. If the suggested plan is implemented, even under worst rainfall conditions, this deficit will not be encountered at all. A productive animal husbandry can thus be supported. It has been estimated that on an average a family of 6 members needs about 3.75 tonnes of dry wood fuel per year. At this rate, for one year, the total dry wood required for western Rajasthan will be of the order of 5.37 million tonnes. The exploitation of woody biomass for various uses in arid Rajasthan was around 24 million tonnes during 1971. If the suggested plan is implemented, around 25.83 million tonnes of dry wood can be expected which of course will be available during the felling cycles of the suggested tree species. The region will also be benefited from production of timbers, pods, seeds, etc. from these trees.

The income benefits from the plan are also quite attractive. By implementing the plan, a net annual income of Rs. 7035 crores can be expected under favourable conditions and Rs. 4032 crores under worst conditions (Table 9.19). These estimates thus reflect the usual apprehensions of inadequate financial returns for development projects in the desert. The benefit cost ratios of the technologies to be followed in the suggested plan were estimated to be exceeding one and are thus very sound investment proposals.

The suggested plan will also generate about 389 millions mandays of employment annually. Taking the total workers in agriculture in these districts and assuming that they will work for 300 days in year, total mandays available for work in western Rajasthan will be roughly 593 million. Thus the employment potential of the suggested plan clearly indicates that those who seek work will largely find it.

## REFERENCES


Ahuja, L. D., 1977 : 'Improving range land productivity', in *Desertification and its Control*, Indian Council of Agricultural Research, New Delhi.

Anonymous, 1964 : *Agriculture and livestock in Rajasthan,* National Council of Applied Economic Research, New Delhi.

———, 1974 : *Basic resources of Bikaner district* (Rajasthan), Central Arid Zone Research Institute, Jodhpur.

———, 1978 : *Survey of basic and human resources, Jodhpur district,* Central Arid Zone Research Institute, Jodhpur.

———, 1980 : *Perspective plan of Rajasthan,* 1974-89, Volumes 1 and 2, National Council of Applied Economic Research, New Delhi.

Dhir, R. P., 1977 : Western Rajasthan Soils : Their characteristics and properties, in *Desertification and its Control*, Indian Council of Agricultural Research, New Delhi.

Jodha, N. S., 1974 : A case of process of tractorisation, *Economic and Political Weekly* (Quarterly Review of Agriculture), 9(52).

———, 1979 : *The process of desertification and the choice of interventions*, Progress Report 2, Economic Program, International Crops Research Institute for Semi-arid tropics, Hyderabad.

Kalla, J. C., Joshi, B. R. and Mehta, A. C., 1978 : A strategy to meet energy demands in arid areas of western Rajasthan, *Urja,* February 18, p. 73.

Mann, H. S. and Singh, R. P., 1977 : Crop production in the Indian arid zone, in *Desertification and its Control*, Indian Council of Agricultural Research, New Delhi.

Malhotra, S. P., 1980 : *Role of communities in project implementation*, FAO/DANIDA Training course on Sand dune, *stabilisation, shelterbelt and afforestation in the dry zones,* CAZRI, Jodhpur, March 3 to 30.

———, 1981 : Human aspects of desertification and of measures of combat it in arid Rajasthan, *The Threatened Drylands*, School of Geography, Kensington, Australia.

Mruthyunjaya, 1982 : *Economic evaluation of arid zone technologies*, Department of Economic and Sociology, Central Arid Zone Research Institute, Jodhpur.

Mruthyunjaya, Malhotra, S. P., 1983 : *Technological possibilities for agricultural growth and stability in western Rajasthan*, CAZRI, Jodhpur.

Paranjape, N. R., Purohit, N. B. and Sathe, N. D., 1981 : Social Forestry : A programme for tree pastures, *The Economic Times,* New Delhi, June 6.

Sen, A. K. and Gheesalal, 1977 : *Present land utilisation survey of Jodhpur district,* Rajasthan, Division of Basic Resources Studies, Central Arid Zone Research Institute, Jodhpur (Mimeo).

Sen, A. K., 1978 : *Landuse classification system in Indian arid zone*, Monograph No. 9, Central Arid Zone Research Institute, Jodhpur.

Sen, A. K., Gheesalal and Gupta, K. N., 1979 : Present land utilisation system survey of Nagaur district, *Rajasthan Studies*, Central Arid Zone Research Institute, Jodhpur (Mimeo).

Sen, A. K. and Gheesalal, 1980 : Land utilisation survey in Jaisalmer district, Rajasthan, Division of Basic Resources Studies, Central Arid Zone Research Institute, Jodhpur (Mimeo).

Shankarnarayan, K. A., 1982 : Alternate landuse system for drylands. A lead paper presented at the National Seminar on *A Decades of Dryland Agriculture Research in India and Thrusts in the Eighties*, held at Hyderabad, January 18-20.

Singh, D. V., 1981 : Adoption and problems of adoption of dry land farming technology in Jodhpur district, Western Rajasthan, Dry Farming Scheme, Central Arid Zone Research Institute, Jodhpur.

Vyas, V. S. and Khanna, Inderjit, 1982 : Famine and famine relief : Coping Strategy at the District Level, *Working Paper* (W.P. No. 418), Indian Institute of Management, Ahmedabad, India.

# 10

# TRANSFORMATIONS IN AGRICULTURAL LANDUSE IN POLAND: 1946-1988

WLADYSLAWA STOLA

AFTER WORLD War II the socio-economic development of Poland has been accompanied by changes in the proportion of agricultural land in the country's total area as well as by transformations in the structure and orientations of landuse.

In the first postwar years (until 1950) the total area of agricultural land increased; the increase was particularly great as far as the arable land is concerned which was abandoned during the war, but had been recultivated. In 1947, these idle lands, accounted for about 40 per cent of arable land whereas in 1950 this percentage was under 7. From 1950 onwards the remaining idle lands were put to use. Simultaneously, a decrease of the area of agricultural land (as well as its proportion of the total area) began : either land of little agricultural value was transferred for afforestation, or the growing industrialisation and urbanisation of the country caused the taking over of agricultural land by industry, urban construction, or transport. Poland's agricultural land fell down from 20,440 thousand hectares in 1950 to 18742 thousand hectares in 1988, i.e. by 1698 thousand hectares; the proportion of agricultural land in the total area went down from 65.6 to 59.9 per cent; the respective index per one inhabitant decreased from 0.86 hectare in 1950 to 0.53 hectare in 1988.

The decrease in the area of agricultural land has not affected the spatial picture of its distribution shaped in the past as an effect of the diverse capability of the individual parts of the country for agricultural production and under the impact of demographic pressure. The highest percentage (over 70) of agricultural land is a characteristic to Central Poland, i.e. for Mazovia, Kujavia, Greater Poland, as well as for the fertile soils of the Sandomierz Uplands, Lublin Plateau and the Silesian Lowland. A lower percentage (under 55) appears in northern and western Poland and in the mountains.

The decrease of the area and of the proportion of agricultural land in the country's total was accompanied by transformations in its structure. A relatively high percentage of arable land (over three-fourths of agricultural land) and a low percentage of permanent grassland as well as perennial crops are characteristic features of the agricultural landuse in Poland. A decrease in the area of agricultural land was obtained mainly at the expense of arable land, the area of which fell down from *circa* 16.0 million hectares in 1946 (51.3 per cent of the country's total and 78.2 per cent of the area of agricultural land) to 14.5 million hectares i.e. 46.2 per cent and 77.2 per cent respectively in 1988. A marked decrease of arable land, similarly as that of agricultural land, could be observed between 1960 and 1965, as well as between 1970 and 1978, when almost 60 per cent and 30 per cent respectively i.e. a total of almost 90 per cent of all arable land transferred to other uses between 1946 and 1988, were taken out of agriculture.

However, the proportion of sown area in arable land increased systematically (in 1988 to almost 100%). Significant changes occurred also in the proportion of crops cultivated on arable land; their spatial picture was formed according to natural and other conditions influencing the development of agriculture, as well as according to the social and ownership characteristics of farms, the degree of their intensity and production orientations. As arable land prevails among agricultural land almost everywhere in Poland, transformations occurring in the structure of arable landuse had a far-reaching effect on changes taking place in the whole agricultural landuse.

In 1946-1988 both the area as well as the share of permanent grassland in the area of agricultural land did not change much: from 4,156 thousand hectares (20.3%) in 1947 to 4,040 thousand hectares (21.4%) in 1988. But the structure of grassland changed because the proportion of own meadows increased (from 11.7 per cent to 13.3 per cent of agricultural land), mainly at the expense of little-productive pastures, as an effect of their improvement and a higher standard of agricultural techniques (Table 10.1).

TABLE 10.1: Agricultural land

| Sl. No. | Agricultural land | Area in thousand hectares in the years | | | | | Per cent of the country's area | | | | | Per cent of the area of agricultural land | | | | |
|---|---|---|---|---|---|---|---|---|---|---|---|---|---|---|---|---|
| | | 1950 | 1960 | 1970 | 1980 | 1988 | 1950 | 1960 | 1970 | 1980 | 1988 | 1950 | 1960 | 1970 | 1980 | 1988 |
| 1. | Arable land | 15986 | 15961 | 15088 | 14621 | 14464 | 51.3 | 51.2 | 48.4 | 46.8 | 46.2 | 78.2 | 78.2 | 77.2 | 77.2 | 77.2 |
| 2. | Perennial crops | 299 | 262* | 238 | 280 | 238 | 1.0 | 0.8 | 0.8 | 0.9 | 0.8 | 1.5 | 1.3 | 1.2 | 1.5 | 1.3 |
| 3. | Meadows | 2398 | 2390 | 2523 | 2503 | 2502 | 7.7 | 7.7 | 8.0 | 8.0 | 8.0 | 11.7 | 11.7 | 12.9 | 13.2 | 13.3 |
| 4. | Pastures | 1758 | 1790 | 1694 | 1543 | 1538 | 5.6 | 5.7 | 5.5 | 4.9 | 4.9 | 8.6 | 8.8 | 8.7 | 8.1 | 8.2 |
| 5. | Total | 20440 | 20403 | 19543 | 18947 | 18342 | 65.6 | 65.4 | 62.7 | 60.6 | 100.0 | 100.0 | 100.0 | 100.0 | 100.0 | 100.0 |

* Since 1961 area of perennial crops above 0, 1 hectare.

Arable land and grassland accounted altogether for 98.6 per cent of the total agricultural land. Arable land accounted for the highest percentage (over 80) of agricultural land in East-central Poland (Greater Poland, Kujavia, western Mazovia) and in the western part of the Lublin Upland and the southern Little-Poland Upland, as well as in the Silesian Plain; the lowest percentage (under 60-70) has been found in the north-eastern part of the country, and also in the Carpathians, Sudetes and the Holy-Cross Mountains. The spatial picture of the proportion of grassland was therefore roughly the opposite of proportion of arable land.

The remaining percentage of agricultural land (1.4) was occupied by perennial crops (orchards, fruit shrubs); though their share in the country's agricultural land did not increase in 1946-1988, yet, relatively great changes in the spatial concentration of orchards and a noticeable increase of intensity, specialisation and productivity of fruit production then took place. The principal areas under orchards were in central and south-eastern Poland, in particular in the Warsaw and Sub-Carpathian regions.

Changes in the proportions of the separate forms of landuse, i.e. between arable land, permanent grassland and perennial crops, and in their internal structures are synthetically represented by orientations of agricultural landuse. The orientations determine the system of agricultural use and types of agricultural production.

To identify orientations the method of successive quotients was used, the number of quotients being established at six. This figure is divisible by the number of categories of the main landuses and the principal groups of crops identified based on agro-technical premises (extractive or exacting, intensifying, and structure-forming crops).

The author has investigated orientations of agricultural landuse by powiats* in seven time cross-sections between 1947 and 1971, both for total agriculture as a whole and separately for, i.e. private, State and collective farming (from 1950 onwards). Almost, 1,800 of landuse structures and crop combinations have been singled out. Some of those combinations appeared more frequently, some rarely, and some only sporadically. Subsequently, the identified combinations have been

* The number of powiats (the administrative division of second order) in Poland was in 1950, 1960, 1971 respectively 271, 401, 396; powiats with State farms respectively 271, 310, 312; powiats with collective farms respectively, 187, 280, 254. Data for 1947 referred to agriculture as a whole.

grouped according to their landuse structures and within them according to the proportion of individual crops cultivated on arable land. In consequence, these *circa* 1,800 combinations have been reduced to 70 orientations, among which there are some specific orientations, as well as some more or less closely resembling one another.

In the first postwar years the field system with a secondary proportion of permanent grassland ($P_5+Z_1$)* dominated in a majority of powiats (60%) as well as the highly field system ($P_6$) dominated in about one-fourth of powiats. Idle and fallow lands represented a relatively big share within arable land; they were identified in the orientations of over two-fifths of the powiats situated mainly in the Territories recovered after the World War II and in the areas of former as bridge-heads of the war time. Among the orientations, without idle lands, that of field system, grain crop rye with a secondary proportion of oats, potatoes, structure-forming annuals and grassland — $P_5[E_3(r_2+o_1)+ I_1(P_1)S_1(a)]+Z_1$ dominated.** The next place was occupied by highly field grain crop, oats-rye with a secondary proportion of potatoes and structure-forming annuals — $P_6 [E_4/o_2+r_3/ +I_1/P_1/+S_1/a_1/]$ and by field grain crop, rye with a secondary proportion of oats or oats with a secondary proportion of rye, potatoes and grassland — $P_5[E_4(r_3+o_1 \text{ or } o_3+r_1)+I_1(P_1)]+Z_1$. They appeared mostly in Central Poland.

In 1950 in individual farming, covering about 90 per cent of the

* $P_6$- highly field system
$P_5+Z_1$ field with a secondary proportion of grassland
$P_4+Z_2$ field with grassland
$P_3+Z_3$ field-grassland
$P_2+Z_4$ grassland with field
$P_1+Z_5$ grassland with a secondary proportion of field
$Z_6$- highly grassland system

** *P*-field
*Z*-grassland
*T*-perennial crops
*E*-extractive crops/cereals/
b-barley, o-oats, w-wheat
*I*-intensifying crops/root crops, industrial crops, vegetables/
p-potatoes, s-sugar beet, v-vegetables
*S*-structure forming/leguminous crops, grass cultivated on fields/
c-clover, l-lucerne, g-grass cultivated, a-annuals
*N*-idle lands.

country's agricultural land, the orientations "field with a secondary proportion of grassland" prevailed (in 55% of the powiats); the most common were : field, rye with a secondary proportion of oats or wheat (or barley), potatoes and grassland and rye with a secondary proportion of oats, potatoes, structure-forming annuals crops and grassland. They appeared mostly in western and northern Poland. The next place was occupied by highly fields orientations, represented mainly by oats-rye orientation with a secondary share of potatoes and structure-forming crops, appearing mainly in Greater Poland, rye-wheat or rye with a secondary proportion of other cereals and potatoes, appearing mainly in Lower Silesia and in south-eastern Poland. The orientations were much more differentiated in 1955 and 1960. Those, described above, were supplemented by the expanding orientation : field with grassland, rye-meadows with a secondary share of other cereals and potatoes — $P_4[(E_3(r_2+w_1 \text{ or } o_1)+I_1(P_1)]+Z_2$; moreover, orientations in which an important role was played among intensifying crops in industrial crops, mainly sugarbeets, rapeseed, were identified in Lower Silesia and Kujavia or by vegetables, for example, vegetables with potatoes, or potatoes with vegetables, and a secondary proportion of cereals and grassland, appearing mainly in the suburban zone of Warsaw and occasionally in towns in the various parts of the country.

Between 1960 and 1971 great changes took place in the spatial pattern of separate orientations. The range of field orientations with a secondary share of grassland and a big share of oats and rye diminished, and field orientations with a secondary share of grassland and within arable land with a relatively big share of wheat, barley, industrial crops or vegetables extended their range. Such orientations as wheat-rye or wheat with a secondary share of barley, potatoes and grassland, or wheat with a secondary share of barley sugar beets (or rapeseed), potatoes (sporadically vegetables) and grassland appeared on large areas in Lower Silesia and in the Vistula Fenland (*Zulawy*), whereas the orientation : highly fields, rye or wheat with a secondary share of potatoes, sugar beets (or rapeseed) and clover (or lucrene) was identified in Kujavia. The highly field, rye with a secondary share of potatoes and sugar beets orientations were quite common in the Lublin Plateau and Sandomierz Upland. As a result of transformations in the sowing structure on arable land and a spatial regrouping of the orientations, the areas developed in which relatively intensive orientations prevailed; for example, south-western Poland, territories along the Middle and Lower Vistula, and Lublin Plateau, the

Sandomierz Upland, the Szczecin Fenland.

A comparison of State and individual farms has revealed that the orientations in State farming were more diverse. In 1950, when State farms spread over an area of under 10 per cent of agricultural land, orientations with a share of idle lands still appeared in about one-fifth of powiats. The most common were the following orientations : fields with a secondary share of grassland, rye with a secondary share of oats or potatoes, idle lands and structure-forming crops, mainly annuals. They were identified mainly in north-western, north-eastern and south-eastern Poland and sporadically in Lower Silesia. The orientation : fields with grassland—$P_4[E_3(r_2+o_1)+N_1]+Z_2$; rye-meadows with a secondary proportion of oats and idle lands occurred in the Bieszczady Mountains and Vistula Fenland. Most commonly occurring orientations without idle lands were: fields with a secondary share of grassland, rye or rye-wheat with a secondary share of other cereals, potatoes, structure-forming annuals and grassland (in almost 30 per cent of powiats). The field orientation with grassland, rye-meadows with a secondary proportion of other cereals and intensifying crops — $P_4[E_3(r_2+o_1 \text{ or } b_1)+I_1(P_1)]+Z_2$ and the orientation — $P_4[E_2(r_1+o_1)+I_1(P_1)]+S_1(c_1)]+Z_2$ with grassland orientations with rye (or wheat), a secondary share of potatoes — $P_3[E_2(r_2 \text{ or } w_2)+I_1(P_1)]+Z_3$ and grassland orientation with a secondary share of cereals (or potatoes) or structure-forming crops, represented by system of agricultural-landuse, mainly in north-eastern Poland and in the Carpathians (they occurred altogether in 30 per cent of powiats). Orientations with a share of industrial crops or vegetables were relatively few. In the 1950s the number of orientations did not increase, whereas a numerical and spatial concentration of certain orientations took place; for example, this refers to such orientations as — $P_6[E_3(r_2+w_1) +I_1 (p_1)] + S_2 (c_1+I_1)]$, $P_6 [E_3(w_2+r_1) + I_2(p_1+s_1)] + S_1(c_1)]$, highly field, with wheat and rye out of extractive crops with potatoes or industrial crops out of intensifying crops, and of clover or lucerne out of structure-forming crops, which extended its range from 8 to 20 per cent of powiats, mainly in Greater Poland, Kujavia and the Silesian Plain.

The range of occurrence of orientations with a relatively big share of orchards developed sporadically.

In 1960 – 1971 the area of agricultural land increased in State farms 12 to 15 per cent of the country's total. The orientations with a share of idle lands which had occurred occasionally in the eastern

Mazurian powiats still in 1960, disappeared. Though no significant changes took place in the proportions of the main forms of landuse, they were quite substantial in the proportions of certain crops and their groups in the sowing structure on arable landuse. As the number of powiats with orientations including industrial crops (sporadically vegetables), among which new orientations were relatively many, went up (to 30%), changes also occurred in the spatial distribution of the orientations. The orientations with a big share of grassland, like $P_4$ $[E_2(w_1+r_1) + I_1(p_1) + S_1(c_1)]+Z_2$, cereal-meadows with a share of intensifying and structure-forming crops, became relatively more frequent in the Mazurian Lakeland, in the Sudetes and Bieszczady Mountains. In Greater Poland, however, orientations with a share of industrial crops were partly replaced by less intensive orientations. Orientations with a more intensive use of arable land shifted to central and south-eastern Poland, i.e. an area where State farms were scarce.

Collective farms have developed in Poland since 1948. In the peak period of their growth, that is until 1956, they occupied 11 per cent of the country's agricultural land and were characterised by a great variety and changeability of the orientations of agricultural-landuse. Orientations with a lesser or greater share of idle lands occurred among them, especially in north-eastern Poland. Orientation with a share of intensifying crops, and in particular of industrial crops, vegetables and more exacting cereals, were relatively few. The highly field system $(P_6)$ was quite frequent (50 per cent of the powiats with the collective farms in 1950); the most common were three orientations — $P_6[E_5(r_3+o_2)+I_1(p_1)]$, $P_6$ $[E_4(r_2+o_1+b_1) + I_1(p_1+S_1(a_1)]$ and $P_6[E_4(r_3+o_1) +I_2 (p_2)]$. Rye and oats dominated among extractive crops, potatoes among intensifying crops, and annuals among structure-forming crops. They appeared in about one-third of the powiats, mainly in Greater Poland and Kujavia, i.e. the area where collective farms were quite numerous. The field orientations with a secondary share of grassland $(P_5Z_1)$ with a structure of groups of crops cultivated on arable land — $E_3 + I_1 + S_1$ and $E_4 + I_1$, occurred quite frequently (in one-fourth of powiats). Those orientations had dominated until 1955. The number of collective farms went down in the subsequent five years, while the number of orientations went up. Their spatial distribution was mosaic-like; the orientations with a big share of grassland appeared mainly in the south-eastern and western parts of the country (lying along the frontier). Arable land was relatively more intensively utilised mainly in Lower Silesia and in southern Greater Poland. In the 1960s the area with highly field orientations decreased (down to 23%) to the benefit of field orientations with a secondary share of grassland (up to 40%), grassland-field and grassland orientations, occurring mainly in south-eastern and north-western Poland. The number of powiats with orientations with a share of industrial crops significantly increased (to over 30 %), mainly in western Poland and Lower Silesia.

The division into powiats was cancelled when a new, two-level administrative division of the county voivodship commune, was introduced in 1975 to replace the former three-level division (voivodship, powiat, commune); thus, comparative research by powiats could no longer be continued.

The orientations of agricultural landuse in private farming (Table 10.2) is presented in a general way by 49 voivodships in Fig. 10.1. The combinations of the structural land and crops were determined (by R. Szczesny) also by means of the method of successive quotients and subsequently grouped in the same way as the orientations in 1947-1971. On the 72% of the country's agricultural land (this percentage of land was covered by private farming in 1988) 17 orientations identified on the voivodship scale, namely :

- highly field ($P_6$), occurring only in 10 per cent of the voivodship, represented by the relatively intensive wheat-rye-barley-potatoes-sugar beet-clover orientation, appearing in Kujavia, and potatoes-rye with a secondary share of wheat and barley, appearing in the Lublin Plateau and Greater Poland;
- field with a secondary share of grassland ($P_5 + Z_1$), very diversified as regards arable-landuse, occurred in over 70 per cent of voivodships. A relatively extensive arable-landuse is represented by cereal orientations with a secondary share of intensifying crops ($E_4 + Z_1$), rye with a share of oats, barley and potatoes, characteristic of the Sub-Carpathian region and Pomerania, and a similar orientation though slightly more intensive, as oats or barley are replaced by wheat, appearing in north-western Poland.

The subsequent two cereal orientations with a secondary share on intensifying crops and structure-forming crops ($E_3 + I_1 + S_1$), wheat with a secondary share of rye or barley, potatoes, and clover — $P_5[E_3(w_2 + r_1 \text{ or } b_1) + I_1(p_1) + S_1(c_1)] + Z_1$, and another one with a

TABLE 10.2: **Structure of arable landuse**

*(percentage)*

| Sl. No. | Crops | 1950 total | 1988 Total | 1988 Private farms | 1988 State farms | 1988 Collective farms |
|---|---|---|---|---|---|---|
| 1 | 2 | 3 | 4 | 5 | 6 | 7 |
| I | Extractive crops (cereals) | 63.6 | 58.4 | 61.2 | 47.1 | 59.6 |
| 1. | wheat | 9.9 | 15.2 | 14.7 | 16.8 | 16.7 |
| 2. | rye | 33.8 | 16.2 | 18.4 | 7.8 | 13.1 |
| 3. | barley | 5.6 | 8.7 | 7.9 | 11.2 | 12.4 |
| 4. | oats | 11.3 | 5.9 | 6.1 | 5.0 | 6.6 |
| 5. | Buckwheat, millet | 1.2 | 0.4 | 0.5 | 0.2 | 0.4 |
| 6. | mixed cereals | 1.8 | 8.2 | 10.3 | 1.0 | 3.8 |
| 7. | wheat-rye | — | 3.8 | 3.3 | 5.1 | 6.6 |
| II | Intensifying crops | 24.9 | 25.2 | 26.5 | 21.1 | 18.6 |
| 1. | potatoes | 17.4 | 13.0 | 15.6 | 4.5 | 3.4 |
| 2. | fodder root crops | 1.6 | 1.5 | 1.7 | 0.6 | 0.4 |
| 3. | vegetables | 0.8 | 1.8 | 2.2 | 0.6 | 0.7 |
| 4. | industrial crops | 4.1 | 6.8 | 5.2 | 12.7 | 9.8 |
| | a.magar beet | 1.9 | 2.9 | 3.1 | 2.3 | 2.3 |
| | b.oleaginous | 1.1 | 3.3 | 1.4 | 10.4 | 7.5 |
| | c.fibrous | 0.9 | 0.2 | 0.3 | 0.0 | 0.0 |
| | d.onthers industrial crops | 0.2 | 0.3 | 0.4 | 0.0 | 0.0 |
| 5. | onthers crops | 1.0 | 2.1 | 1.8 | 2.7 | 4.3 |
| III | Structure forming crops | 11.5 | 13.8 | 11.6 | 23.4 | 14.3 |
| 1. | papilionaceous (clover, lucrene) | 3.0 | 6.1 | 6.4 | 5.6 | 2.9 |
| 2. | annuals papilionaceous | 4.7 | 5.4 | 4.1 | 11.6 | 4.3 |
| 3. | leguminous fodder grains | 3.1 | 2.0 | 0.8 | 5.7 | 6.4 |
| 4. | edible leguminous | 0.7 | 0.4 | 0.3 | 0.5 | 0.7 |
| IV | Maize | — | 2.6 | 0.7 | 8.3 | 7.5 |
| I to IV | Total | 100.0 | 100.0 | 100.0 | 100.0 | 100.0 |

lesser proportion of wheat to the benefit of other cereals are of a slightly different character. They both appeared in south-eastern Poland, the Carpathian territories excluded. As regards arable-landuse cereal-intensifying crops or intensifying crops-cereal $(E_3 + I_2)$ orientations are much more intensive and relatively numerous. Starting with rye with potatoes orientations characteristic of the field system in southern Greater Poland and the Czestochowa Upland, going through potato orientations with a secondary share of wheat, rye, oats or barley (Upper Silesia, northern Mazovia), we arrive at potatoes-wheat with a secondary share of rye or mixed cereals or wheat with potatoes, sugar beet and barley orientations in lower Silesia. Because of a significant proportion of vegetables the orientation : rye with a secondary share of oats, potatoes and vegetables, and grassland — $P_5[(E_3)r_2 + o_1(+I_2)p_1 + (V_1)] + Z_1$, appearing in the Warsaw voivodship, should be mentioned.

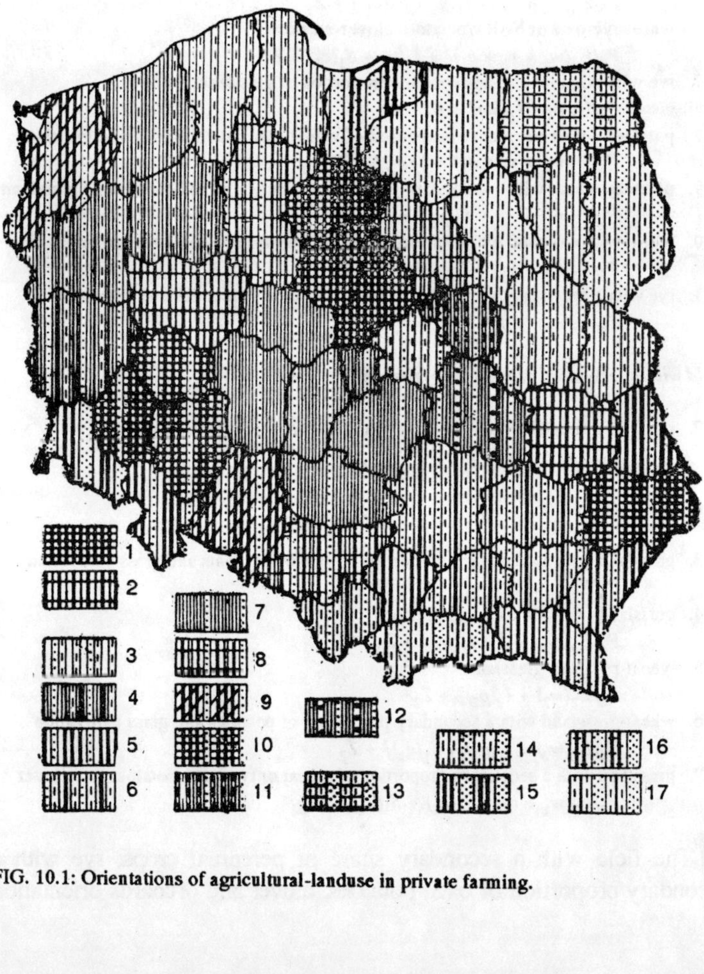


FIG. 10.1: Orientations of agricultural-landuse in private farming.

*Highly Field Orientations ($P_6$)*

1. wheat-rye-barley-potatoes-sugar beet-clover
   $P_6[E_3(w_1 + r_1 + b_1) + I_2(p_1 + s_1) + S(c_1)]$
2. potatoes-rye with a secondary proportion of wheat and barley
   $P_6[E_4(r_2 + w_1 + b_1) + I_2(p_2)]$

*Field with a secondary proportion of grassland ($P_5 + Z_1$)*

3. rye with a secondary proportion of oats, barley, potatoes and grassland
   $P_5[E_4(r_2 + o_1 + b_1) + I_1(p_1)] + Z_1$
4. rye with a secondary proportion of wheat, oats or barley, potatoes and grassland
   $P_5[E_4(r_2 + w_1 + o_1) + I_1(p_1)] + Z_1$
5. wheat with a secondary proportion of rye or barley, potatoes, clover and grassland
   $P_5[E_3(w_2 + r_1) + I_1(p_1 + S_1(c_1)] + Z_1$
6. wheat-rye-oats or barley/potatoes clover-meadows
   $P_5[E_3(w_1 + r_1 + o_1)] + I_1(p_1 + S_1(c_1) + Z_1$
7. rye with potatoes and a secondary proportion of grassland
   $P_5[E_3(r_3 + I_2(p_2)] + Z_1$
8. potatoes with a secondary proportion of wheat, rye, oats/or barley/and grassland
   $P_5[E_3(w_1 + r_1 + o_1) + I_2(p_2)] + Z_1$
9. potatoes-wheat with a secondary proportion of rye/or mixed cereals and grassland
   $P_5[E_3(w_2 + r_1) + I_2(p_2)] + Z_1$
10. wheat with a secondary proportion of barley, potatoes, sugar beet and grassland
    $P_5[E_3(w_2 + b_1) + I_2(p_2 + s_1)] + Z_1$
11. rye with a secondary proportion of oats, potatoes, vegetables and grassland
    $P_5[E_3(r_2 + o_1) + I_2(p_1 + v_1)] + Z_1$

*Field with a secondary proportion of perennial crops ($P_5 + T_1$)*

12. rye with a secondary proportion of oats, potatoes, clover and orchards
    $P_5[E_3(r_2 + o_1) + I_2(p_1) + S_1(c_1)] + T_1$

*Field with grassland ($P_4 + Z_2$)*

13. grassland with a secondary proportion of rye, barley, oats and grass cultivation
    $P_4[E_3(r_1 + b_1 + o_1) + S_1(g_1)] + Z_2$
14. grassland-rye with oats and potatoes
    $P_4[E_3(r_2 + o_1) + I_1(p_1)] + Z_2$
15. wheat-potatoes-grassland
    $P_4[E_2(w_2) + I_2(p_2)] + Z_2$
16. wheat-grassland with a secondary proportion of potatoes and grass cultivation
    $P_4[E_2(w_2) + I_1(p_1) + S_1(g_1)] + Z_2$
17. grassland with a secondary proportion of wheat or rye, oats, potatoes and clover
    $P_4[E_2(w_1 + r_1) + I_1(p_1) + S_1(c_1)] + Z_2$

The field with a secondary share of perennial crops, rye with a secondary proportion of oats, potatoes, clover and orchards orientation was identified only in the Random voivodship, into which the principal group orchard area of Poland was incorporated — $P_5\{E_3(r_2 + o_1) + I_1(p_1) + S_1 + (c_1)] + T_1$.

Field orientations with grassland $(P_4 + Z_2)$, characteristic of the north-eastern areas and the Carpathian and Sudetes mountains, show a different system of agricultural-landuse. Of them less intensive are : grassland with a secondary share of rye, barley, oats and grass cultivation — $P_4[E_3(r_1 + b_1 + o_1) + S_1(g_1)] + Z_2$, as well as grassland-rye with a share of oats and potatoes : whereas more intensive are: $P_4[E_2(w_2) + I_2(p_2)] + Z_2$ wheat-potatoes-grassland, $P_4[E_2(w_2) + I_1(p_1) + S_1(g_1)] + Z_2$ — wheat-grassland with a secondary proportion of potatoes and grass cultivation, and — $P_4[E_2(w_1 + r_1) + I_1(p_1) + S_1(c_1)] + Z_2$, grassland with a secondary proportion of wheat or rye, oats, potatoes and clover (Fig. 10.1).

The orientations of agricultural landuse, identified for private farming, can be recognised as representative also of agriculture as a whole in the areas where the proportion of socialised farming is little, that is mainly in the south-eastern halt of the country, starting with the Bialystok voivodship in the north, the Konin voivodship in the west, through the Sieradz and Czestochowa voivodships up to the Bielsko Biata voivodship, the eastern area of the Zamosc, Przemysl and Krosno voivodships excluding. In the remaining area, north-western in particular, where socialised farming, mainly State farms, occupied 40-50 per cent of agricultural land, the orientations of agricultural-landuse as regards agriculture as a whole are undoubtedly different, especially as far as the orientations of arable landuse are concerned : this is, of course, caused by the different ways of husbandry in private and in socialised farming.



11

# LANDUSE IN THE EASTERN HIMALAYAN REGION

SULEKHA LAHIRI

CORRECT APPRAISAL of the nature of the utilisation of land in any region helps better understanding of the problems and prospects of that region. Mountainous region with their high altitude, steep slopes, narrow valleys with fast flowing streams offer certain resources which are different from those of the low-lying regions. The study of land utilisation of a small unit area in the Lesser and Outer Himalayas of the Eastern Himalayan Region reveals the dominating control of a physical set-up, on the one hand and certain political events in the last century on the other.

The area of study comprises the three hill sub-divisions of Darjeeling district viz., Darjeeling Sadar, Kurseong and Kalimpong. They cover an area of 239,298 hectares out of a total of 307,000 hectares for the whole district. The region lies between Latitude 26° 48′N — 27° 13′30″N and Longitude 87° 57′ — 87° 59′E. This hill territory was annexed to Bengal only in 1835 after the Titalya Treaty with Nepal, Sikkim and British Government. Kalimpong Sub-division was acquired even later from Bhutan after the war with her in 1865-66.

Dense forests clothed these majestic mountains. Hunting and primitive migratory cultivation by the local inhabitants — Lepchas were the earliest form of landuse in this mountainous territory. Later on, a sanatorium was established for the British soldiers. This laid the foundation of this famous hill resort of the Eastern Himalayas. A depot for the recruitment of soldiers for the Gorkha Regiment was also established here.

Estate farming, a typical form of commercial agriculture was introduced here by the end of 19th Century. Several political and social factors operating at the end of this Century were responsible for the rapid expansion of plantation acreage throughout the Western half of the district. An accumulation of capital in Great Britain during the early and middle part of the 19th Century, coupled with the loss of British tea trade monopoly in China in 1833, encouraged the idea of introducing tea, cinchona and indigo in India. As indigo plantation was not successful, more stress was given to other plantation crops. The opening of the Suez Canal in 1869 also facilitated the export of plantation products. Because of its limited use the area under cinchona remained very limited. Rapid expansion of area under tea estates resulted in the fast clearing of virgin forests, and rapid increase in population. But as the workers of the tea gardens were provided with staple food articles, the area under indigenous cultivation did not increase with the increase in population density. But in the eastern part of the district in the Kalimpong Sub-division, large tracts were kept as reserved forests and pastures. Only a few tea estates were allowed to be set-up there. Kalimpong served as a entrepot for the trade between Tibet and Bengal at that time. The areas remaining outside the reserved and protected forests and tea estates were known as Khasmahals, where indigenous farming with other subsidiary activities are carried on.

Thus the history of development of these hill sub-divisions of Darjeeling shows a distinct trend in landuse pattern set by early administrators. Few other additions were made later on, but the main trend remains the same, even today — that is, tea plantation, siviculture and agriculture with terraced cereal and vegetable cultivation supplemented by animal husbandry.

TABLE 11.1: Landuse patterns

*(Area in Hectares)*

| Landuse/Sub. Div. | Darjeeling | | Kurseong | | Kalimpong | |
|---|---|---|---|---|---|---|
| | Area | % age | Area | % age | Area | % age |
| Total reported area | 91,118.82 | — | 42,529.55 | — | 105,650.00 | — |
| Forest | 31,042.25 | 34 | 21,221.18 | 49.8 | 48,725.00 | 46.12 |
| Cultivated Area | 16,125.69 | 17.50 | 5,562.15 | 13.00 | 21,301.21 | 18.16 |
| Tea Estates | 23,258.71 | 25 | 14,675.00 | 34.5 | 3,029.55 | 2.87 |
| Area not available for cultivation | 19,972.7 | 16.76 | 4,169.41 | 0.9 | 34,689.88 | 29.96 |
| Current fallow | 721.9 | 0.7 | 589.0 | 0.79 | 720.0 | 0.75 |

*Source*: District Rural Development Agency, Darjeeling.

A brief description of the natural environment including the surface relief, rock character, soil and climate would help us to grasp the nature of control of environment on the utilisation of land in this area.

**Physical Environment**

These hill tracts are divided into two distinct parts, Darjeeling and Kalimpong divisions by the North-South valley of the river, Tista. The Darjeeling division comprises the Sadar Sub-division, and the Kurseong Sub-division, and the Kalimpong Sub-division, and the Kalimpong Division covers the whole area of Kalimpong Sub-division.

The Darjeeling division is traversed by a long ridge entering the area from the north. This is a continuation of the Singalila Range, which is a part of the Lesser Himalayas. This range, known as, Senchal-Mahaldhiram Range here, runs first in west-east direction, then turns southwards to taper off gradually below Kurseong town. The average elevation of this range, varies from 1,875m. to above 2,000 m. Highest point in this range is reached at Sandakpu Peak, 3727.8 m. Ridge tops are too high and precipitous for any other use but silviculture. Elevation decreases gradually on all ridges from this range. A narrow belt of highlands with an average altitude between 1250m — 1875m encircles this high range. Below this level lies a broad belt with an average elevation varying from 625m to 1250m. These two lower levels with comparatively less steep gradients have been thoroughly utilised by the tea planters. Although the relative relief of this high altitude region is above 625m, there are local variations in gradient. Land slips, debris slips, soil-slips, sheet-wash and gully-erosion have often modified the steep slopes into moderate ones. These areas with moderate slopes often accommodate the tea gardens and other cultivated areas. Areas with altitude varying between 312m to 625m lie mainly near the valleys of the main rivers. Relative relief varies from 312m to 625m, but slopes near the main valleys are usually very steep. Tea and cinchona plantations as well as few patches of cultivated blocks are found here.

This whole division is drained by two major rivers and their tributaries, viz., the Tista and the Great Rangeet river system and the Mahananda-Balason river system. The Senchal-Mahaldhiram range which forms the major water-parting of this area is being dissected on all sides by the fast flowing streams. Long and narrow spurs separated by the steep sided young valleys containing torrential streams characterise the landscape of this mountainous region. Besides these major streams, innumerable 'jhoras' and 'kholas' originating from the rapid run off resulting from heavy monsoon rains have also made the surface more rugged with excessive soil erosion.

The Kalimpong Division, to the east of the river Tista shows a broad mass of a highland in its north-central part. This is a continuation of the high ranges of Bhutan. Highest point in this division is reached at Rishila (above 3152m), which forms the nodal point from which spurs radiate on all directions, separating deep mountain valleys. Over most of the section, the central range has an altitude between 1875m and 2500m. Only in a small section in the north central part, the elevation exceeds 2500m. Next to this high altitude belt is another highland zone, with an elevation varying between 1250m to 1875m. Finally there is a broad belt with an average altitude ranging from 625m to 1250m. Areas lying between 312m and 625m form a comparatively wide belt on the outer margin of this division, bordering the narrow valley floors. There are small patches of lowlands (156m to 312m) occurring at the zone of contact with the Dooars plains in the south. The relative relief of the highest area shows the highest value—that is above 625m. This region is consequently covered with thick forests. Over rest of the area of this division the relative relief varies from 312m to 625m. Most of the cultivated areas are located at this level. Southern most fringe of this division has got minimum slope ranging from 78m to 312m. This area also remains under forests probably due to heavy rain and dissected topography.

Kalimpong division is drained mainly by the river Tista and its tributaries. Only a small portion in the eastern part forms a part of the river Jaldhaka basin. Heavy rainfall on the southern and the western flanks of the central highland mass is wearing away the land surface at a very rapid rate. A large number of parallel streams are dissecting this area into an intricate pattern of narrow valleys and spurs. Northern flanks get less rainfall and consequently the drainage density is less here.

## Geology and Soil

Geologically, three different types of rock zones could be identified in this area. Low frontal hills of Siwalik formations lying below 937m, are composed of pebbly sandstone and conglomerates at upper levels and sandstones in the lower levels. Shales, claystone and siltstone are found in the lower part of the sequence. Next group is the narrow belt of Gondwanas lying between the Siwaliks and the higher hills above 1000m. It is composed of coal seams, shales, clays and sandstones.

The hill ranges beyond this Gondwana belt is composed of metamorphic rocks. The outer zone of these mepamorphites is composed of slates, phyllites and low grade schists with subordinate quartzites. Similar group of rocks known as Buxa series occur in a narrow land from Neora river to Jaldhaka river in Kalimpong Sub-division. The Buxa group contains dolomite in addition to slates, phyllites and schist. This dolomite has great value in ameleorating the acidity of the soil of this area. Soil is grey porous sandy loam.

Overlying the Dalings are the Darjeeling gneiss consisting of schists and gneisses of higher rank with subordinate quartzites. At upper level gneisses ultimately give place to garnetiferous silimanite gneiss/schist. A stiff brownish loam results from this gnessic rock but few areas also show almost pure sandy soil or stiff red clay. Under conditions heavy rainfall and low summer temperature there has been much podzolisation, but not uniformly throughout the area. The surface layer is often found to be shallow. But soil-cover was thicker over the heavily wooded areas. Removal of plant cover has often resulted in complete washing away of the soil cover. Thin and light coloured soils found over steep slopes are less productive than the dark coloured soils found over gentler slopes.

On the whole soil of the hilly region are grouped under the name 'grey brown forest soils'. An analysis of soil samples collected from different parts of Darjeeling district shows the availability of the plant nutrients as well as the pH level of the soil.

## Climate

Altitudinal variation brings a quick change in climate within a short distance in this hill region. Climate changes from tropical in the foot hills region through sub-tropical in the middle hills to temperate in the upper hills region.

Temperature below freezing point is experienced every year in higher areas. Snow falls occasionally above 2187m. Frost is very frequent in areas above this altitude. Frost assists in disintegration of rocks and forming soil. Gradual melting of snow helps plant growth and also feeds the springs with water in the dry season.

TABLE 11.2: Range of nutrient status and pH in different types of soil

| Subdivisions | Name of Block | N | AV.$P_2O_5$ kg/ha. | AV.$K_2O$ kg/ha. | pH | m.mhos (E.C.basis) | Predominant Type | % of total area | Other type | % of total area |
|---|---|---|---|---|---|---|---|---|---|---|
| Darjeeling | Bijonbari | 0.40-2.20 | Trace-261.62 | 79.5-820.96 | 5.0-6.0 | 0.10-0.25 | | 95 | | 5 |
| | Sukhiapokhri | 0.83-3.08 | Trace-309.1 | 134.4-985.6 | 3.8-5.9 | 0.04-0.27 | Sandy loam | | Loam | |
| | Takdah | 0.47-1.21 | 2.73-184.75 | 61.6-896.0 | 5.0-6.9 | 0.10-0.30 | | | | |
| Kurseong | Kurseong | 0.26-0.90 | 17.53-370.0 | 87.36-483.84 | 5.5-6.0 | N.A. | | 70 | Sandy and clay loam | 30 |
| | Mirik | 1.05-1.46 | 13.16-161.5 | 192.64-560.0 | 5.6-6.2 | 0.10-0.17 | Sandy loam | | | |
| Kalimpong | Kalimpong I | 0.5-1.35 | 5.1-29.9 | 64.0-554.3 | 4.6-6.5 | 0.05-1.90 | | | | |
| | Kalimpong II | 0.37-1.87 | 3.9-52.0 | 120.0-720.0 | 4.0-6.2 | 0.04-0.55 | Sandy Loam and | 67 | Loam | 5 |
| | Gorubathan | 0.82-1.66 | 25.0-192.3 | 80.0-1120.0 | 4.7-6.3 | 0.05-0.23 | Sandy | 28 | | |

*Source*: District Annual Plan on Agriculture for Darjeeling, 1985-86.

TABLE 11.3: Variation in temperatures at different altitudinal levels

| Stations | Altitude | Mean Temperature Coldest Month | | Mean Temperature Warmest Month | |
|---|---|---|---|---|---|
| Darjeeling | 2200m | 6.2 C | January | 17.4 C | August |
| Kalimpong | 1224m | 11.4 C | January | 23.7 C | June |
| MO* | 1562m | 11.7 C | February | 21.1 C | June |
| Lava* | 2120m | 5.7 C | February | 17.3 C | June |
| Nagri Farm | 1158m | 10.15 C | February | 25.8 C | September |

* Data is available only for 1972. It has been collected from the Forest Office, Darjeeling.

*Source*: (a) District Census Handbook, Darjeeling, 1961.
(b) Project Report for the development of under-developed forests in North Bengal through the agency of West Bengal Forest Development Corporation Ltd. Pt. 1. Darjeeling and Kalimpong Divisions, Directorate of Forests, Government of West Bengal, 1974.
(c) Annual Scientific Report 1972-76, Tea Research Association, Calcutta.

This region receives heavy to very heavy rainfall from the south-west monsoon during the period extending from June to September. The mean annual rainfall exceeds 610 cm. at Paren (altitute 1500m) located on the eastern flank of the North-Central range of Kalimpong division. On the whole rainfall exceeds 350 cm in all the stations located on the southern flanks of the central ranges of both the divisions. But a decrease in the amount is noticed on the ridge tops as well as on the lower slopes.

Table 11.4 illustrates the varition in the amount of rainfall at different attitudinal levels. High amount of rainfall in such a short period often aggravates soil erosion and leaching, particularly on steeper slopes. A small quantity of rain is also received in winter and it helps the farmers to grow vegetables. Pre-monsoon rains in March and April bring fresh flush of leaves to the tea bushes after the dormant winter period. Quality tea is often made from these spring flushes. Again too heavy rain in later period often deteriorates the quality of tea. Humidity is very high and thee sunshine is deficit. Thick mist prevails over these hill tracts throughout the greater part of the year. The mist however, is one of the important factors influencing the production of the quality tea in Darjeeling.

Under this distinct nature of physical set up, enumerated above, the landuse pattern in this area has assumed certain characteristic features which will be discussed below.

TABLE 11.4: Variation in the amount of rainfall at different altitudinal levels

| Sub-Division | Stations | Altitude | Total Annual Rainfall in cm.* |
|---|---|---|---|
| Darjeeling | Sandakphu | 3727.8 m | 330.73 |
| | Darjeeling | 2200 | 243.13 |
| | Sukhia Pokhri | 1900 m (approx.) | 309.43 |
| | Takdah | 1607.8 m | 237.56 |
| | Namring | 1280 m | 357.55 |
| Kurseong | Kurseong | 1875 m | 480 |
| | Mahaldhiram | 1718 m | 480 |
| | Latpanchor | 1171 m | 396 |
| | Pankhabari | 578 m | 368 |
| | Sukna | 166 m | 380 |
| Kalimpong | Labha | 2120 m | 301.4 |
| | Kalimpong | 1224 m | 205.6 |
| | Paren | 1500 m | 610.9 |
| | Kuapani | 1400 m | 543.2 |
| | Samsing | 625 m | 516.1 |
| | Gorubathan | 609 m | 505.9 |

* Data available only for 1972.

TABLE 11.5: Monthly rainfall average for 15 years

*(Rainfall in Centimetres)*

| Months | Darjeeling | Kalimpong | Kurseong |
|---|---|---|---|
| January | 1.61 | 0.2 | 1.2 |
| February | 0.46 | 2.5 | 1.3 |
| March | 3.38 | 2.5 | 4.3 |
| April | 9.86 | 3.0 | 5.8 |
| May | 17.12 | 20.0 | 18.3 |
| June | 49.79 | 70.0 | 65.0 |
| July | 74.71 | 110.0 | 90.8 |
| August | 54.0 | 66.6 | 37.9 |
| September | 38.73 | 69.0 | 55.3 |
| October | 27.68 | 18.0 | 13.3 |
| November | 0.37 | 0.3 | 2.1 |
| December | 0.16 | — | 0.2 |

*Source*: District Annual Plan for Agriculture for Darjeeling, 1985-86.

## Forests

The percentage of land kept under forests should be considerably high in a rugged region with very heavy rainfall concentrated within a short period of the year. Forests, managed properly not only gives the highest monetary returns in hilly terrains at high altitude, but also provides the best safeguards for the maintenance of the ecological balance of the whole region.

Forests extend from 312 m to the highest levels of this highland region. Consequently a large number of physiographic, climatic, edaphic and biotic factors influence the character of vegetation. The variation in rainfall and aspect even over comparatively smaller areas has considerable effect on the distribution of species and the composition of tree crops. Thus in the valley forests, shorea Robusta (Sal) grow at more or less same altitudes mainly on southern aspects. Similarly, in the Rangeet Valley at Badamtam where the rainfall is low Pinus Longifolia (chirpine) appears on the ridges and on the southern aspects, whereas this species is absent on the northern aspect. Again at Rammam where rainfall is less than occurring further east Tsuga (Tengra Salla) brunoniana appears in compact patches, whereas it is absent in areas with the same elevation further east. As a result of these variations, a large variety of subtypes of forests are found in this area. Table 11.6 gives the classification of forests of this region according to Champion and Seth.

TABLE 11.6: **Classification of forest**

| Type No. | Type of Forest | Altitudinal Belts of Occurrence (Approximate) |
|---|---|---|
| 14/C2 | East Himalayan Alpine Birch/Fir. | Above 2812 m |
| 12/C3a | East Himalayan Mixed coniferous | 1875-2812 m |
| 11B/C1/C1c | High Level Oak Forest | 1875-2500 m |
| 11B/C1/C1b | Buk-Oak Forest | 2187-2500 m |
| 116/C1/C1a | Laurecous Forest | 1875-2187 m |
| 8b/C1 | East Himalayan Sub-tropical Wet hill forest | 1561-1875 m |
| 3c/3cb | East Himalayan moist mixed deciduous | 625-1250 m |
| 3c/C3b | East Himalayan Lower Bhabar Sal | 250-625 m |
| 2b | Assam Valley Tropical Semi-Evergreen | 625-1875 m found in Kalimpong division only |
| 3c/C1b(1) | Upper Bhabar Sal | 625-1250 m |
| C3/C1a(1) | East Himalayan Sal | 312-625 m |
| 5/1S2 | Khair-Sisso forest | 156-312 m |

*Source*: Directorate of Forests, Government of W.B., Darjeeling Division.

Darjeeling forest division was formed as early as 1874 and Kurseong and Kalimpong forest divisions were formed little later for the better management of the forests of this area. For an all round development and utilisation of the forest areas the West Bengal Forest Development Corporation was entrusted with the management of the forest divisions of Darjeeling and Kalimpong.

The main objectives of forest management in this area are the following:

(a) To develop commercial lumbering by large scale harvesting and replanting of suitable species.
(b) To encourage mechanisation of logging and extraction operations to minimise wastage and to develop a network of all-weather trackable roads.
(c) To market various forest produce and to encourage and establish forest based industries.
(d) To develop recreational facilities and to create infrastructure for development of tourism.
(e) To promote, develop and manage wildlife and fishing.
(f) To prevent soil erosion and increase water supply.
(g) To convert natural forests which have degenerated through old age into new plantations with suitable species.
(h) To train up persons for taking up more advanced management of forests, flora and fauna.

TABLE 11.7: **The area under clear felling and replantation**

| Forest Divisions | Area felled in hectares | Plantations raised in hectares | | |
|---|---|---|---|---|
| | 1983-84 | 1982-83 | 1983-84 | 1984-85 |
| Darjeeling | 111 | 230 | 356 | 244 |
| Kalimpong | 479 | 733 | 265 | 303 |
| Kurseong* | 121 | 484 | 684 | 626 |
| Kalimpong Soil Conservation | | 147 | 22 | 20 |
| Kurseong Soil Conservation | | 105 | 100 | 134 |

* Figures include parts of Siliguri Sub-division.
*Source*: West Bengal Forest Statistics, 1984.

It is clear from Table 11.7 that Forest Directorate is planting more than it is felling. Blank areas in reserved forest and the degraded areas (resumed forests) from the tea estates and Khasmahals are converted into valuable plantation areas. Entire felling is being undertaken strictly as per working plan schedule laid down by the directorate of forests. The middle and the upper hill areas are planted with a mixture of indigeneous broad leaved species like birch, pipli, and maple together with quick growing coniferous species like cupressius, cashmerinca, pinus patula and hemlock. In the foothills teak, sal, champ, gokul, mudane, chikrasi, panisaj, chilaone and other valuable miscellaneous species are raised.

Natural forests, where trees like oak, maples and champ have attained great age are rapidly deteriorating. Here natural regeneration is virtually absent. So clear felling system followed by artificial regeneration was evolved. Clear felling is not confined in big patches. It is rather distributed in suitable areas with average of 15 hectares or so in one place. This process of conversion of the natural forests is taking place in the area for the last 100 years and so far only 27 per cent of the natural forests have been converted into planted forests. Protected forests near the source of streams, streambanks, over the steep slopes and catchment areas are always preserved. Over 33 per cent of the forest areas of Darjeeling hills are strictly kept under total protection.

Commercial forestry depends to a great extent on the accessibility of the forests. According to a report prepared by the Directorate of Forests, Government of West Bengal 1974, 44 per cent of the total area of the forests in Darjeeling and Kalimpong Subdivisions is accessible (within 2 kms. from the nearest road). 40 per cent is inaccessible (at least 5 kms. away from the road), and 15 per cent is partially accessible (2.5 kms. away from the road). Accessibility has improved significantly in only a few areas where sky-line felling is done. In Kurseong Sub-division large parts of forest are accessible, but some areas in lower hills and foothills are either inaccessible or partially so. Many valuable timber trees growing in this area take 30-60 years to mature. The time for maturity varies from species to species (Table 11.8).

Minor products like, medicinal plants such as Aconitum Spicatum, Siverita-Chirota, Aluridum and Apalmatum are obtained from Singalila Range. Bamboo is also produced in considerable quantity. Fodder is in great demand and at present, permits are given to all, especially to Khasmahal tenants for grazing in the forests in accordance with the directives of the Divisional Forest Officer.

TABLE 11.8: **Availability of timber per utility classes**

| Type of timber/sub. div. | Darjeeling | Kurseong | Kalimpong |
|---|---|---|---|
| Constructional timber | 16.01 | 10.61 | 8.88 |
| Ply-wood | 6.32 | 15.23 | 8.52 |
| Box-wood | 1.64 | 8.08 | 2.31 |
| Furniture wood | 0.91 | 3.59 | 0.90 |
| Match wood | 0.59 | 2.44 | 0.02 |
| Pulp wood | 0.04 | 0.29 | 0.42 |
| Fuel wood | 0.03 | 0.02 | 0.06 |
| In Lakh cubic meters | 27.55 | 30.91 | 17.50 |

* Figures for the whole Kurseong Forest Division which includes parts of Siliguri Subdivision of Darjeeling district. Separate figures for Kurseong sub-division not available.

*Source*: Project report for the development of underdeveloped forests in North Bengal through the agency of West Bengal Forest Development Corporation Limited Part I Darjeeling and Kalimpong Forest Divisions. Prepared by Directorate of Forests, Government of West Bengal, 1974.

A study of Forest Resources of Kurseong Forest Division — P. Guhathakurta. Siviculture (North) W.B.

At present North Bengal has only few wood based industries, viz., 60 units of saw mills, 6 units of ply-wood factories making tea chests, blackboards and commercial ply-wood. All these units consume much of the forest produce of this hill area but a part of the produce is also sent to Calcutta area. Sal sleepers are supplied to the Railways. Cryptomaria Japanica has got great demand for house building in the hills. There is a heavy demand for firewood in the hill region. So fuel wood is often converted into charcoal and finds a ready market in the vicinity of the producing areas.

In the recent years there has been a rapid degeneration of forest areas in this hill region. Actual deforestation has taken place in the tea garden and Khashmahal forests and large gaps have been created in the reserved and protected forests. Encroachment on the forest areas by the local farmer, excessive exploitation for fuel and fodder, and illegal felling of timber have resulted in this degradation. Excessive soil erosion, frequent landslides, drying up of the streams, frequency of floods in the lower reaches of the rivers are few of the problems which are being faced by the people today.

As a remedial measure a special working circle styled as firewood working circle has been formed for certain identified areas in and around human habitations with the sole objective of meeting firewood and fodder needs of people as part of social forestry programme.

Several quick growing varieties of trees are being planted for fuel and fodder. The names of few of such species are : Alnus nepalensis (*utis*), Fecus Hookeri (*nebharo*), Eribotrya Petobata (*Maya*), Macaranga spp. (*Malata*), Sauraia spp. (*gogun*). The Uttis tree has got the added quality of fixing nitrogen to the soil.

Of late, people have started taking active interest in social forestry. In few urban and Khasmahal areas seedlings are being planted. Of course, people should be made aware of the grave danger of deforestation. Motivation to plant and protection of the trees would come partly through the knowledge of environmental processes and partly through the assurance of some economic gain.

### Tea Estates

Tea estates in the hill region provide us with good examples of commercial plantation farming in the tropics. A large number of rural population of the two hill subdivisions are entirely dependent on this type of farming. According to a recent estimate the tea industry employs approximately 50 thousand persons on a permanent basis and 30 thousand more are engaged during the plucking seasons. However, about 80,000 workers constituting roughly 18 per cent of this total rural population are engaged in the Tea Gardens of the three hill subdivisions.

### Location of Tea Gardens

Physical factors particularly of the relief are mainly responsible for localising the tea estates in certain zones. A small northern zone is located north of the Senchal-Mahaldhiram waterparting and a larger one is situated to the south of this divide. Uptill now 1877 m is considered to be practically the upper limit for the cultivation of tea in this area. Steepness of the terrain and the lack of right temperature above this level are the causes of this limitation.

The smaller northern tea block extends from the little Rangeet Valley in the west to little above the Teesta Valley in the east with few gaps resulting from the southward extension of two spurs — Lebong and Takdah. Though the crests of these spurs are avoided, the flanks are covered with tea bushes. Except in the Teesta Valley which is covered with protective forests the bushes extend right up to the edge of the mountain streams which are vitally important for the water supply.

The larger southern tea block is practically hemmed in by the S.M. waterparting in the north and east. Except where the off shoots from the higher northern ridges intervene, the tea growing areas here is compact. This area is drained by the Balason-Mahanadi river system which has eroded away much of the outstretching spurs of the waterparting.

On the whole the majority of the tea gardens are located in areas with relative relief ranging from 78m to 625m. This is because of the essentially rugged nature of this tract.

TABLE 11.9: **Proportion of tea areas under different relative relief categories**

| Relative relief in meters | Approximate areas of tea in hectares | Per cent of the total tea area |
|---|---|---|
| 78-312 | 3798 | 17.1 |
| 312-625 | 11841 | 53.3 |
| Above 625 | 6586 | 29.6 |
| Total | 22,225 | 100.0 |

*Source:* **From field work and computation of data.**

Out of the three tea gardens established earlier in Kalimpong sub-division only two are operating now. These are located on one of the low lying spurs coming out of the high range in the north. Defective management is responsible for the closure of tea estates.

Production of tea in gardens located in different localities varies from 5.5 quintals per hectare in the N.W. part of the northern block to 8 quintals in the eastern part of the same block. Relief, climate, and edaphic factors are responsible for such variations. China variety of tea bushes of this hill region gives low yield but highly flavoured quality tea. Assam variety of tea bushes planted in lower altitudes and in the plain areas give much higher yield.

The low productivity of tea is also attributed to the great age of the tea bushes and the decreasing fertility of the soil. More than 90 per cent of the total bushes are more than 50 years old. Pests and diseases which often attack the old bushes lower the yield. Replantation is proceeding here at a very slow rate. Because the growers have to wait for several years before the plant is ready for plucking. Recently a new high yielding variety of clones have been developed at Tea Research

Association's Proving Station at Ging Tea Garden in Darjeeling. The high yield from these clones will easily make up for the loss of crop resulting from replantation.

Soil erosion resulting from deforestation is another factor contributing to low productivity of tea. Establishing a good plant cover over the exposed slopes, maintaining drains along the contours would check rapid run off and thus impede soil erosion. Proper care of the bushes, efficient manner of plucking and pruning, irrigation during the dry season, and application of proper doses of fertiliser would help the healthy growth of tea bushes.

After the steady withdrawal of the established foreign tea companies in the late forties, very few of the new owners allowed a normal period of gestation for the tea bush to yield its optimum. Very little attention was paid to the well-being of the tea bushes. Condition deteriorated gradually and about two dozen gardens were closed down from 1950 to 1970. At last in 1985 a scheme was launched by the Tea Board to start the developmental activities in the Tea Estates. A promotional campaign for only Darjeeling tea has also been started to create a consumer consciousness about the quality aspect of Darjeeling Tea. Now-a-days first flush quality Darjeeling tea has a good market in Western Europe and Japan.

### Landuse in the Tea Estates

Considerable portion of the land in a tea estate is planted with tea, but a large proportion is also kept under forest. Steep slopes are always kept under forest cover, because the thin soils of steeper slope are not suitable for the development of tea roots (Fig. 11.1). Plucking and pruning operations are also difficult in areas with higher degree of slope. A small fraction is also used for factories, manager's bungalow, labour colonies etc.

### Khasmahal Lands

Indigenous cultivation in the hill region is carried on mainly in Khasmahal lands which belong to the state. Mixed farming with more stress on maize, potato and vegetables is noticed here. Animal husbandry often supplements the farm income. Compared to the low lying areas, the rural density in the hills is not very high. But owing to the steepness of the slope and the thinness of the soil a large percentage of the area of the village is kept outside the farming activity, although fuel and fodder are collected from these areas. Thus the actual farm density is very high in these villages. The average density of inhabited villages per km² is 329, 233, and 196 in Darjeeling, Kurseong and Kalimpong sub-divisions respectively.

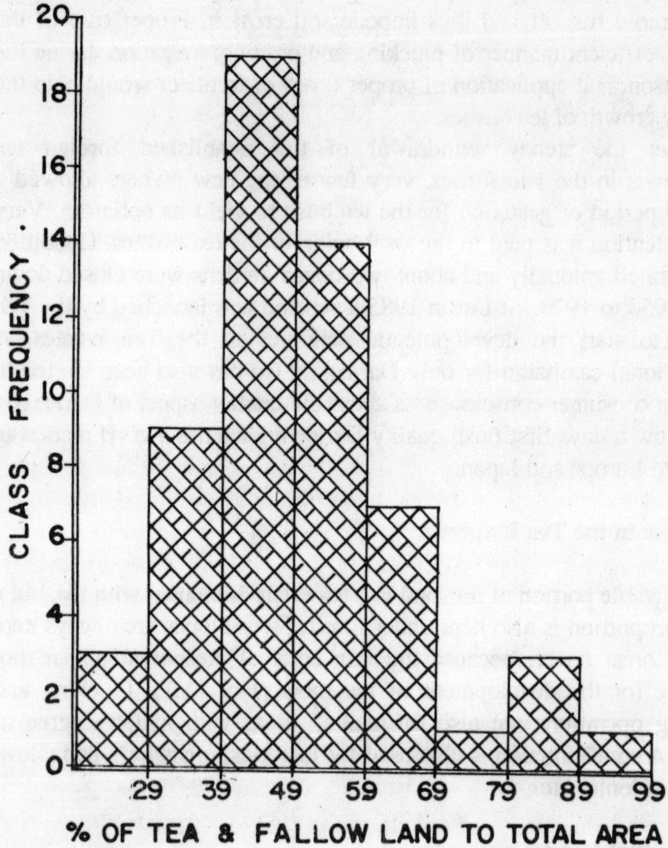


FIG. 11.1: Landuse in the tea estates.

The problem of land alienation or non-personal cultivation is not very acute here. Because these areas have been settled only a little over 100 years ago and were never under permanent settlement. Table 11.11 indicates the ratio between cultivators owning land and the agricultural labourers. Although the number of landless labourers is not very high the actual size of land holding for individual farmer is also not very high. The size of land holding of maximum number of farmers in all the subdivisions is below one hectare and yield per hectare is quite low.

TABLE 11.10: **Proportion of tea area to the total area of the estate**

| Subdivisions | Area under Tea Estates (hectares) | Area under Tea (hectares) 1987-'88 | % of Tea to total Tea Estates |
|---|---|---|---|
| Darjeeling | 23,258.71 | 110041.75 | 47.4% |
| Kurseong | 14,675.00 | 5,734.42 | 33.2% |
| Kalimpong | 3,029.55 | 707.92 | 23.3% |

*Source*: Darjeeling Tea Planters' Association.

Extra effort is needed in the area to make proper terraces and to maintain them. The fields lying at lower elevations, with longer growing seasons are being utilised more intensively with the increase of area under irrigation in recent years.

Field observations show that unirrigated dryfields are often used for maize or millets. Pulses are found to be growing round the edges of the main plots. Paddy fields are irrigated. Small vegetable plots near the homesteads are often provided with irrigation water diverted from nearby streams through bamboo channels.

Field observations have also shown that these dry fields are usually unirrigated and are used for growing maize or millets. Old fallow lands are not cultivated and often lie on steeper areas. Small vegetable gardens near the homesteads are often provided with irrigation water diverted from nearby jhoras through bamboo channels.

Kalimpong sub-division shows the highest percentage of total area under crops, the reason being the virtual absence of tea estates. In Darjeeling and Kurseong cultivated lands are found in areas which have not been taken by tea estates. In the Sadar sub-division, few non-contiguous cultivated patches are found north of the main water parting on the basins of the rivers Great and Little Rangeet. Few cultivated areas are also present just to the south of the water parting. Flanks of the Takdah and Lebong spurs accommodate several farming villages. The cinchona plantations of Mungpoo and Latpanchor are located on the eastern flanks of this water divide. Kurseong sub-division has got very little area left by the Reserved Forests and the Tea Estates.

TABLE 11.11: The size and ownership of land holdings

| | Darjeeling | Kurseong | Kalimpong |
|---|---|---|---|
| *Size up to one hectare:* | | | |
| Number of holding | 5973 | 1019 | 6126 |
| Percentage total holding | 51.25 | 43.38 | 45.62 |
| *Size upto 1-2 hectares:* | | | |
| Number of holding | 2874 | 855 | 3315 |
| Percentage of total holding | 24.66 | 36.40 | 24.68 |
| *Size upto 2-4 hectares:* | | | |
| Number of holding | 2200 | 350 | 2892 |
| Percentage of total holding | 18.87 | 14.90 | 21.54 |
| *Size above 4 hectares:* | | | |
| Number of holding | 608 | 125 | 1097 |
| Percentage of total holding | 5.22 | 5.32 | 8.16 |

*Source*: District Annual Plan on Agriculture for Darjeeling 1985-86.

TABLE 11.12: Number of agriculture labour in all the three sub-divisions

| Subdivisions | Cultivators | Agricultural Labourers | % of Ag. Lab. to the Cultivators | Total Area under Agricultural Land Holding |
|---|---|---|---|---|
| Darjeeling | 26,094 | 4,402 | 5.92 | 14,778 |
| Kurseong | 4,953 | 1,088 | 4.55 | 3,150 |
| Kalimpong | 30,457 | 3,522 | 8.64 | 18,170 |

TABLE 11.13: Irrigated and double cropped area

| Subdivision | Net Sown Area | Net Irrigated Area | % of Irrigated area | Gross Cropped Area | Double Cropped Area |
|---|---|---|---|---|---|
| Darjeeling | 14,524 | 2769.25 | 23.98 | 22,585 | 7,656.5 |
| Kurseong | 3,395.6 | 972.06 | 30.78 | 5,453.8 | 2,058 |
| Kalimpong | 19,191.09 | 6034.0 | 30.30 | 25,110.5 | 12,995 |

*Source*: District Rural Development Agency.

In Kalimpong sub-division, larger blocks of cultivated area are found in the upper basins of the western and southern river valleys.

Smaller areas of cultivation are also found north of the main range in the upper basin of the river Rishi Khola.

On the whole the cropping pattern in this area shows considerable variations due to altitude and the resulting climatic changes. An ideal crop rotation schedule suitable for this region has been evolved by the Department of Agriculture, (Government of West Bengal).

TABLE 11.14: **Variable crop rotation to be followed at different elevation levels**

| Levels of Altitude | January-June | June-October | October-February |
|---|---|---|---|
| 1718 m - 2185 m | (i) Summer potato<br>(ii) Maize | Radish, carrot, peas, beans, turnip | Cauliflower, cabbage, peas |
| 1250 m - 1718 m | (i) Maize<br>(ii) Summer potato<br>(iii) Winter potato | Radish, carrot, peas beans squash Soyabeans, chilly | (i)Cauliflower, wheat, cabbage.<br>Winter potato |
| 625 m - 1250 m | Maize, ladies finger | Paddy, vegetables | (i) Wheat, winter potato, beans, tomato<br>(ii) Cauliflower, cabbage<br>(iii) Winter potato, cabbage<br>(iv) Beans, cauliflower, wheat |

*Source*: District Agriculture Office, Darjeeling.

Although this ideal cropping pattern implies farming activity throughout the year, in practice multiple cropping is carried on only where the irrigation facilities are available.

TABLE 11.15: **Area under different crops, 1984-85**

*(in hectares)*

| Sub-divisions | Paddy | Wheat | Maize | Other cereals | Potato | Pulses | Rape & Mustard | Til |
|---|---|---|---|---|---|---|---|---|
| Darjeeling | 1600 | 50 | 10,000 | 3,150 | 2570 | 200 | 90 | 90 |
| Kurseong | 130 | 50 | 2,480 | 300 | 470 | 35 | 20 | 20 |
| Kalimpong | 7200 | 300 | 11,940 | 6,550 | 675 | 265 | 90 | 90 |

*Source*: Principal Agricultural Officer.

TABLE 11.16: **Area under different crops, 1987-88**

*(in hectares)*

| Subdivisions | Ginger | Cardamom | Orange | Pineapple |
|---|---|---|---|---|
| Darjeeling | 329.96 | 1,041.30 | 508.50 | — |
| Kurseong | 176.52 | 388.26 | 205.55 | 22.27 |
| Kalimpong | 666.80 | 633.60 | 187.45 | — |

*Source*: District Rural Development Agency.

TABLE 11.17: **Area under different crops, 1987-88**

| Subdivisions | Mandarin Orange | Temperate Fruits |
|---|---|---|
| Darjeeling | 772 hectares | N.A. |
| Kurseong | 370 hectares | N.A. |
| Kalimpong | 335 hectares | N.A. |
| Total | 1777 hectares | 480 hectares |

*Source*: District Office of Horticulture, Darjeeling.

Four major plantations of cinchona are in Mungpoo (Darjeeling Sub-division), Mansong and Rongo in Kalimpong sub-division and Latpanchor in Kurseong sub-division. A part of the plantation area is also used for growing Ipecac which is used for making medicines for ameobiosis.

Orchard farming with orange at lower levels (800m-1600m) and temperate fruits like apple, pear, plum, peach and apricot at higher levels is gaining importance. Acidic soils of Darjeeling hills need lime treatment to improve the quality of fruit. Farmers are provided with better quality seedlings from the nurseries set up in different parts of the hills. Horticultural research stations have been set up at Kalimpong and Sonada (near Darjeeling) for the improvement of horticulture in this area. Farmers are being advised to keep more areas of their farms under tree crops. Because the menace of soil erosion increase when slopes over 33 ° are planted with maize and millet.

Encouragement is being given to the small and marginal farmers to take up different types of subsidiary occupations like bee keeping, floriculture, sericulture, animal husbandry etc. Co-operative societies have been set up in recent years to organise the production and marketing of different farm products. Six milk chilling plants, operating in the hill area, are sending the milk from the house dairies to the Central Dairy at Matigara near Siliguri. 101 primary bee keepers co-operative societies are gradually showing the way for the use of least known resources like the flowers of oranges, cardamom, squash (a vegetable), mustard etc.

## Conclusion

Main trend in the landuse pattern has remained same in this area. Excessive and unwise use of some of the vital resources, like forest have resulted in the dwindling of potentialities of this region. The shrinking of the forest lands, in its wake have resulted in excessive soil erosion and drying up of the springs and headwaters in the catchment areas. All these again has increased the incidence of land slides and debris-slips in this area. People first have to be made aware of this environmental degradation. Plans should be made in such a way that the resources could be utilised without disturbing the ecological balance.

## REFERENCES

Chatterjee, A. B., et al, (eds), 1970 : *Geological Framework of the Himalayan Foothills*, West Bengal, Presidency College, Calcutta.

Census, 1981 : *District Census Handbook*, Darjeeling.

Lahiri, S., 1957 : "Influence of Physiography on the location and landuse of Tea-gardens of the Darjeeling District". *The Indian Geographer*, Vols. 1&2, Number 2&1 August, 1957.

(a) Lama, M. P., 1986 : *The Tea Industry of Darjeeling: A Productivity Analysis.*

(b) Sarkar, R. L. and Lama, M. P., (eds), 1986 : *Environment and Economy*. Publisher, Atmaram and Sons, Delhi, Lucknow.



## Conclusions



Banerjee, A. B. et al. (ed.), 1970 *Socio-economic survey of the Himalayan Region, West Bengal*, Presidency College, Calcutta.



(a) Lama, M. P., 1988 *The Darjeeling Himalaya*, Darjeeling : A Preliminary Analysis.



# 12

# MICROLEVEL MONITORING OF LANDUSE AND AGRICULTURE IN HIMALAYAS

A. R. KHAN AND R. B. SINGH

AGRICULTURE HAS been one of the earliest anthropogenic activities which tried to make use of natural bounty according to its own wish and satisfaction. The constant interference and nature's course of yielding, did not for all the time hovered around the perfect balance but the insatiable need often ushered man to the states where it had to pay heavy toll for the "misdeed". The situation of certain areas where the balance of natural ecosystem is rather delicate and the human pressure and interaction go on increasing unabated, more often than not, the activities turn counter-productive. The natural calamities such as famine, flood and landslides are less frequent though, the slow depletion of vital productive nutrients from the soil is going to affect the productivity in the long run if the monitoring of the whole activity is not taken up.

Monitoring of landuses and agriculture is of vital importance, because land is finite and its uses restricted (to the stretch of technology), which in most normal circumstances has been put to agricultural activities; supplying the life source to the mankind. The importance of monitoring for landuse and agriculture becomes vital because both the factors have direct bearing on the sustenance of man. The correct status, mode, rate, direction, and resultant ramifications; if so there is any, is possible, when the watch on the activity undertaken is detailed one. The monitoring of larger areas in this regard is rather difficult with the level of statistics available and analysis becoming operationable. Therefore, there is great need for the development of data base for smaller areas with larger number of variables and compare them with the other spatial units for monitoring to serve the needs like achieving the status evaluation of its development and prognosticate the future course of development.

Himalayan landuse and agricultural practices present a different scenario if they are compared with the same activities carried out in the plain regions. If a closer look is taken, a marked variation within the Himalayan regions can also be traced which merits micro level efforts to explain the factors better. The Himachal Himalayas has many major river systems and the region falling mainly between the river Beas and Sutlej Systems, also includes an administrative unit as Mandi Tahsil in the Mandi district of Central Himalayas. The old occupance of the Mandi Tahsil because of its location on the trade route from the permanent snow covered ranges and the plain makes it a special case to study the long time landuse practices and agricultural uses.

The analysis of the resource base and the occupational structure highlights agriculture as the essential ingredient and the mainstay of the people of Mandi Tahsil. It has been the main attractive and dependable source of revenue for the princely states of Mandi and Suket in older times and consequent governments alike.

Two major breakthrough in the immediate agricultural past made with the induction of the agricultural programme with the help of Federal Republic of Germany with the prime objective to improve the existing agricultural conditions and to change the direction of agricultural production in order to improve the production of cash crops more suited to the area. The programme was spread over the period of 12 years, (1962-1974) and was adjudged as one of the best composite and integrated agricultural projects in the country. Under this project improved equipments of various types, farm machinery, improved breeds of cattle, sheep and poultry etc., were made available to the farmers and a well equipped soil testing laboratory was set up at Sundar Nagar. The project has visibly brought a big change in the living standard of the rural people of the district, resulting in a change in their social and cultural behaviour.

**Landuse System**

The landuse of an area, to a great extent, speaks about the economic status of a region. The various uses the land is put to, in the district and Mandi Tahsil can be seen from the Table 12.1 and Fig. 12.1.

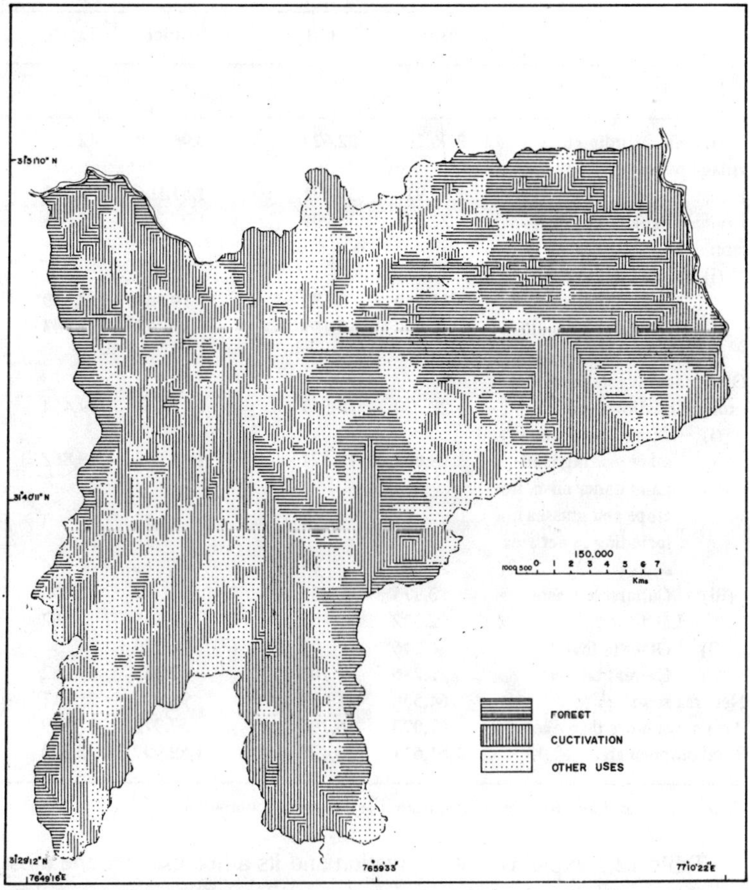


FIG. 12.1: Landuse pattern (Mandi tahsil).

TABLE 12.1: **Changing landuse system**

*(Area in hectares)*

| Item | 1984-85 | | 1986-87 | |
|---|---|---|---|---|
| | District | Tahsil | District | Tahsil |
| 1 | 2 | 3 | 4 | 5 |
| Total area according to village papers | 3,96,163 | 82,627 | 3,96,320 | 82,627 |
| Forest | 1,53,918 | 35,071 | 1,57,506 | 35,449 |
| Land available for agriculture: | 22,937 | 5,299 | 23,356 | 5,621 |
| (i) Land put to non-agricultural use | 11,940 | 3,135 | 10,864 | 2,838 |
| (ii) Barren and uncultivable land | 10,997 | 2,164 | 12,492 | 2,783 |
| Other uncultivable land excluding fallow land | 1,22,238 | 20,519 | 1,17,908 | 19,464 |
| (i) Permanent pastures and other grazing land | 1,18,062 | 19,621 | 1,14,249 | 18,862 |
| (ii) Land under misc. tree crops and grasses not including in net area sown | 201 | 43 | — | 46 |
| (iii) Culturable waste | 3,975 | 855 | — | 556 |
| Fallow Land | 2,532 | 105 | 2,525 | 282 |
| (i) Other fallow | 246 | 5 | 140 | — |
| (ii) Current fallow | 2,286 | 100 | 2,385 | 282 |
| Net area sown | 64,538 | 21,633 | 95,025 | 21,811 |
| Area sown more than once | 67,073 | 16,795 | 57,972 | 11,509 |
| Total cropped area | 1,61,611 | 38,428 | 1,52,997 | 3,320 |

*Source*: District Land Record Office, Mandi and personal computation.

Table 12.1 explains that cultivation and its allied uses are the most important activities that the land is put to. The area sown more than once occupy about two-third of the total area in Mandi Tahsil. The second most important landuse is that of forestry. However, there is a significant increase in forest area over the period '84-87' in the district of Mandi. About 93% population is living in rural areas whose main occupation is cultivation. Owing to mountainous topography and varied cultivation practices, different crops are grown. The net area sown per agricultural worker is one of the highest among the Tahsils of the district. In the district of Mandi a total of 95,025 hectare was sown in 1986-1987. It engaged 2,20,087 people that is, 0.43 hectare per person on an average. The number of agricultural worker in the calendar year 1986-87 in Mandi Tahsil was 35,828 persons and this gave every agricultural worker an area of 0.60 hectare on an average, which was 0.43 hectare each agricultural worker in the district. This gave 0.17 hectare land on an average extra to each agricultural worker in the Mandi Tahsil than to those of the district agricultural worker taken as a whole.

**Agricultural Characteristics**

Agriculture in the hills of the study area is the prime activity. It is also the most important of landuses whose monitoring is of vital significance. A brief account of it reveals that there are three types viz., Katil cultivation, Upraon cultivation and Talaon cultivation; the latter two being more popular. Katil cultivation is of the primitive form of hill agriculture and considered as the first stage of terracing. This is comparable of Jhuming in north eastern India. The shrubs and forests are cleared and burnt and the land is broken with hoe. The seeds are then broadcast over the ashes which are the only form of manure.

Upraon cultivation is pursued on upland slopes which are permanently terraced and superior in quality of those of katil lands. The qualities vary according to the available soil nutrients, the evenness of the fields, their proximity to the forests and accessibility of the slopes, nearness to homesteads, transport facilities and their position in Sunny and Shady aspect of the ridges. However, they are at a relatively higher level without any means of irrigation and are adaptable mostly to the dry crops. Among Kharif crops are Mandua, Jhangora, Kanin and Bhat etc., while those of rabi are wheat, barley, mustard etc. grown through this method.

Lands irrigated and irrigable, generally at the foot of the hills are known as Talaon lands. They are characterised by the proximity to the rivers and their tributaries, mild climate, broad level valleys and low lying fields, high fertility and development, existence of permanent irrigational facilities and possession of alluvium. Paddy is the main crop of Kharif season, while wheat barley, mustard are grown in rabi season (Table 12.2).

TABLE 12.2: **Cultivation types and their characteristic features in Mandi Tahsil**

| Cultivation type | Location | Crop type | Terracing | Gradient | Irrigation source | Nutrient type |
|---|---|---|---|---|---|---|
| Katil | Distributed at all heights | Mandua, Chua, Gohat Turmeric | First stage of terracing | varied | Nil | |
| Upraon | Relatively higher levels | Mandua or ragi, Jhangora; Kanin Bhat, Rabi Wheat, barley mustard | Permanently terraced | Upland slopes | Nil | Composite |
| Talaon | Foot of the hills near water bodies | Paddy, wheat barley, mustard | Relatively flat stairs | Low lying field | Permanent irrigational facilities | Chemical fertiliser |

*Source*: Personal survey.

## Terracing

The steep and rugged mountainous terrain of Mandi has scarcity of tracts of level grounds, small proportions of cultivable land and the frequent soil erosion have all necessitated resort to terracing in the region. Mountain slopes have been cut into hand-made fields rising in succession like gigantic flights of steps, ascending from the banks of river Beas and Uhl towards the summits of the hills in almost all directions as river at times, twists to its left and right. The usual width varies between three and six metres, length between five and twenty-five metres, height between a metre and half to three metres and area between fifty and two hundred square metres as observed in Mandi Tahsil. Terraces, three to six metres are common in Upraon lands and in certain parts of the west of Mandi Tahsil, at places more than two hundred terraces can be counted in continuous flight. The maintenance and repair is a very tough job as nearly forty per cent of the terraces in the upland break each year during the rainy season.

Apart from terracing and irrigation, the manuring aspect is the other factor which makes the hill cultivation unique. In the Katil cultivation region the burnt ashes are the only source of manure to the fields. In Talaon that is mainly near and around the valleys of river Uhl and Beas, the manure is not required and the silt brought down by the small channel is supposed to be enough nourishment added year after year. The region where the requirement of manuring is felt most is Upraon lands. Cow and buffalo dung are most common manure in the region. Many farmers collect the dung, fifth leaf in a ditch which they use in subsequent year when they turn into manure.

## Irrigation

While the small channels taken out from brooks were much common till the last few decades, rainfall has been the single major source of irrigation in present time. There was an excess rainfall in 1988 but prior to that northern and north eastern parts of Mandi did not receive much rainfall for more than seven seasons in succession. Despite the fact that rainfall is one of the lowest and irrigational systems are insufficient, Kharif has dominated the crops. The Tahsil of Mandi in the year 1986-87 had 21,811 hectare net area under cultivation, out of which 2,946 hectare was irrigated through different sources. This leaves only 13.5 per cent area irrigated to net area sown in the same period. This figure, however, is much close to the district average where 14.1 per cent net area is irrigated to net area sown in the year 1986-87. As far as the sources of irrigation are concerned, canals and tanks cover a very insignificant part of the irrigated land. Wells and tubewells irrigate a net area of 101 hectare and other sources irrigated 2,845 hectare land during cropping season of 1986-87.

Table 12.3 indicates the percentage of area irrigated in Mandi Tahsil under each crop in three different years.

TABLE 12.3: **Changing pattern of irrigated area by crop**

*(Hectare)*

| Crop | Year | | | | | |
|---|---|---|---|---|---|---|
| | 1984-85 | | 1985-86 | | 1986-87 | |
| | Total area | %irrigated area | Total area | %irrigated area | Total area | %irrigated area |
| Wheat | 16420 | 17 | 12481 | 18 | 11375 | 18 |
| Rice | 8232 | 42 | 6297 | 44 | 7799 | 43 |
| Maize | 9438 | 5 | 11116 | 4 | 7799 | 5 |
| Barley | 468 | 5 | 442 | 4 | 1097 | 4 |
| Pulses | 5714 | 3 | 1348 | 3 | 902 | 2 |
| Oil seeds | 93 | 46 | 59 | 50 | 58 | 50 |

*Source*: Land Records Office, Mandi and Statistical Abstract Report of Mandi District,1987 and personal computation.

Table 12.4(a, b) indicates that there is a general rise in the percentage of area under irrigation. The increase is more fast among the popular crops. The irrigation practices are mainly confined to most popular crops like wheat, rice and oil seeds. Oil seeds which had the maximum area under irrigation also recorded maximum per cent increase (4%) during the period of three years.

### Monitoring Crop System

A comparative study of total cropped area shows a marked variation in the cropping pattern of Mandi district as well as the Tahsil (Table 12.4 a and b). Wheat occupies the largest area under cropping with 43 per cent of gross cropped area in the district and 44.6 per cent in the Tahsil. Rice and Maize are also one of the major crops of Mandi. Other food crops like Potato, Sugarcane, Ginger and Chillies are also grown albeit in small quantities. In non-food crops rape, mustard and tobacco are significant.

In Mandi, the area under each crop has not shown any significant change over the years but a slight increase is evident in the area under rice which was 16 per cent for the district in 1984-85 and increased to 17.3 per cent in 1986-87 and for the Tahsil it increased from 22.3 per cent in 1984-85 to 28.2 per cent in 1986-87.

The intensity of cropping in Mandi Tahsil, however, is one of the lowest among other Tahsils. Despite the fact that the agricultural worker to total population is one of the highest among other Tahsils, Mandi has a very little difference in the net area sown and gross area sown. According to Revenue Survey (1986-87) of Blocks in district, the two Blocks, Mandi Sadar and Rewalsar taken together recorded 33,320 hectare land as gross cropped area. However, a major proportion of the gross cropped area was net sown area. The figure for net sown area in the agricultural calendar year 1986-87 was 21,811 hectare. The cropping intensity of the Mandi Tahsil in the year 1986-87 was 1.52. This intensity figure is lower and also a shade lower than the cropping intensity figure of the district which is 1.61 (Table 12.4 a and b).

### Horticulture

The agro-climatic conditions occurring in Mandi district are quite suitable for propagating the tropical and sub-tropical fruits. Since late sixties, the district has made a considerable stride in the field of horticulture. The apple orchards are located predominantly in Karog and Chachyot tahsils while plum and apricot orchards have come up in Mandi and Sundar Nagar Tahsils. Large number of dwarf root stocks have been imported from the Federal Republic of Germany under the Indo-German Agricultural Project and multiplied in government progeny orchards not only in the Mandi Tahsil but also in the whole of Himachal Pradesh.

TABLE 12.4a: Comparative study of total cropped area (ha) and different crop production

*(tonnes)*

| Crop | 1984-85 | | | | 1986-87 | | | |
|---|---|---|---|---|---|---|---|---|
| | Area under each crop | | % of gross cropped area | | Area under each crop | | % of gross cropped area | |
| | District | Tahsil | District | Tahsil | District | Tahsil | District | Tahsil |
| 1 | 2 | 3 | 4 | 5 | 6 | 7 | 8 | 9 |
| I. *Food Crops* | | | | | | | | |
| *cereals:* | | | | | | | | |
| Wheat | 67904 | 16420 | 43.0 | 44.6 | 60782 | 11375 | 42.1 | 38.2 |
| Rice | 26226 | 8232 | 16.0 | 22.3 | 25067 | 7799 | 17.3 | 26.2 |
| Maize | 24025 | 2438 | 27.0 | 25.6 | 40002 | 7799 | 27.7 | 26.2 |
| Barley | 4536 | 468 | 2.9 | 1.2 | 4672 | 1029 | 3.2 | 3.4 |
| Millets & other cereals | 1613 | 47 | 1.0 | 0.1 | 1571 | 37 | 1.0 | 0.1 |
| Ragi | 3497 | 716 | 2.2 | 1.9 | 3212 | 21 | 2.2 | 2.3 |
| II. *Pulses:* | | | | | | | | |
| Gram | 114 | 86 | 0.1 | 0.2 | 191 | 13 | 0.1 | 0.1 |
| Other Pulses | 5600 | 1145 | 3.6 | 3.1 | 5266 | 889 | 3.6 | 2.9 |
| III. *Other Food Crops:* | | | | | | | | |
| 1. Potato | 1934 | 48 | 1.2 | 0.1 | 1773 | 46 | 1.2 | 0.2 |
| 2. Sugarcane | 19 | 4 | 0.1 | 0.1 | 15 | — | 0.1 | — |
| 3. Ginger | 7 | — | 0.1 | — | 13 | 1 | 0.1 | 0.1 |
| 4. Chillies | 159 | 54 | 0.1 | 0.1 | 149 | 51 | 0.1 | 0.2 |
| *Non-Food Crops:* | | | | | | | | |
| I. *Oil Seeds:* | | | | | | | | |
| Rape & Mustard | 360 | 23 | 0.2 | 0.2 | 320 | 58 | 0.1 | 0.2 |
| II. *Other Non-Food Crops:* | | | | | | | | |
| Tobacco | 58 | 1 | 0.2 | 0..1 | 30 | — | 0.1 | — |

*Source*: Land Record Office, Mandi and Statistical Abstract Report of Mandi District, 1987 and personal computation.

TABLE 12.4b: **Comparison of changing productivity of principal crops**

| Crops | Year | | | | | |
|---|---|---|---|---|---|---|
| | 1984-85 | | | 1986-87 | | |
| | State | District | Tahsil | State | District | Tahsil |
| 1 | 2 | 3 | 4 | 5 | 6 | 7 |
| Wheat | 0.7 | 0.66 | 0.65 | 1.21 | — | — |
| Rice | 1.24 | 1.00 | 1.07 | 1.10 | — | — |
| Maize | 1.86 | 3.00 | 1.10 | 1.85 | — | — |
| Barley | 0.83 | 0.70 | 0.77 | 1.00 | — | — |
| Millets & other cereals | 0.43 | 0.40 | 0.46 | 0.60 | 0.38 | 0.24 |
| Gram | 0.10 | 0.10 | 0.11 | 0.27 | 0.17 | 0.20 |
| Other pulses | 0.15 | 0.20 | 0.28 | 0.20 | 0.26 | 0.009 |
| Sugarcane | 1..41 | 1.30 | 1.25 | 1.00 | 1.86 | — |
| Ginger | 0.67 | 1.00 | — | — | 1.00 | 1.00 |
| Chillies | 0.11 | 0.30 | 0.005 | — | 0.18 | 0.10 |
| Rape & Mustard | 0.08 | 0.005 | 0.004 | 0.13 | 0.11 | 0.008 |
| Tobacco | 0.50 | 0.008 | 1.00 | — | 0.16 | — |

*Source*: Land Record Office, Mandi and Statistical Abstract Report of Mandi District, 1987 and personal computation.

With a view to utilise the unmarketable surplus fruits of the Tahsil and other parts of the district of Mandi as well as of the adjoining areas, two canning-cum-fruits reservation plants at Panarba and Jarol (Sundar Nagar) have been established. The apple juice processing plant at Jarol is providing a boon to the horticulturists in providing ready market for these surplus apples. The plant protection centres are located at Mandi with sub-centres at each block headquarters which render technical guidance to the orchardists and supply pesticides through the agriculture department. The development in horticultural activity is more in Mandi Sadar block than in Rewalsar block despite the fact that topographical advantages are more with Rewalsar block.

## Livestock and Poultry

Apart from agriculture, livestock and poultry are other two major mainstay of the rural inhabitants of Mandi Tahsil. The availability of pastures on hills has been an immediate advantage to the masses who keep cattle. The out-migration of people from the remote areas has created a dearth of manpower for agricultural operation as most of the 'stayed back ones' were either very young or old, both unable to carry out cultivation on the terraces. This led people to take refuge in the pastoral economy as they did not want to leave their ancestral land. The other factor which led to the shrinking up of the agricultural activities and livestock development, is the money-order economy. The out migrated people sent back money to old ones who invested those money in buying cattle as the avenues for the expenses are very small and their age checked them to cultivate on hills. The whole system made a cumulative impact and the fragile mountain ecosystem got disturbed and again it had further snow falling effect on several other activities, ultimately boomeranging on the economic status of the rural masses.

The idea about the number of livestock in Mandi Tahsil can be had from the Table 12.5. The decrease in total number of livestock shows that the land available for their grazing is shrinking. However, there has been a significant increase in cattle population over the five years (1977–82).

The milching cattle is most preferred. Buffalo, as it requires lots of fodder and the pastures have shrunk considerably in number is becoming unpopular. Horses, ponies and donkeys have lost their popularity as they no more are taken to be the cheapest mode of transportation. Buses and other vehicles have taken their place as they are fast and economical.

## A Micro Level Monitoring of Village Bhatwar

Primary activities involving land in most of the cases in the Himalayan rural areas are carried out at a local level, not larger than the village. An effective monitoring of landuse and agricultural practices must also be given due importance at micro level for better understanding of the phenomena. The selection of the village Bhatwar under micro level monitoring was guided by many factors of which the main factor is that it has many of its characteristic figures very close to the Tahsil average. To get a clear picture of the status of the uses of resources in the Tahsil it was better thought to study some village which lies closer to the average than on one of its extremes.

The village Bhatwar, under study, is located about 13 kms. west of the Tahsil headquarters and almost at the same distance from its block headquarters at Rewalsar. The village is situated at a height of about 1500 mts. from the mean sea level. There is no metalled road connecting the village but a well demarcated trek path of a three and half kilometres joins the village with the main district town.

TABLE 12.5: Livestock change in Mandi Tahsil

| | Year | |
|---|---|---|
| | 1977 | 1982 |
| Cattle | 84,108 | 87,604 |
| Buffaloes | 23,960 | 23,512 |
| Horses and Ponies | 567 | 572 |
| Donkeys | 13 | 30 |
| Mules | 171 | 20 |
| Goats | 39,518 | 32,022 |
| Sheep | 37,121 | 38,626 |
| Pigs | 118 | 206 |
| Total Livestock | 1,85,576 | 1,82,772 |
| Poultry | 11,746 | 10,184 |

*Source*: Community Development Block, Mandi Sadar, Rewalsar.

The slope of the village is rather flattened by the cultivators and one can count hundreds of terraces in one flight. The rocks are sandy and soil, and in terms of broad classification, fall under the category of Inceptisols. The general slope is towards east and in the west a hill records the highest point in the village. There are two small seasonal *Nullahs* passing through the village from west to east and meeting Scodhy Khad, a tributary or River Beas.

## Land Resources

The forest cover of the village is around 24 per cent which is very low for a hilly area. There can be observed some plantations undertaken during last seven years. The per cent land under forest in the village was very high till mid-sixties. The rising market demand in the Tahsil forced the local poor people to cut down the trees for sale in the market. The economic backwardness of the region ultimately made this naturally rich and endowed village devoid of their common wealth — forest. The plantation undertaken by the forest department can be commendable as one can see some newer trees in the area.

Water availability due to many environmental factors got badly affected and as there is no perennial river passing through the village, people have very tough time during summer.

The area of the village is 784 bighas of which maximum land is put to agricultural uses. Cultivation is practiced on 50.8 per cent of the area which is quite high and also indicates that how much the pressure the land carries. The allied activities to cultivation which also indicate pressure of population on its carrying capacity is pasture. About 16.2 per cent of the village is put to use under pastures. Forest accounts for 23.9 per cent of land and other uses include those of horticulture and uncultivated summing to a total of 9.1 per cent of the total land. A look at the Fig. 12.2 shows that how localisation of the landuses have taken place.

Forest is confined to the eastern most boundary of the study region with a small stretch extending towards north. The area adjacent to forest is put to other uncultivated uses. The other uses of land in terms of horticulture can be found along the western margin of the village. The centre of the village have terraces where most of the agricultural practices are carried out. The reason for cultivable land to be in the centre is be its fertility and nearness to the settlement.

The condition of the village in terms of the agriculture is not very bright. The soil of the village is not very fertile and many farmers are too poor to afford to use fertilisers and more so a condition where there is no certainty of monsoon. If rain is not adequate, the farmers suffer a great loss. The main crops of the village include maize, paddy, pulses etc. There are a few vegetables also grown in the village and some of them find their place in the Tahsil market also.

The village Bhatwar has about 16.2 per cent land under pastures and that gives the villagers a good chance to generate some resource through cattle. The commercial use of cattle is also made as quite a few families supply milk to the houses in the Tahsil headquarters. There is not a single person who entirely depends on cattle as his income source rather he undertakes it as one of his part time jobs. Ox, even today, remains the only source for turning soil through plough. It is also the cheapest available option to the farmers. There are 46 sheep and goats in the village which are used for wool and meat.

**Concluding Remarks**

The land resource study of Mandi Tahsil indicates that the region makes use of it in primary sector and mostly in primitive form. Cultivation practices dominate the scene, while pastoral and horticultural activities have a very small share of landuse. The area under different crops in terms of percentage is nearlv same as that of

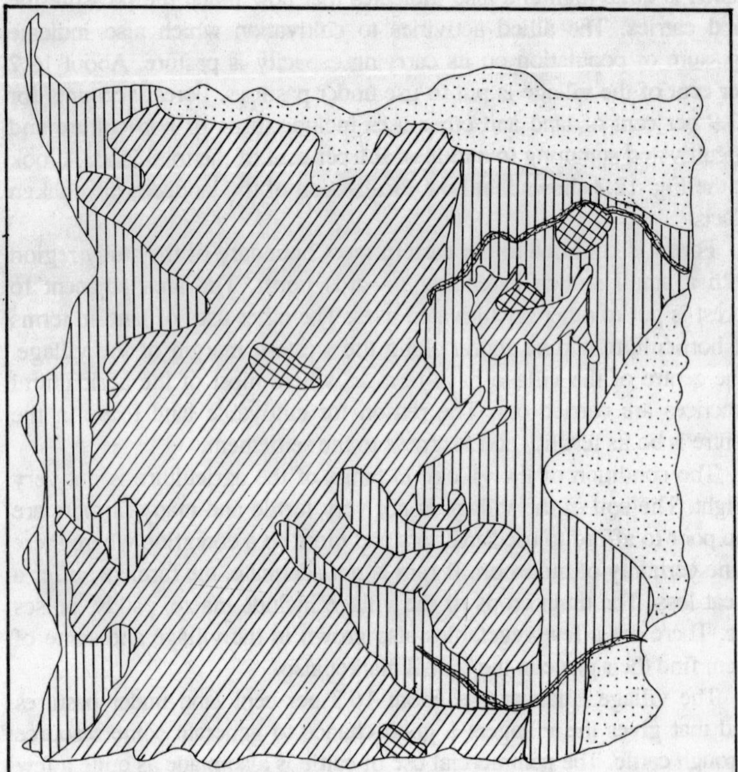


FIG. 12.2: Landuse map (Bhatwar).

the district average. Mandi Tahsil, despite being so close to the district headquarters, does not make much use of market availability. This also indicates that Mandi Tahsil has subsistence type of cultivation. The changing consciousness of people, however, holds hope that people will shift to horticultural activities as was evident from several villagers case in Bijni and Bhatwar villages. However, there is a great need to go into the details of the agriculture and landuse systems. The consideration of their sub-systems such as their historisity, interactive mechanism, different activities involved, popular strategies, input-output analysis, human involvement, livestock and fisheries development etc. hold great responsive role in the elaboration and extension of their monitoring to have a balanced and sound strategy formulation for the effective utilisation of the resources.

## REFERENCES


Arora, R. C., 1976 : *Development of Agriculture and Allied Sectors,* New Delhi, S. Chand and Co.

Census of India, *District Census Hand Book,* Mandi District, 1981.

Chattopadhya, B. C., 1979 : Emerging Concept of Rural Development Planning, *Decision,* 6, p. 4.

Director of Economics and Statistics, Himachal Pradesh, Shimla, *Blockwise Indicator of District Mandi,* 1978-88.

Director of Economics and Statistics, Himachal Pradesh, Shimla, *Economic and Statistical Bulletin,* 1989, April-June.

Director of Economics and Statistics, Himachal Pradesh, Shimla, *Important Statistics of Himachal Pradesh,* 1989.

The Directorate of Economics and Statistics, Government of Himachal Pradesh, Shimla, *Statistical Abstract of Mandi District,* 1987.

Directorate of Economics and Statistics, Himachal Pradesh Shimla, Mandi, *Statistical Report,* 1987-88.

District Statistical Office, Mandi District, *Economy in Figures,* 1978-88.

Economics and Statistics Department, Government of Himachal Pradesh, *An Economic Classification of the Himachal Budget,* 1987-88.

Government of India, Department of Rural Development, 1976, *Annual Report,* New Delhi, Ministry of Agriculture and Irrigation.

Kayastha, S. L. and Khurana, V. K., 1987 : Geographical Environmental: Approach for Integrated Rural Development, *The Journal of Scientific Research,* Vol. XXIX(2).

Mukherjee, A. B., 1962 : Agricultural Regions and Geographical Planning for Indian Agriculture, *National Geographer,* December.

Planning Department, the Directorate of Economics and Statistics, Himachal Pradesh, *Economic Review of Himachal Pradesh,* 1988.

Singh, R. B., 1986 : *Geography of Rural Development: The Indian Micro-Level Experience,* New Delhi.

13

# IMPACT OF OUTMIGRATION ON PATTERNS OF AGRICULTURAL LANDUSE IN A HILL AREA

NOOR MOHAMMAD AND SURESH KUMAR BANDOONI

THE MOVEMENT of population is inherent in the process of socio-economic development. The causes of population movement are varied, but its basic objectives have primarily been to seek better economic opportunities in order to enhance the individuals income, leading to material advancement and enhanced social status. Outmigration has, by and large, brought prosperity by way of flow of funds, technology and cultural contacts in the region of origin.

Outmigration affects both the areas of origin as well as the areas of destination in a variety of ways, which could broadly be assessed within techno-economic and socio-institutional framework. It influences the economic base of the region specially in the region of origin. Rural outmigration remains a potential mechanism for change in the rural areas through diffusion process (Goldscheider, 1984). The ideas and money brought back by migrants have given encouragement to the villagers to change their methods and techniques of farming. They make a gradual shift from food crops to commercial crops. In several areas, most of the technology and capital needed for commercial farming is supported by migrants.

This phenomenon is more noticeable in the mountanous areas, specially in the Himalayan region, where due to variety of reasons, by and large, male population migrates to the plains. It is more so in Uttrakhand Himalaya, where agriculture, although it is the main activity of the people, can provide food to its people only for 4 to 6 months and employment to its inhabitants only for 4 to 5 months. The other sectors of economy are also not capable enough to provide job opportunities to a large force of unemployed people of the region. Consequently, a large number of unemployed workforce is compelled to migrate in different areas in search of better economic opportunities. In due course of time, the flow of capital, ideas and technology takes place, which brings a number of socio-economic changes. Since, this is a predominantly an agricultural area, the changes are more prominent in this sector of economy. An attempt has therefore, been made to study and explain the impact of outmigration on agricultural landuse pattern in Pauri Garhwal District of Uttrakhand Himalaya.

**Study Area**

Pauri Garhwal district lies in the south-eastern part of the Garhwal zone of the Uttrakhand Himalaya. It is located between 29° to 30°15′ North latitude and 78°10′ to 79°10′ East longitude and occupies 5,540 sq. km. area (Fig. 13.1). Except a narrow belt in the south, the whole district comes under mountains with the surface height ranging from 350 metres in south-west to 3100 metres in north-east. The district has a complex physiographic structure and it may be divided into three physiographic divides viz. The Bhabar and Doon, the Shiwalik, and the Lesser Himalaya. The Bhabar and doons are located in a narrow stretch in south between the foothills of the Shiwalik range and tarai of northern plain. It is of recent origin and is largely composed of the detritus washed away from the Shiwalik hills by swift flowing streams. It is the most potential area for agricultural purposes. The Shiwalik is a zone of low hills (400-1200 mts) and is separated from the Himalaya by the major faults– Main Boundary Fault (MBF). It consists of a series of parallel ridges covered with thick forests, leaving small pockets of agriculture and population concentration. The Lesser Himalaya, being bounded by MBF in the south and Main Central Thrust (MCT) in the north, consists of scattered mountains with numerous spurs running parallel from north-east to south-west direction.

As a result of complex physiography, the region experiences varied climate–sub-tropical to cool, with a complex pattern of distribution of its elements which have direct and indirect bearing on patterns of landuse. However, the micro climatic conditions dominate due to the complicated relief features. The district has a variety of soil ranging from yellow-brown to black soil with varied degree of nitrogen, phosphorous and potassium content. It presents an extremely mosaic pattern of distribution according to different ecological conditions. Landuse of the district is dominated by the forest cover (61%) which provide fuel, fodder, timber and raw materials. Agriculture, although occupies only 23 per cent of the total area, is the main occupation of the people, engaging around 90 per cent of the total workers. There are two main agricultural seasons viz. Kharif and Rabi. The major crops of the Kharif season are rice, maize, kauda, jhungora, chillies, soyabean and pulses, while wheat, barely and mustard are the main crops of Rabi season. The district supports a population of 627,617 (1981) with a density of 117 persons per sq. km.

**Data Source and Methodology**

The present study has been carried out at the village level with farmers as unit of analysis for indepth study. It is entirely based on primary data, because the secondary data does not provide complete information about the outmigration and its related consequences in the district. For the collection of primary data, stratified purposive sampling technique has been used. The study area has been divided into three divisions viz. The Bhabar and doon, the river valleys, and the ridge. On the basis of location, accessibility, movement of population etc., fourteen villages have been selected from these divisions. Eight villages have been selected from the river valleys, while five villages have been chosen from the ridges or higher belt of the district. Only one village has been taken from the Bhabar division (Fig. 13.1).

Since, it was not possible to collect data from each and every household of the sample villages, 168 respondents have been selected from them. The respondents have been interviewed according to a well structured questionnaire so as to get the required data. After careful scrutiny of the survey schedules, data have been processed and analysed with simple statistical and cartographical techniques. The overall volume of change in agriculture in 1985-86 over 1975-76 has been worked out with the help of Weaver's modified index of volume of change as given below:

I.V.C. = A/B where
I.V.C. = Index of volume of change
A = is the difference of percentage as landuse categories of increase, and
B = the difference of percentage as landuse categories of decrease for the periods 1975-76 to 1985-86.

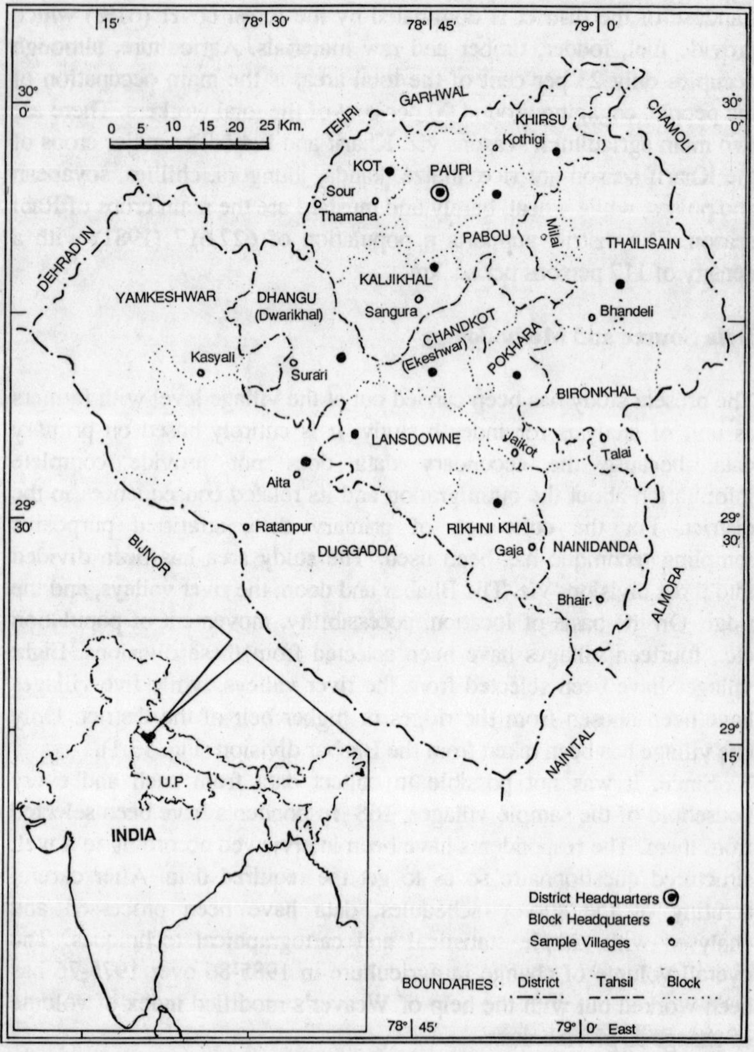


FIG. 13.1: Location of sample villages.

## Outmigration and Agricultural Landuse Pattern

A scrutiny of the questionnaire indicates that seven of the selected villages have more outmigration and 4 have moderate, while the remaining three villages have less outmigration. Theses villages are at different levels of transport connectivity, but generally outmigration is positively related to the transport network. For the assessment of impact of outmigration on landuse pattern, the respondents have been grouped in 5 categories on the basis of number of persons migrated from the family (Table 13.1).

TABLE 13.1: **Classification of respondents on the basis of outmigration**

| S.No. | Category | Outmigration rate per family | No. of Respond-dents | %age to total |
|---|---|---|---|---|
| 1. | More outmigration | 3 and above | 22 | 13.09 |
| 2. | Moderate outmigration | 2 | 44 | 26.19 |
| 3. | Less outmigration | 1 | 72 | 42.86 |
| 4. | Returned migration | — | 18 | 10.72 |
| 5. | No outmigration | 0 | 12 | 7.14 |
| | Total | | 168 | 100.00 |

Before analysing the dynamics of landuse, it would be worthwhile to understand some of the salient characteristics of the outmigrants. The basic reasons for outmigration from the villages are lack of job opportunities, inadequate agricultural land, poor economic conditions and so on. Out of total outmigrants, about 69 per cent came under middle age groups (25-49 years) and 26 per cent belong to young (25 years) while only 5 per cent are more than 50 years of age. It shows that only working population have migrated with some exceptions. The average monthly income of about 8 per cent outmigrants is more than Rs. 1800, while 47 per cent have income ranging from Rs. 1200 to Rs. 1800 per month. The remaining 45 per cent get less than Rs. 1200 per month. The outmigrants remit money to their native place ranging from Rs. 100 to Rs. 800 per month. About 21 per cent of total outmigrants remit on an average less than rupees 100 per month and 17 per cent sent rupees 100 to 200 per month, but majority of the outmigrants (33%) remit rupees 200 to 300 per month. More than one-fifth of outmigrants remit more than rupees 300 on an average, while 4 per cent of total outmigrants do not remit money at all.

The outmigrants remit money at various periodicity. About 31 per cent remit monthly, while one-third sent quarterly. There are 17, 10 and 5 per cent outmigrants, who remit half yearly, yearly and rarely respectively. The remaining outmigrants do not remit the money at all because they have almost lost their contact with their native place. The money thus received, is used for different purposes, but substantial part of it is used for agricultural purposes, specially in the development of infrastructure and purchasing of inputs. They purchase agricultural land as well. Generally, outmigrants visit their native place half yearly, and yearly in order to participate in agricultural activities and to celebrate various socio-religious functions. The outmigrants bring certain agricultural innovations through which they try to improve the agricultural environment of the area. However, the impact of outmigrants at respondent level on patterns of agricultural landuse in kharif and rabi seasons is assessed separately in the following lines.

## Agricultural Landuse Pattern in Kharif Season and Volume of Change

The picture of agricultural landuse (Fig. 13.2) and the volume of change gets reflected at the micro-level. The cereals and pulses occupying around 70 per cent of the cropped land, dominate the scene in all the five categories of respondents. Cash crops come next but it occupies slightly more than 15 per cent of the cropped land. Horticulture gets a third order in sequences, occupying a little more than 10 per cent of the total cropped land.

TABLE 13.2: **Volume of change in agricultural landuse pattern in kharif season in groups of respondents (1975-85)**

| Sl.No. | Respondent Groups | Year | Area in percentage | | | | |
|---|---|---|---|---|---|---|---|
| | | | Cereals and Pulses | Cash crops | Other crops | Horti-culture | Fallow Land |
| 1. | More outmigration | 1975 | 78.88 | 4.73 | — | 1.56 | 14.93 |
| | | 1985 | 67.25 | 16.59 | — | 10.94 | 5.22 |
| 2. | Moderate outmigration | 1975 | 78.46 | 6.65 | — | 3.50 | 11.39 |
| | | 1985 | 63.61 | 19.27 | — | 12.20 | 4.92 |
| 3. | Less outmigration | 1975 | 78.92 | 8.97 | 0.09 | 2.72 | 9.29 |
| | | 1985 | 69.64 | 14.27 | 0.12 | 11.12 | 4.85 |
| 4. | Returned migrants | 1975 | 82.91 | 4.82 | 0.21 | 5.86 | 6.20 |
| | | 1985 | 65.58 | 15.83 | 0.25 | 16.79 | 1.55 |
| 5. | No outmigration | 1975 | 76.55 | 6.85 | 1.05 | 0.53 | 15.02 |
| | | 1985 | 66.09 | 14.30 | 0.99 | 1.97 | 16.65 |
| | Aggregate | 1975 | 79.33 | 6.91 | 0.11 | 3.20 | 10.45 |
| | | 1985 | 77.71 | 16.32 | 0.13 | 11.98 | 4.86 |

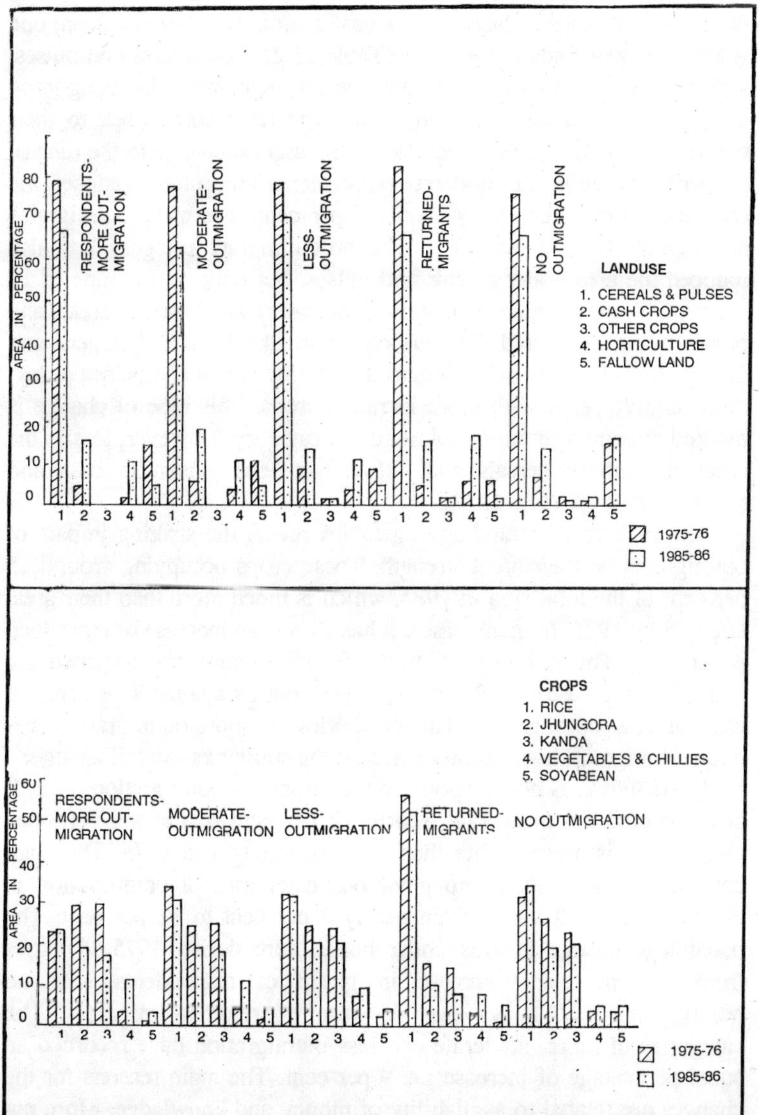


**FIG. 13.2: Changing agricultural landuse pattern in kharif season.**

During the decade under review, there have been a lot of changes in different categories of landuse in kharif season, which varies from one group of respondents to the other (Table 13.2). The cereals and pulses, which have recorded the maximum loss, have shown a declining trend of around 15 per cent. The migrants who have come back to their native land permanently, have shown the highest change to the tune of 17 per cent, while the moderate respondents group have reduced the area under this category by about 15 per cent. The more outmigrants are slightly lagging behind it. The no-outmigrant category has also reduced the area under cereals and pulses, but only to the tune of 10 per cent. There has also been shift in the nature and type of cereals and pulses being cultivated. The course grain like Kauda, Jhungera and Cheena or millets are being replaced with fine grains and pulses or less remunerative crops with more lucrative crops. This type of change is marked maximum in more outmigration category. However, loss in the areal strength of cereals and pulses has been gained in cash and horticulture crops (Fig. 13.2).

The cash crops including vegetables reveal the striking impact of outmigrants on their areal strength. These crops occupying around 15 per cent of the total area in 1985, which is much more than their areal strength in 1975. In many cases, it has shown an increase of more than 4 per cent. The category of moderate outmigrants has recorded the maximum increase of more than 16 per cent as against 7 per cent in case of less outmigrants. The categories of more-outmigration and returned migrants have recorded almost the similar amount of change.

Horticulture, is no exception to the impact of outmigration on it. It has occupied on an average nearly 12 per cent of the total cropped area, which is much higher than its areal strength in 1975. There are considerable variations among various categories of outmigration in this connection. It varies from nearly 2 per cent to 12 per cent. The percentage change in area under horticulture during 1975-85 varies from one per cent increase in group of respondents with no outmigration to 11 per cent in case of returned migrants. The categories of more, moderate and less outmigration have recorded an equal percentage of increase i.e. 9 per cent. The main reasons for the changes are related to availability of money and knowledge. More out and returned migrants have more money and knowledge as compared to groups of respondents with less and no outmigration. Therefore, later groups are unable to adopt the farming of cash crops. In case of fruit farming, 5 to 10 years lapse between the planting of trees and their economically benefits. Poor farmers cannot wait for such a long time. So they do not change their agricultural landuse pattern. The area under other crops has shown, by and large, a declining trend, which shows the more interest towards specialisation.

Another striking feature of landuse pattern during kharif season is the rapid decline in the areal strength of fallow land. It has occupied on an average 15 per cent of the net area sown in 1975, which shows intensification of agriculture as a result of improved infrastructural facilities due to better economic conditions. However, this change has not been uniform among all the categories of the respondents. It is the highest in case of returned migrants (11%) and more outmigrants (9%) categories followed by moderate, and less outmigrants categories to a tune of 7 and 5 per cent respectively. But is has gained around 2 per cent in case of the category of no-outmigration, which shows that they lack in resources. Hence, they cannot bring it under cultivation with the help of various modern agricultural techniques and innovations. The above analysis clearly indicates that the outmigrants have influenced the pattern of agricultural landuse during the kharif season.

The above analysis clearly indicates that the outmigrants have influenced the pattern of agricultural landuse during kharif season. As a result, overall volume of change is found high in categories of returned migrants (21.98%), more outmigration (21.34%) and moderate outmigration (21.29%). Least changes have taken place in categories of less outmigration (13.76%) and no-outmigration (10.52%).

### Agricultural Landuse Pattern in Rabi Season and Volume of Change

The rabi crops are usually sown in the months of October-November and are harvested in the months of March to May. The major crops of the season are wheat, barley, mustard, pulses and potatoes. About half of the total cropped area in rabi season is covered by cereals and pulses in 1985-86. They occupy significant proportion in all the categories and it varies from 42.05 to 50.55 per cent. Cash crops account a sizeable proportion in groups of respondents with more and moderate out-migration and returned migrants, while these crops account only 3.95 per cent of total area in category of no-outmigration. Although, fallow lands occupy nearly one-forth of total area in all the groups, proportion of fallow land is lowest in group of respondents with returned migrants (23.47%) and highest in no-outmigration category (43.53%).

But, there have been considerable changes in all these categories of landuses of rabi season with reference to different groups of outmigrations (Table 13.3). The area under cereals and pulses has shown a declining trend varying from 5 to 17 per cent among all groups of outmigrants (Fig. 13.3). The maximum decline is noticed in case of returned outmigrants, who have, perhaps, come back with adequate financial resources and knowledge of agricultural innovations. The group of more outmigration comes next with 12 per cent decrease and is followed by other groups. This declining trend is mainly due to the shift of area towards cash and fruit crops, which is generally possible with sound economic base.

Cash crops including vegetables and potatoes have shown a rapidly increasing trend ranging from one per cent in case of the group of no-outmigration to more than 7 per cent in almost all the groups of respondents; but the rate is uniform among various groups. Cropwise analysis of volume of change shows the increasing trend in area under vegetables and potatoes. Due to progressive nature of migrants, more increase in area under vegetables and potatoes have occurred in groups of respondents with more outmigration, returned migrants and moderate outmigrants. Pulses have shown both the trends — positive as well as negative, but changes in area under pulses is less and insignificant. On the other hand, due to more returns from vegetables and fruits, area under wheat and barley has declined. Percentage of decrease has occurred maximum in group of respondents with returned migrants (10.08%) and more outmigration (8.40%), as compared to group of respondents with no outmigration (0.84%). It indicates the lack of will power of respondents of last group to take risk in changing their cropping pattern (Fig. 13.3).

Likewise, horticulture has also shown an increasing trend ranging from one per cent to about 12 per cent. The group of returned migrants has recorded the maximum increase followed by the groups of more moderate and less outmigrants, whereas it is only one per cent in the group of no-outmigration. It shows that families of last group are either unable or not interested to change their agricultural landuse pattern. It is the result of lack of money, knowledge of new ideas, which are found in abundance in the case of families of out and returned migrants.

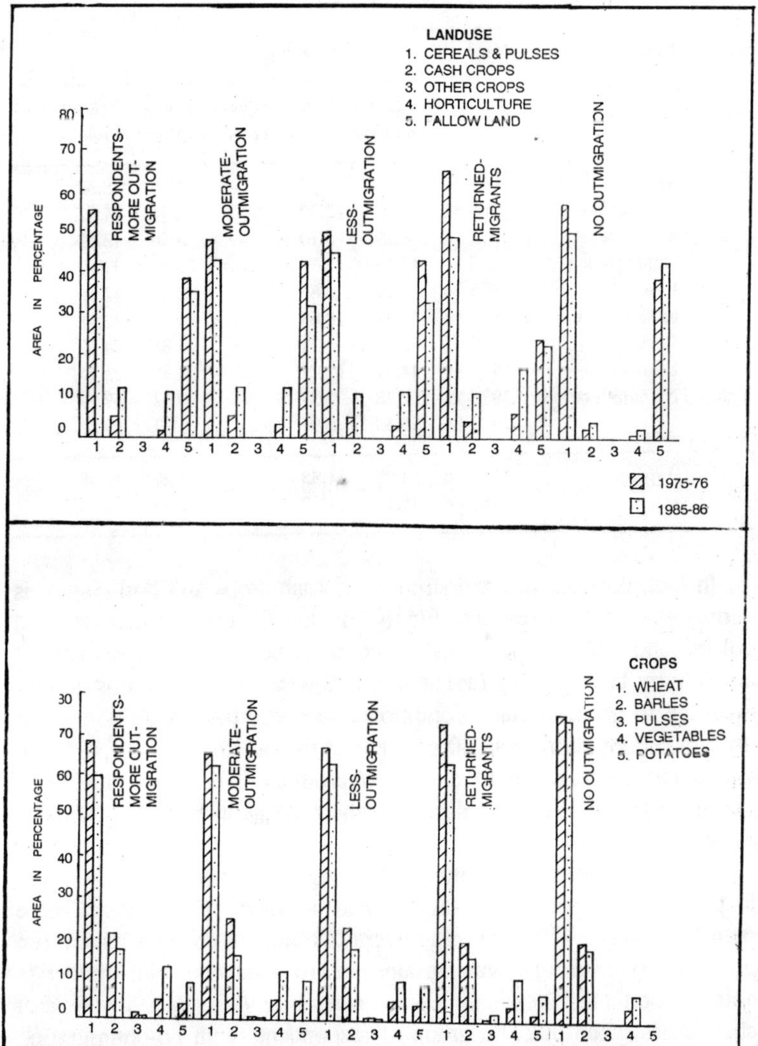


**FIG. 13.3: Changing agricultural landuse pattern in rabi season.**

TABLE 13.3: **Volume of change in agricultural landuse pattern in rabi season in groups of respondents (1975-76 – 1985-86)**

| Sl.No. | Respondent Groups | Year | Area in percentage | | | | |
|---|---|---|---|---|---|---|---|
| | | | Cereals &Pulses | Cash crops | Other crops | Horti-culture | Fallow Land |
| 1. | More | 1975-76 | 54.58 | 5.54 | – | 1.56 | 38.32 |
| | outmigration | 1985-86 | 42.05 | 12.30 | – | 10.94 | 34.71 |
| 2. | Moderate | 1975-76 | 48.29 | 5.10 | – | 3.50 | 43.11 |
| | Outmigration | 1985-86 | 43.67 | 12.16 | – | 12.20 | 31.97 |
| 3. | Less | 1975-76 | 50.02 | 4.89 | – | 2.73 | 42.36 |
| | outmigration | 1985-86 | 45.32 | 10.65 | – | 11.12 | 32.91 |
| 4. | Returned | 1975-76 | 65.20 | 4.31 | – | 5.86 | 24.63 |
| | migrants | 1985-86 | 48.57 | 11.17 | – | 16.79 | 23.47 |
| 5. | No outmigration | 1975-76 | 57.18 | 2.50 | – | 0.53 | 39.79 |
| | | 1985-86 | 50.55 | 3.95 | – | 1.97 | 43.53 |
| | Aggregate | 1975-76 | 52.97 | 4.85 | – | 3.20 | 38.98 |
| | | 1985-86 | 45.15 | 11.12 | – | 11.98 | 31.75 |

In fact, the gain in areal strength of cash crops and horticulture is mainly due to two reasons. Firstly, decline in area of cereals and pulses and fallow land and secondly, horizontal expansion of agricultural land. During rabi season fallow lands occupy considerable area due to harsh climatic conditions, specially the severe winters. It used to occupy more than 40 per cent of the net area sown (1975-76), which has come down to a tune of around 30 per cent (1985-86), except in group of no-outmigration where it has increased by about 4 per cent.

However, the overall volume of change have been observed more in groups of respondents with returned migrants (13.85%) and more outmigration (13.42%). Average overall volume of changes have taken place in two groups viz. groups of respondents with moderate outmigration (12.22%) and less outmigration (10.12%), while least change has taken place in group of respondents with no-outmigration (3.05%).

## Conclusion

The foregoing study reveals many interesting facts about outmigration and its impact on patterns of landuse. The rate and volume of outmigration is much more in hilly areas as compared to the plains. A lot of people have migrated from Pauri Garhwal district in different directions, especially in the plains of U.P. and the neighbouring states including Delhi. The volume of outmigration is positively related with the transport connectivity. These outmigrants not only remit money of varying amount at frequent intervals but also visit their native place, depending upon their nature of employment and job. Thus moneyorder remittance together with the new agricultural technology affects the pattern of landuse in the district. It helps in improving agricultural infrastructure and adoption of agricultural innovations. The substantial amount of the moneyorder remittance is used for agricultural purposes.

Consequently, structural changes have taken place in agriculture. The net cropped area has adequately increased as a result of horizontal expansion and intensification of agriculture. The outmigrants have not only intensified their agriculture, but also have changed their crops and cropping system. The course and less renumerative crops like Kauda, Jhungora and Cheena have been replaced by fine and more renumerative crops such as rice, soyabean and potatoes. The most striking change which has taken place in the agriculture in general and cropping pattern of the outmigrants in particular, is the shift towards horticulture. The area under horticulture has considerably increased among the more, returned and moderate migrant categories of respondents as compared to the other categories of respondents i.e less and no-outmigrants. It shows the trend of commercialisation and specialisation of agriculture. Thus, outmigrants act as agent of agricultural change in the place of origin and contribute to promote further rural development.

## REFERENCES

Bandooni, S. K., 1988 : *Impact of Outmigration on Patterns of Agricultural Landuse in Pauri Garhwal District*, Unpublished M.Phil. Dissertation submitted to Department of Geography, Delhi University.

Bisht, N. S., 1984 : 'Strategy for integrated agricultural development of Uttrakhand Himalaya', in S.C. Joshi et. al (Eds.), *Rural Development in the Himalaya: Problems and Prospects*, Gyanodaya Prakashan, Nainital, pp. 131-142.

Bora, R. S., 1987 : 'Extent and Causes of Migration from the Hill Regions of U.P.', in Vidyut Joshi (Ed.), *Migrant Labour and Related Issues*, IBH Publishing Company, New Delhi, pp. 167-210.

Census of India, 1981 : *Migration Tables: Series 22*, U.P., Directorate of Census Operation, Lucknow, Part V, A and B.

David, L., and Hambros, S. O., 1962 : 'Emigration and depopulation: Some neglected aspects of population geography', *Geographical Review*, American Geographical Society, New York, Vol. LII, No. 2, pp 195-210.

District Census Handbook, 1971 : *Primary Census Abstract: District Pauri Garhwal*, Lucknow, Series 21, U.P. Part XB.

Dobhal, G. L., 1987 : *Development of the Hill Areas: A Case Study of Pauri Garhwal District*, Concept Publishing Co., New Delhi.

Kaur, J., 1985 : *Himalayan Pilgrimages and the New Tourism*, Himalayan Books, New Delhi.

Mohammad, N., 1978 : *Agricultural Landuse in India: A Case Study*, Inter-India Publications, Delhi.

Negi, P. S., 1981 : 'Impact of moneyorder economy on socio-economic development of Garhwal Region', *National Geographer*, Allahabad Geographical Society, Allahabad, Vol. 16, No. 1, pp 71-76.

Oberai, A. K., and Singh, H. K. M., 1983 : *Causes and Consequences of Internal Migration: A Case Study in Indian Village*, Oxford University Press, Delhi.

Rawat, P. S., 1987 : 'Migration and Development in U.P. Himalaya', in V. P. S. Pangtey and S. C. Joshi (Eds.), *Western Himalaya*, Vol. II, Gyanodaya Publication, Nainital, pp. 305-309.

Singh, J., 1974 : *An Agricultural Atlas of India: A Geographical Analysis*, Vishal Publications, Kurukshetra.

State Planning Commission, 1987 : *Shankikya Patrika of Pauri Garhwal District (Hindi)*, Statistical Department , Lucknow.

# 14

# THE NON-LINEAR DYNAMICS OF SECTORAL SHIFTS

EMILIO CASETTI AND KAVITA PANDIT

DURING A country's transition from a premodern to a postindustrial economy the proportion of its labour force in agriculture tends to decline, that in manufacturing increases at first and then declines, and that in services tends to increase. These trends have been empirically validated and theoretically justified. However, the literature has recognised two distinct sub-patterns within these trends. Specifically, while some countries experienced a surge in their manufacturing employment at relatively early stages of development, others such as the presently developing countries have seen rapid early increases in service employment with labour absorption in manufacturing remaining low.

This paper suggests that these sub-patterns can be related to the existence of multiple equilibrium partitions of the labour force. More specifically, it will be shown that a model based on conventional economic presuppositions produces two stable equilibria at intermediate levels of development. Of these equilibria, one is characterised by the dominance of manufacturing and the other by the dominance of services. The results obtained explain at the qualitative level of resolution the empirical patterns observed and provide a theoretical rationale for policies aimed at promoting industrialisation as means to more rapid economic development.

## Background

The sectoral shifts that occur during the process of development

---

*Source*: *Economic Geography*, Vol. 63, 1987, pp. 241-258.

constitute a cluster of theoretically justified empirical regularities that have attained a quasi-law status. The early impetus to the investigation of sectoral shifts was provided by the work of Clark (1940), Fisher (1939) and Fourastie (1949). Canonical explanations for these shifts hinge upon the interplay of sectoral differentials in income elasticities of demand and differential growth of labour productivities. The low elasticity in the demand for agricultural products and the rising agricultural productivity cause relative employment in this sector to fall during development. At the same time, an increasing demand for manufactured goods causes relative manufacturing employment to rise. In later stages of development, productivity increases in manufacturing along with the high elasticity of demand for services cause transfers of labour from manufacturing to services.

The pioneering research by Clark, Fisher and Fourastie was followed by a number of empirical studies Bairoch and Limber, 1966; Chenery, 1960, 1968, 1975 and Kuznets, 1966, 1971. A considerable uniformity was noted in the tendency of the agricultural share of the labour force to decline. As regards the non-agricultural sectors, two subpatterns were identified. During the development of the industrialised countries, the decline in agricultural labour was accompanied by a substantial growth in the manufacturing sector. This pattern of sectoral shifts has been noted for instance by Clark (1940), Kuznets (1966), Bairoch and Limber (1968), Sorrentino (1969) and UNECE (1977).

In contrast, in the presently developing countries the shift of labour out of agriculture has been accompanied by an unprecedented growth in services, with manufacturing labour shares growing only at modest rates. This pattern has been observed for instance by ECLA (1965), Ramos (1970), Bairoch (1975), Berry (1978) and Kirsch (1973) in post-World War II Latin America; by Mehta (1961) and Oshima (1971) in Asia; and by Bairoch and Limber (1963), Turnham (1971) and Sabolo (1975) for other less developed countries.

Several theories have been proposed to explain these service dominated sectoral shifts. Bairoch (1975) and Mouley and Costa (1974) posit that in the developing countries, high rates of natural increase in population together with rapid rural to urban migration allowed the supply of non-agricultural labour to outstrip the absorptive capacity of the industrial sector. This excess non-agricultural labour joins the "informal" service sector, made of low-skilled, low-wage, individual and family enterprises (Hart, 1973). In fact, it has been noted that the "traditional" or informal services are the ones undergoing the most rapid expansion in the contemporary less developed countries (Emi, 1969; Kuznets, 1971 and Berry and Sabot, 1984).

Other explanations of the service-dominated sectoral shifts in the less developed countries emphasised the inadequate growth of the industrial sector due to social and institutional obstacles and/or to neo-colonial structures (Harris and Todaro, 1970; Mazumdar, 1976; and McGee, 1967 and Katouzian, 1970), and the tendency to the adoption of capital intensive techniques in manufacturing (Singer, 1964; Baer and Herve, 1966; Dore, 1974; I.L.O., 1976 and Berry, 1978).

In Figs. 14.1 and 14.2 examples of the manufacturing dominated sectoral shifts in the now-industrialised countries are contrasted to examples of service dominated shifts that are occurring in the contemporary less developed countries. In Fig. 14.1, ratios of service employment to manufacturing employment, *S/M*, are plotted against percentages of labour force in agriculture. The ratios for the Philippines, Brazil, and Mexico are clearly higher than those for Germany, United Kingdom, and Belgium and remain higher as the percentage labour force in agriculture declines. The same data are plotted in the triangular diagram in Fig. 14.2. In this diagram, every point in the triangle maps one tripartition of the labour force between agriculture manufacturing and services. Here, too, the structural shifts of the two groups of countries form visually distinct patterns. Later, it will be shown that these patterns closely resemble the ones generated by the analyses that constitute the focal point of this paper.

In this paper, the manufacturing dominated and the service dominated sectoral shifts are explained in terms of multiple systemic equilibria that materialise at intermediate development levels. This explanation extends an earlier investigation (Casetti, 1980) on equilibrium population partitions between the urban and rural sectors. In it Casetti suggested that in pre-modern times there is only one equilibrium partition with most of the population in the rural sector. With development, two stable equilibrium partitions appear, separated by an unstable equilibrium. In mature economies, a single stable partition exists, with most of the population in the urban sector. The relation between this two-sector model and the conventional sectoral partition into agriculture, manufacturing, and services was investigated by Casetti and Pandit ( Casetti and Pandit, 1984).

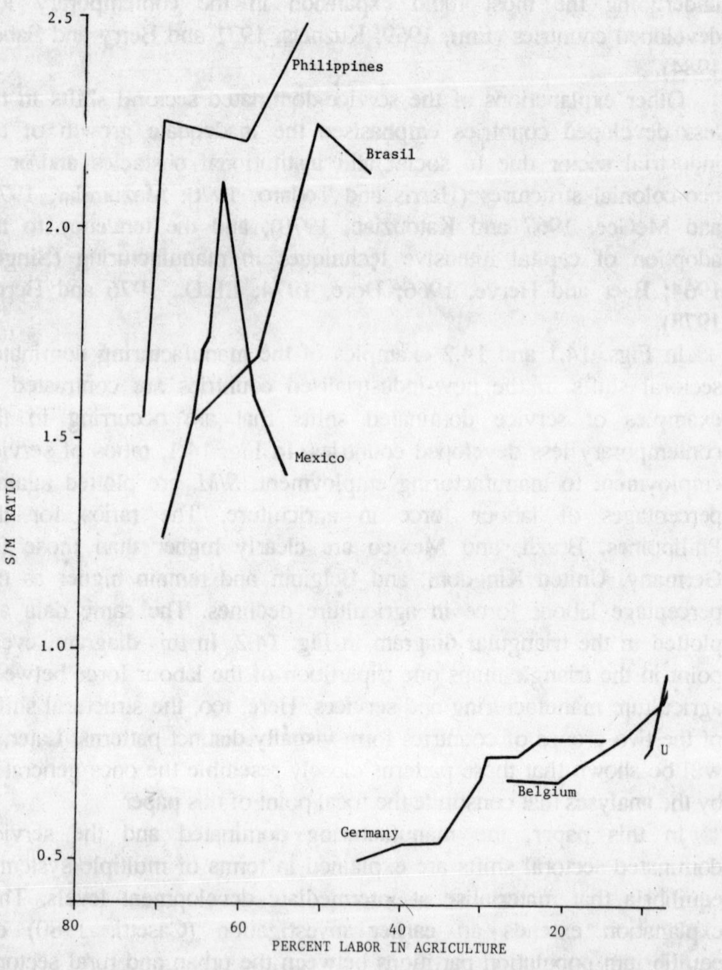


**FIG. 14.1: Relationship between the service to Manufacturing Labour Ratio and the Agricultural Labour Share for selected countries.**

*Source:* **Sabolo (44, pp. 6-9).**

The broad contours of the investigation reported in this paper can be described as follows. First, an initial model of sectoral dynamics at a fixed level of development is formulated. The model consists of differential equations relating to rates and levels of sectoral labour forces. This "sub-system" model reflects the presupposition that for a given capital stock and technological level, agriculture and services are characterised by diminishing returns to scale; whereas in manufacturing, scale economies operate initially but are eventually replaced by scale diseconomies and diminishing returns.

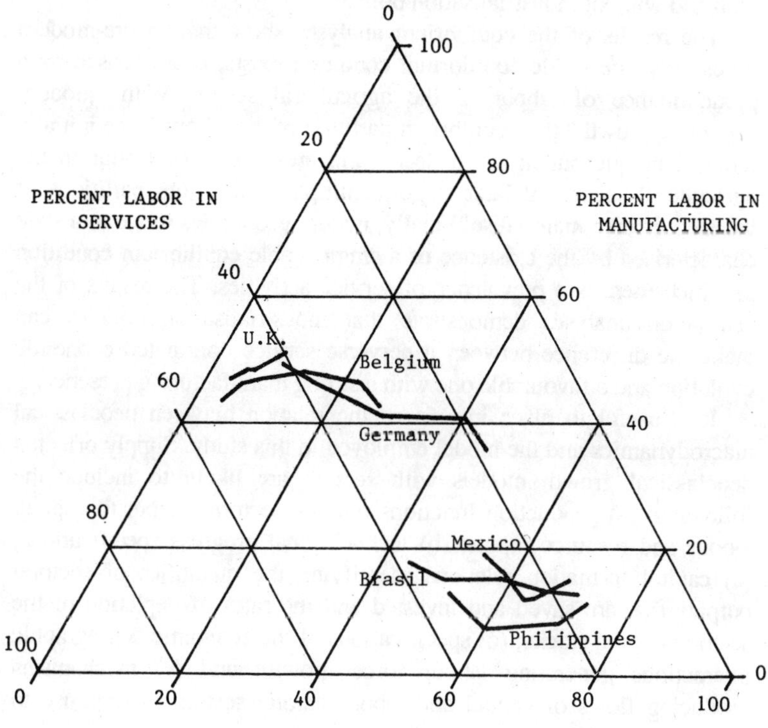


FIG. 14.2: **Patterns of Labour Force Shifts between Agriculture, Manufacturing, and Services for selected countries.**
*Source:* Sabolo (44, pp. 6-9).

The parameters of this sub-system map aspects of the economic system, such as prices, capital stock and technological levels, that change over the course of development. The evolution of the sectoral dynamic sub-system during the course of development is modelled by "expansion equations" redefining some of these parameters into functions of time.

To investigate the implications of these formalisms, (i) plausible numerical values of the subsystem's parameters and of the expansion equations attached to it are specified; (ii) equilibrium partitions of the labour force and equilibrium wages are calculated over 200-years-time interval, and (iii) the entire dynamic system is solved for scenarios with and without industrialisation policies.

The results of the equilibrium analyses show that in pre-modern times only one stable equilibrium condition exists, characterised by a predominance of labour in the agricultural sector. With "modern economic growth" the equilibrium partition of the labour force initially remains unique but involves decreasing proportions of labour in the agricultural sector. At later stages, multiple equilibrium partitions of the labour force materialise. Finally, mature economies move to a state characterised by the existence of a single stable equilibrium condition in which there is a prevalence of service activities. The results of the simulation analyses demonstrate that industrialisation policies can make the difference between a perverse service dominated economic evolution and a favourable one with a strong manufacturing presence.

It is useful to place into focus the relation between neoclassical macrodynamics and the model employed in this study. Supply oriented neoclassical growth models with *sectors* are likely to include the following: (a) production functions relating sectoral output to capital, labour and resource inputs; (b) technological progress specifications; (c) capital formation relations identifying the quantities of sectoral outputs that are saved and invested and the rates of depletion of the sectoral capital stocks; (d) specifications of the economic demographic interactions generating labour force growth; and (e) mechanisms producing flows of capital and labour among sectors in response to remuneration differentials.

The building blocks of neoclassical growth theory are indeed essential for understanding the development processes, and many interesting contributions have been constructed upon them (Hamberg, 1971; Wan, 1971; Burmeister and Dobell, 1970 and Tinbergen and Box, 1972). The strategy employed here consisted in formalising explicitly only some neoclassical building blocks and mechanisms while leaving others in the background. In fact, the equations in this study specify only a labour sub-system and the changes of some of its parameters during development, which are focal points of our analyses. These equations, however, were conceptualised in terms of economic relations and variables, some of which do not appear explicitly in our formalisms.

## The Model

The model used in this paper was arrived at by combining a sectoral shifts sub-system model with expansion equations specifying the manner in which the sub-system changes with development.

The modelling strategy employed implements the Expansion Method (Casetti, 1972 and 1976), a technique for model construction that allows for the controlled introduction of complexities into simple initial structures. Basically, the expansion methodology involves: (a) defining an "initial model"; (b) specifying the modalities of its contextual variation by "expansion equations" that relate some or all of the initial model's parameters to relevant variables; and (c) replacing the expansion equations into the initial model to produce a more complex "terminal model".

The model employed in this paper was designed to investigate the manner in which the dynamics of the sectoral shifts change during the course of economic development. To construct it, first an initial model of the sectoral shifts sub-system was defined. This initial model consists of a set of differential equations producing stable equilibrium partition(s) of the labour force between agriculture, manufacturing and services. The model's parameters reflect economic structures and level of development. Selected parameters of the initial model were then expanded into functions of time representing development induced changes. Finally, the expansion equations were inserted into the initial model to produce a complex "terminal" formulation encompassing both a sectoral dynamics mechanism and the change of the mechanism over time.

In the sections that follow, the initial model, the expansion equations and the terminal model will be presented. Then the terminal model is investigated to determine the equilibrium partitions of the labour force, it would produce over the course of development and the time paths of sectoral partitions with and without forced industrialisation policies.

## The Initial Model

Assume an ideal country with a labour force of $N$. Denote by $N_A$, $N_M$ and $N_s$ the portions of $N$ in the agricultural, manufacturing and service sectors, so that

$$N = N_A + N_M + N_S \quad (1)$$

Here and hereafter, the subscripts *A*, *M* and *S* stand, respectively, for agriculture, manufacturing and services. Let $w_A$, $w_M$ and $w_S$ be the wage rates in the three sectors, and assume that intersectoral labour flows are proportional to wage differentials. Then, a flow of $F_{AM}$ persons per year from/to agriculture to/from manufacturing is produced by the wage differential $w_M$-$w_A$ multiplied by a constant $z_{AM}$ with dimension "persons per dollar":

$$F_{AM} = z_{AM}(w_A - w_M) \quad (2)$$

Assume $F_{ji} = -F_{ij}$ for any sectors *i* and *j*. Let the ideal country be a closed system with no labour movements across its boundaries. Then, the instantaneous rates of change of its sectoral labour force are given by the following:

$$dN_A/dt = F_{AM} + F_{AS} = z_{AM}(w_A - w_M) + z_{AS}(w_A - w_S) \quad (3)$$

$$dN_M/dt = F_{MA} + F_{MS} = z_{AM}(w_M - w_A) + z_{MS}(w_M - w_S) \quad (4)$$

$$dN_S/dt = F_{SA} + F_{SM} = z_{AS}(w_S - w_A) + z_{MS}(w_S - w_M) \quad (5)$$

Assume that for any sector *i*, the wage $w_i$ is proportional to the marginal productivity of labour, defined as the product of sectoral prices, $p_i$, times physical marginal labour productivities $M_i$:

$$w_i = vp_iM_i \qquad i = A, M, S \quad (6)$$

where $v$ is a proportionality coefficient.

The physical marginal productivities of labour are taken to be functions of the sectoral labour forces. These functions play a key role in this investigation and need to be justified at some length.

The technological levels and capital stock increase during development, but may be regarded as fixed at a given development level. The sectoral dynamics sub-system employed here is defined for a fixed development level. At any given development level, when a sectoral partition of the labour force is specified, it will produce a redistribution of the capital stock among sectors, which in turn will determine the marginal labour productivities. We did not specify explicitly the engines of sectoral capital flows and the sectoral production functions. However, they are the conceptual backdrop of the "black box" relations between sectoral labour forces and sectoral marginal labour productivities that are modelled explicitly.

For agriculture and services, it was assumed that the physical marginal productivities decrease monotonically with increases in labour. However, in the case of manufacturing, it was assumed that the physical marginal productivity first increases and then declines with rising $N_M$, attaining a local maximum at some intermediate level of employment. Figures 14.3 through 14.5 portray graphically these assumptions.

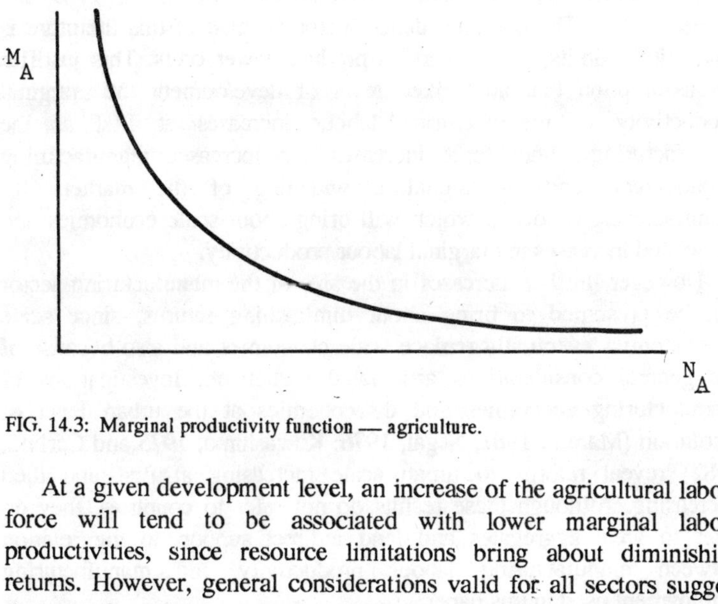


FIG. 14.3: **Marginal productivity function — agriculture.**

At a given development level, an increase of the agricultural labour force will tend to be associated with lower marginal labour productivities, since resource limitations bring about diminishing returns. However, general considerations valid for all sectors suggest that at a given development level increases in size and labour force will eventually bring about diminishing returns. In fact, the best productive opportunities will be the first to attract entrepreneurs, capital and labour. Therefore, any sectoral expansion will tend to be realised by increasingly suboptimal productive opportunities. Furthermore, since the labour force is made of qualitatively different units, it can be presumed that as the labour force share of any one sector grows larger it will tend to disproportionately large fraction of lower quality labour.

All these arguments suggest that at a given level of development, increases in sectors' sizes and labour forces will eventually bring about lower marginal labour productivities.

The notion that division of labour, scale economies, agglomeration economies and external economies bring about lower costs when some size and market thresholds are exceeded has a long history and has been referred especially to manufacturing. This notion has been theorised and investigated empirically at levels of resolution encompassing plants, firms, urban centres, regions, countries, and economic sectors (Fisher, 1983; Sawyer, 1981; Pratten, 1971; Gould, 1972; Nourse, 1978; Denison, 1967; Jones, 1933; Emi, 1969 and Stigles, 1951). The common denominator of most of this literature is that, within limits, larger markets produce lower costs. This justifies the assumption that at a fixed level of development the marginal productivity of manufacturing labour increases at first as the manufacturing labour force increases. An increased manufacturing employment tends to signal a widening of the markets for manufacturing products, which will bring about scale economies and associated increases in marginal labour productivity.

However, further increases in the size of the manufacturing sector can be presumed to bring about diminishing returns, since scale diseconomies eventually replace scale economies and also because of the general considerations articulated earlier on. Investigations of manufacturing economies and diseconomies at the urban level of resolution (Marelli, 1981; Segal, 1976; Kawashima, 1975 and Carlino, 1982) reveal returns to urban scale increasing at first and then decreasing. Although these results do not refer to countries, they do refer to areal aggregates and lend indirect support to the relation between manufacturing labour productivity and manufacturing employment used in this paper.

The following specifications of the $M_i(N_i)$ functions were selected:

$$M_A = exp(a - bN_A) \tag{7}$$

$$M_M = exp(m - n(N_M - h^2)) \tag{8}$$

$$M_S = exp(q - rN_S) \tag{9}$$

where *a, b, m, n, h, q* and *r* are positively valued parameters. Equations (7) through (9) imply that the $M_i$'s approach zero when their $N_i$'s grow larger. Equation (8) allows a local maximum at some positive value of $N_M$. The parameters *a*, *m* and *q* reflect the effects of technology and capital on productivities. If they increase, everything else being equal, the marginal labour productivities increase. Consequently, these parameters correspond to a given fixed development level.

The sectoral price functions were specified as follows:

$$p_i = exp(H_i/N_i + P_i) \qquad i = A, M, S \tag{10}$$

These functions imply that the prices of the sectoral outputs approach a constant value of $exp(P_i)$ when a sector grows large but would increase rapidly if a sector's size shrinks too much. This latter effect is designed to represent the increase in prices associated with decline in production in a fashion that is mathematically tractable in this context. Specifically, the *H* parameters determine the sensitivity of sectoral process to an "excessive" contraction of the sectoral employments and indirectly, outputs. The *P* parameters are logarithmic transformations of "normal" sectoral prices and are exogenous to the labour dynamics sub-system. The impact of demand shifts on equilibrium sectoral partitions of the labour force can be investigated by varying the *P*'s.

The sub-system identified by equations (1) through (10) is in equilibrium if

$$w_A = w_M = w_S = w^* \tag{11}$$

where $w^*$ depends upon the subsystem's parameters. In fact, if the wage rates in the three sectors are equal, no labour flows between any two sectors take place and consequently $dN_i/dt = 0$ for all *i*'s.

Equation (11) is not only a sufficient but also a necessary equilibrium condition. To show it, let us suppose that some values $w_A \neq w_M \neq w_S$ could produce $dN_i/dt = 0$ for all *i*'s. These ($w_i$'s should produce mutually compensating circular flows, so as to keep the labour force circulating through the system without altering the sizes of the sectors. Suppose, for instance, that labour flows occur from agriculture to manufacturing and from manufacturing to services. Then, a compensating flow from services into agriculture should take place, if the sizes of the sectors are to be preserved. However, a flow from agriculture to manufacturing and from manufacturing to services. Then, a compensating flow from services will take place only if $w_M > w_A$ and $w_S > w_M$. These inequalities imply $w_S > w_A$. But service wages higher than agricultural wages would produce a manpower flow from agriculture into services, rather than the compensating flow from services to agriculture required to keep the labour force circulating through the system without altering the size of the sectors.

The same reasoning would apply to any conceivable circular flow across two or three sectors. Which proves that no $w_A \neq w_M \neq w_S$ triplet can exist that would produce $dN_i/dt = 0$ for all $i$'s. Consequently, equation (11) specifies a necessary and sufficient condition for the sub-system's equilibrium. However, for any given set of parameter values, the condition (11) will not necessarily produce a single equilibrium wage. The occurrence of multiple equilibrium wages and of matching multiple equilibrium partitions of the labour force constitute the focal point of this enquiry.

Equations (1) through (9) constitute an initial model defining the sectoral shifts dynamics at a level of development specified by the model's parameters. Different development levels are associated with different values of these parameters. By expanding pertinent parameters into suitable functions of development levels and by replacing the expanded parameters into the initial model, a terminal model is generated that represents the sectoral dynamics during the transition of a society from premodernity to maturity.

## The Expansion Equations

Time, as a proxy for development, is used to expand a subset of the initial model's parameters. It could be argued that all the aspects of the model and, consequently, all of its parameters drift over time to some degree as a result of development processes. However, the development induced changes in the sub-system environment considered here are prices and labour productivities. The size of each sector's labour force should be presumed to change in response to the demographic transition and also to labour participation trends. In this investigation, however, these aspects of development will not be dealt with in order to place in focus the implications of price and productivity changes.

Here we will assume that the agricultural prices decline exponentially over time; that the prices of the service sector outputs increase exponentially over time; that the manufacturing prices remain constant; and that the "productivity" parameter *a*, *m* and *q* increase linearly over time. These temporal changes and their numerical specifications were selected so as to portray in a general way trends in productivities and relative prices occurring in the course of economic development.

It should be noted, however, that alternative specifications of the time paths of sectoral prices and productivities could be used to investigate the implications of many diverse real or hypothetical development histories. Let us clarify this point by an example. The service sector includes high technology, high price, higher productivity services and the traditional low price, low productivity activities in the so-called informal service sector. The patterns of growth of the service sector, characterised by a preponderance of the informal services noted in many contemporary less developed countries, translate into lower levels and rates of growth in service prices and productivities. In contrast, hypothetical development patterns characterised by massive manpower transfers from agriculture to modern services would be associated with much higher levels and rates of growth of service prices and productivities. In our formalisms, the effects of these and other scenarios on the sectoral dynamics sub-system can be explored using appropriate alternative specifications of the time paths of productivities and prices. However, in this paper we confined ourselves to try trends in prices and productivities that reflect the development experience in an abstracted and unspecific fashion.

Industrialisation policies and their formalisation are discussed more fully in a later section of this paper. Here it will suffice to note that the effects of industrialisation policies successful in influencing the intersectoral labour flows can be represented by the $z_{AM}$, $z_{AS}$, and $z_{MS}$ parameters in equations (3), (4) and (5). These parameters specify the number of workers who move between any two sectors for every dollar wage differential. In the absence of industrialisation policies these three parameters should be assumed equal, signifying that labour responds to wage differentials equally, irrespective of which sectors are involved. In contrast, industrialisation policies, would facilitate the labour movements into manufacturing and discourage other intersectoral flows, which in the context of the model employed here translates into differential values of the $z$ parameters. Industrialisation policies can be represented by time paths of the $z$ parameters. In order to generate a model capable of accommodating alternative industrialisation policies the $z$ parameters were expanded into functions of time.

The expansion equations corresponding to these assumptions are:

$$P_A = P_{A0} + P_{A1}t \tag{12}$$

$$P_S = P_{S0} + P_{S1}t \tag{13}$$

$$a = a_0 + a_1 t \tag{14}$$

$$m = m_0 + m_1 t \tag{15}$$

$$q = q_0 + q_1 t \tag{16}$$

$$z_{AM} = z_{AM0} + z_{AM1}t \tag{17}$$

$$z_{AS} = z_{AS0} + z_{AS1}t \tag{18}$$

$$z_{MS} = z_{MS0} + z_{MS1}t \tag{19}$$

**The Terminal Model**

By replacing the right-hand-sides of equations (12) through (19) into the initial model the following terminal model is arrived at:

$$dN_A/dt = (z_{AM0}+z_{AM1}t)\,(w_A-w_M) + (z_{AS0}+z_{AS1}t)\,(w_A-w_S) \tag{20}$$

$$dN_M/dt = (z_{AM0}+z_{AM1}t)\,(w_M-w_A) + (z_{MS0}+z_{MS1}t)\,(w_M-w_S) \tag{21}$$

$$dN_S/dt = (z_{AS0}+z_{AS1}t)\,(w_S-w_A) + (z_{MS0}+z_{MS1}t)\,(w_S-w_M) \tag{22}$$

$$ln(w_A) = ln(v) + H_A/N_A + P_{A0} + a_0 + (P_{A1} + a_1)t - bN_A \tag{23}$$

$$ln(w_M) = ln(v) + H_M/N_M + P_{M0} + m_0 + m_1 t - n(N_M-h)^2 \tag{24}$$

$$ln(w_S) = ln(v) + H_S/N_S + P_{S0} + q_0 + (P_{S1} + q_1)t - rN_S \tag{25}$$

$$N_A + N_M + N_S = N \tag{26}$$

Equations (20) through (26) constitute the model that will be investigated in the remaining portion of this paper. They form a system of non-linear differential equations that is too complex for an analytical solution. Consequently, numerical analyses based on plausible and representative parameter values will be employed. Two analyses will be reported. The first one is dynamic equilibrium oriented and focuses upon the change in the equilibrium partitions of the labour force during the course of development. This analysis shows that for plausible parameter values the system produces two stable equilibrium partitions of the labour force characterised respectively by manufacturing dominance and by service dominance.

In the second analysis, equations (20) through (26) are solved over a 200-year time horizon to place into focus the impact of industrialisation policies. The results show that in the absence of industrialisation policies the system may be trapped by the perverse service dominated equilibrium until a late stage of development. Instead, industrialisation policies may bring it much sooner into the domain of the positive attractor (the manufacturing dominated equilibrium) that is associated with much larger values of product per capita.

**Dynamic Equilibrium Analysis**

Equations (20) through (26) specify the dynamics of sectoral shifts during development. In order to investigate the temporal change in the equilibria of the sectoral shifts sub-system, the condition (11) was imposed on the terminal model above to obtain the following:

$$ln(w^*) = ln(v) + H_A/N_A + P_{A0} + a_0 + (P_{A1} + a_1)t - bN_A \tag{27}$$

$$ln(w^*) = ln(v) + H_M/N_M + P_{M0} + m_0 + m_1t - n(N_M - h)^2 \tag{28}$$

$$ln(w^*) = ln(v) + H_S/N_S + P_{S0} + q_0 + (P_{S1} + q_1)t - rN_S \tag{29}$$

$$N_A + N_M + N_S = N \tag{30}$$

The $N$'s and the $w^*$ appearing in equations (27) through (30) identify dynamic equilibrium partitions of the labour force and dynamic equilibrium wages. For appropriate parameter values, one set of equilibria is defined for every point in time, namely, for every value of $t$.

In the next sections, first numerical values are assigned to the parameters of the system, and then the changes in the dynamic equilibria of the system are investigated over a time horizon of 200 years, assumed to encompass the transition from premodernity to maturity.

Only the salient aspects of the process by which plausible parameters were arrived at will be presented here. Let us consider first the parameters appearing in equations (27) through (30). Some of the parameters were set to entirely arbitrary values. Parameter $v$ was set to a value of one. The total labour force was assumed to be a constant 10 million people. The price of the manufacturing output $P_M$ was set equal to one. At time zero, the prices of all three outputs are assumed to be equal; $P_{A1}$ is set equal to -0.025, signifying a 2.5 per cent annual price decline in agricultural products relative to manufacturing products; $P_{S1}$ is set equal to 0.0020, representing a 0.2 per cent increase in service output prices relative to manufacturing. The "hyperbolic" constants $H_A$, $H_M$ and $H_S$ were given an arbitrary value of .30, large enough to prevent the occurrence of non-positive equilibrium $N$'s but such that they would not render the price functions overly sensitive to the hyperbolic term $H_i/N_i$. A rate of technological progress of 1.5 per cent per year was locked into the model by assuming $a_1 = m_1 = q_1 = .015$.

The numerical values of the remaining parameters in equations (27) through (30) were arrived at so that they would produce: (a) marginal productivity functions qualitatively similar to the ones shown in Figs. 14.3, 14.4, and 14.5; and (b) single (as opposed to multiple) equilibria at times zero and 200. The parametrised equations employed in the analysis are as follows:

$$ln(w^*) = .30/N_A + 8.1347 - 0.10t - .3946N_A \qquad (31)$$

$$ln(w^*) = .30/N_M + 4.2517 + .015t - .1159(N_M - 4.5)^2 \qquad (32)$$

$$ln(w^*) = .30/N_S + 4.7667 + .017t - .3228N_S \qquad (33)$$

$$N_A + N_M + N_S = 10 \qquad (34)$$

The system of equations (31) through (34) was evaluated at ten-year intervals over a 200-years time horizon. The analysis was carried out by a FORTRAN program calling the ZSPOW subroutine from the IMSL library. Each solution to the system consists of an equilibrium partition of the labour force and an associated equilibrium wage rate. Each equilibrium was tested for stability. To do so, the equilibrium values were perturbed in order to determine whether the system of differential equations (20) through (26) would set the perturbed equilibrium values in motion toward the equilibrium or away from it.

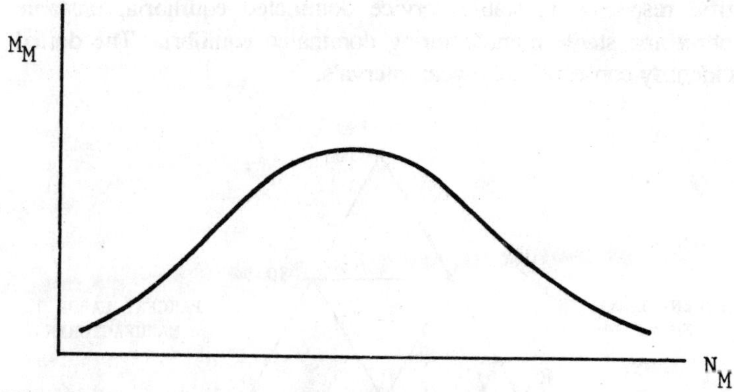


FIG. 14.4: Marginal productivity function — manufacturing.

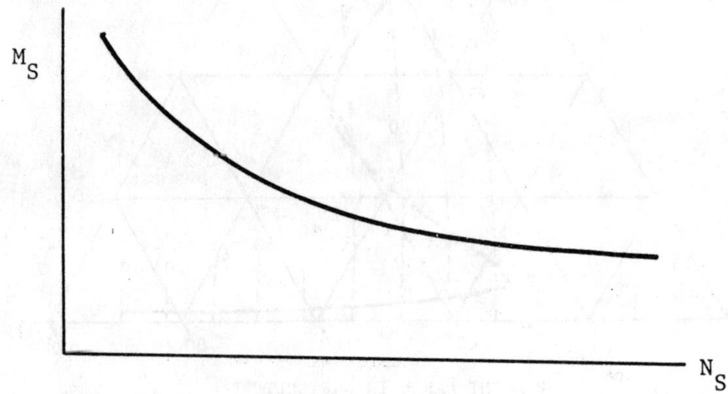


FIG. 14.5: Marginal productivity function — services.

The results of the analysis are shown in Figs. 14.6 and 14.7. In the triangular diagram in Fig. 14.6 the equilibrium partitions of the labour force have been connected to form a continuous curve that shows the equilibrium states of the sub-system over the 200 years. Each point in the triangular diagram corresponds to a partition of the labour force among the three sectors. The segments *AB*, *BC* and *CD* in the diagram identify, respectively, stable service dominated equilibria, unstable equilibria and stable manufacturing dominated equilibria. The dotted lines identify consecutive 20-year intervals.

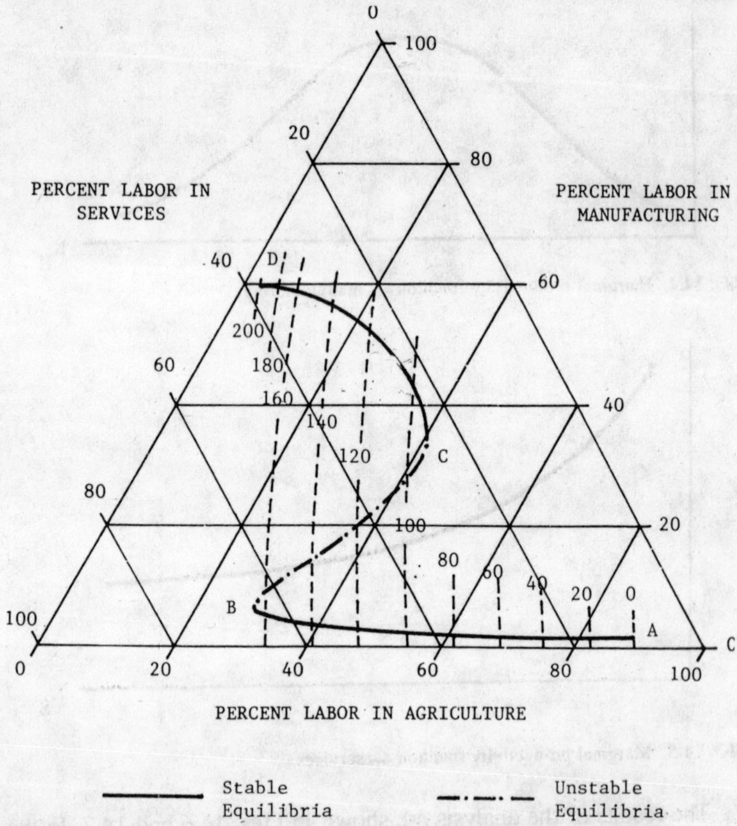


FIG. 14.6: **Equilibrium partitions of the labour force over a 200-year time horizon.**

Clearly, at the beginning and at the end of the time interval only one stable equilibrium exists. In the intermediate period the system possesses three equilibrium states, two stable and one unstable. The two stable equilibria are, respectively, characterised by a prevalence of services and by a prevalence of manufacturing in the non-agricultural segment of the labour force. The service dominated stable equilibrium is perverse in the sense that it is associated with an equilibrium wage level lower than that corresponding to the manufacturing dominated stable equilibrium. This is apparent from Fig. 14.7 that shows the equilibrium wages over time. The lower branch of the curve marked *AB* corresponds to the perverse service dominated stable equilibria. The manufacturing dominated equilibria produce the equilibrium wages depicted by the upper branch of the curve *CD*. The unstable equilibria correspond to the *BC* segment of the curve.

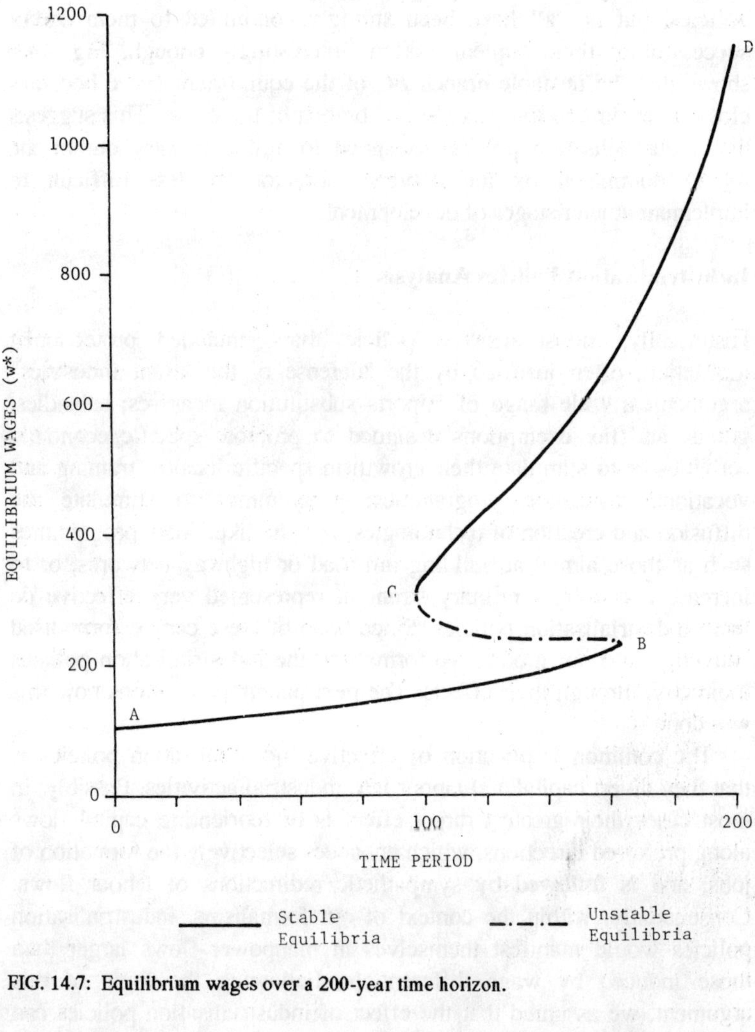


FIG. 14.7: Equilibrium wages over a 200-year time horizon.

These results are congruent with the empirical findings that the decline of the labour force in agriculture during development is associated with a stronger growth of the service sector in some countries and with a stronger growth of manufacturing in other countries.

The higher equilibrium wages associated with manufacturing dominance provide a strong rationale for industrialisation policies. In fact, virtually all developing countries adopted industrialisation policies, but not all have been strongly committed to them and/or successful in their implementation. Interestingly enough, Fig. 14.6 shows that the unstable branch *BC* of the equilibrium curve becomes closer over time to the perverse *AB* branch of the curve. This suggests that industrialisation policies designed to pull a country out of the region dominated by the perverse attractor are less difficult to implement at later stages of development.

**Industrialisation Policies Analysis**

Historically, industrialisation policies have included protectionist legislation, often justified by the "defense of the infant industries" argument: a wide range of imports substitution measures; subsidies, grants, and tax exemptions designed to promote specific economic activities or to stimulate their growth in specific locales; training and vocational education programmes; programmes to stimulate the diffusion and creation of technologies; and the like. Also, programmes such as those aimed at building rail-road or highway networks, or to increase a country's military strength, represented very effective de facto industrialisation policies. Since none of these can be formalised "directly" into our model, we formalised the industrialisation policies indirectly, through their effects. The next paragraph explains how this was done.

The common implication of effective industrialisation policies is that they divert capital and labour into industrial activities. Possibly, in most cases their greatest direct effect is of reorienting capital flows along preferred directions, which enhances selectively the formation of jobs, and is followed by sympathetic redirections of labour flows. Consequently, within the context of our formalisms, industrialisation policies would manifest themselves in manpower flows larger than those induced by wage differentials. Following the logic of this argument, we assumed that the effect of industrialisation policies can be simulated by manipulating the $z$ parameters that control the intersectoral manpower flows.

In order to explore the implications of industrialisation policies, differential equations (20) through (26) were solved over a 200-year time horizon for two sets of $z$ parameters corresponding to scenarios with and without such policies. In the no policies scenario $z_{AM0}$, $z_{AS0}$ and $z_{MS0}$ were set equal to .002, while $z_{AM1}$, $z_{AS1}$ and $z_{MS1}$ were set equal to zero. These parameters imply time invariance of the $z$'s and a between-sectors movement of 2000 persons per year for every dollar wage differential. The second scenario involved temporally phased out industrialisation policies. The expansion equations implementing it are:

$$z_{AM} = (5E - 6) - (2.45E - 8)t \tag{35}$$

$$z_{AS} = (1E - 9) + (4.9E - 10)t \tag{36}$$

$$z_{MS} = (1E - 9) + (4.95E - 10)t \tag{37}$$

These expansions produce enhanced flows from agriculture to manufacturing and reduced flows into services earlier on, and they restore the equality of the $z$ parameters at the end of the 200-year time horizon. The same initial conditions were employed in both simulations. Specifically, it was assumed that at time zero 80 per cent of the labour force was in the agricultural sector, and that the remaining 20 per cent was equally divided between manufacturing and services.

Figure 14.8 shows the two trajectories $T_1$ and $T_2$, that emerge from the simulations. In the case of $T_1$ the partition of the labour force moves over time toward the *AB* branch of the equilibrium curve until this sequence of stable equilibria disappears at *B*. At that point the system becomes attracted by the *CD* branch of the equilibrium curve that is characterised by a prevalence of manufacturing in the non-agricultural labour force. This trajectory identifies the type of sectoral shifts qualitatively similar to the service dominated evolutionary patterns. However, service dominated trajectories can only exist until the stable, equilibria in the *AB* branch of the curve in Fig. 14.8 exists. When this stable equilibrium vanishes at *B* the system will tend to industrialise at an accelerated rate. Consequently, in terms of our model, the predominance of services in the non-agricultural labour force observed in the real world is a perverse condition that eventually is exited. The countries with a service dominated, non-agricultural labour force have not escaped it yet.

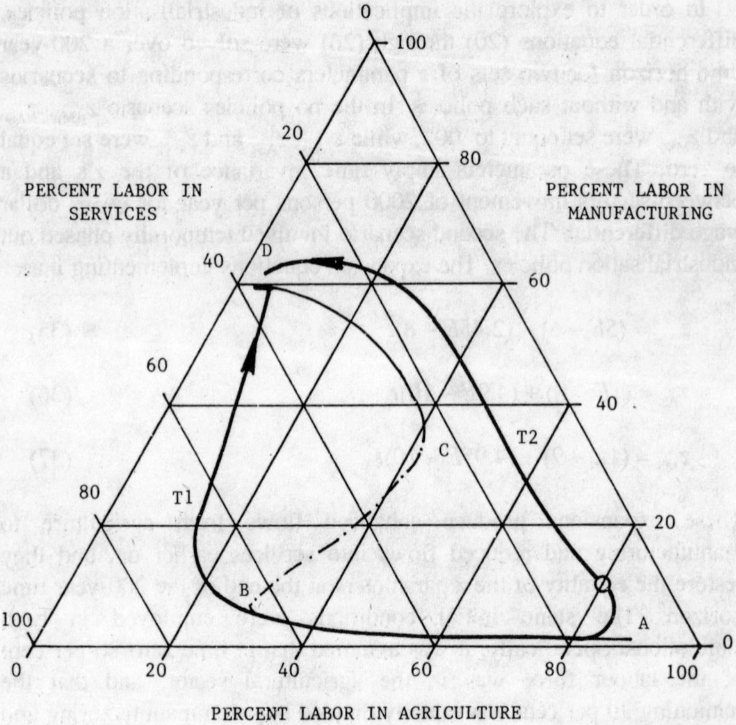


FIG. 14.8: Simulated Services dominated ($T_1$) and Manufacturing dominated ($T_2$) time paths.

A manufacturing dominated evolution is exemplified by the $T_2$ trajectory in Fig. 14.8. In the case of $T_2$ the system is attracted by the upper *CD* branch of the equilibrium curve. The $T_2$ trajectory represents, in an idealised manner, the historical experiences of countries that, via industrialisation policies, attained a stronger manufacturing sector and much higher equilibrium levels of product per capita.

## Conclusion

The literature has recognised two distinct development patterns, one characterised by lower incomes and by a strong service sector, and the other by higher incomes and by a strong manufacturing sector. The investigation reported in this paper explains these patterns as the manifestation of distinct dynamic equilibria. In a capsule, we show that conventional economic presuppositions imply that the single stable equilibrium path in existence at low development levels is eventually complemented by a second stable dynamic equilibrium. Of these, the latter is manufacturing dominated and is favourable, and the former is service dominated and is perverse.

The frame of reference articulated in this paper provides a theoretical rationale for many government interventions aimed at promoting the growth of the manufacturing sector. These interventions may be construed as attempts to pry a country loose from a "perverse attractor" (the services dominated development path) and bring it into the domain of the "positive attractor", represented by the manufacturing dominated development path.

This frame of reference also can explain why the laissez faire ideologies advocating non-interference with market mechanisms played a much greater role in the West than in Third World countries. In Western countries progressing along a manufacturing dominated development path, the market forces produced outcomes by and large favourable, and the laissez faire ideologies and policies reflected this fundamental fact. Instead, in Third World countries trapped in a perverse, service sector dominated development path, the market forces tend to preserve perverse economic structures and trends. It is not surprising that in these countries interventionist views advocating forced industrialisation policies reflect collective perceptions and aspirations.

Perhaps the perspective that we suggest with respect to industrialisation policies has a more general validity. It could be argued that many demands for government intervention arise from a collective perception that some phase of society is trapped in a vicious equilibrium condition, and that government interventions are required to switch into existence a dynamic moving it toward a collective welfare optimum.

## REFERENCES

Baer, W. and M. E. A. Herve, 1966 : "Employment and Industrialisation in Developing Countries", *Quarterly Journal of Economics*, 80, pp. 88-107.

Bairoch, P., 1975, *The Economic Development of the Third World Since 1900,* Berkeley: University of California Press.

Bairoch, P. and J. M. Limber, 1968 : "Changes in the Industrial Distribution of the World Labour Force, by Region. 1880-1969", *International Labour Review*, 98, pp. 311-36.

Berry, A., 1978 : "A Positive Interpretation of the Expansion of Urban Services in Latin America, with Some Columbian Evidence", *Journal of Development Studies*, 14, pp. 210-31.

Berry, A. and R. H. Sabot, 1984 : "Unemployment and Economic Development", *Economic Development and Cultural Change*, 33, pp. 99-116.

Burmeister, E. and A. R. Dobell, 1970 : *Mathematical Theories of Economic Growth.* Macmillan, London.

Carlino, G. A., 1982 : "Manufacturing Agglomeration Economies as Returns to Scale: A Production Function Approach", *Papers of the Regional Science Association*, 50, pp. 95-108.

Casetti, E., 1972 : "Generating Models by the Expansion Method: Applications to Geographic Research", *Geographical Analysis*, 4, pp. 81-91.

———, 1980 : "Equilibrium Population Partitions Between Urban and Agricultural Occupations", *Geographical Analysis*, 12, pp. 47-54.

———, 1986 : "The Dual Expansion Method: An Application to Evaluating the Effects of Population Growth on Development, *IEEE Transactions on Systems, Man and Cybernetics*, 16, pp. 29-39.

Casetti E. and K. Pandit, 1984 : "A Catastrophe Model of Urbanisation", *Modeling and Simulation*, 15, pp. 201-06.

Chenery H. B., 1960 : "Patterns of Industrial Growth", *American Economic Review*, 50, pp. 624-54.

Chenery, H. B. and M. Syrquin, 1975 : *Patterns of Development 1950-1970,* London: Oxford University Press.

Chenery, H. B. and L. Taylor, 1968 : "Development Patterns: Among Countries and Over Time", *Review of Economics and Statistics*, 50, pp. 391-416.

Clark, C., 1940 : *The Conditions of Economic Progress*, Macmillan, London.

Denison, E. F., 1967 : *Why Growth Rates Differ: Postwar Experience in Nine Western Countries*, Washington, D. C. : Brookings Institution.

Dore, R., 1974 : "The Labour Market and Patterns of Development in the Wage Sector of the LDC's: Implications for the Volume of Employment Generated", *World Development*, 2, pp. 1-7.

Economic Commission for Latin America, 1965 : "Structural Changes in Employment within the Context of Latin America's Economic Development", *Economic Bulletin for Latin America,* 10, pp. 163-87.

Emi, K., 1969 : "Employment Structure in the Service Industries", *The Development Economies*, 7, pp. 133-57.

Fisher, A. G. B., 1939 : Production: Primary, Secondary and Tertiary", *Economic Record*, 15, pp. 24-38.

Fisher, D., 1983 : *Macro-Economic Theory: A Survey,* New York: St. Martin's Press.

Fourastie, J., 1949 : *Le Grand Espoir du XX Siecle,* Paris: Presses Universitaires de France.

Haldi, J. and D. Whitcomb, 1967 : Economies of Scale in Industrial Plants", *Journal of Political Economy*, 75, pp. 373-85.

Hamberg, D., 1971 : *Models of Econcmic Growth,* New York, Harper and Row.

Harris, J. R. and M. P. Todaro, 1970 : "Migration, Unemployment and Development: A Two-Sector Analysis", *American Economic Review*, 60, pp. 126-42.

Hart, K., 1973 : "Informal Income Opportunities and Urban Employment in Ghana", *Journal of Modern African Studies*, 11, pp. 61-89.

International Labour Office, 1976 : *Urban Development and Employment in Sao Paulo*, Geneva: International Labour Office.

Jones, G. T., 1933 : *Increasing Return*, Cambridge: Cambridge University Press.

Gould, J. D., 1972 : *Economic Growth in History — Survey and Analysis*, London: Methuen, 1972.

Katouzian, M. A., 1970 : "The Development of the Service Sector: A New Approach", *Oxford Economic Papers*, 22, pp. 362-82.

Kawashima, T, 1975 : "Urban Agglomeration Economies in Manufacturing Industries", *Papers of the Regional Science Association*, 34, pp. 157-75.

Kirsch, H., 1973 : "Employment and the Utilisation of Human Resources in Latin America", *Economic Bulletin for Latin America*, 18, pp. 46-94.

Kuznets, S., 1966 : *Modern Economic Growth*, New Haven: Yale University Press.

————, 1971 : *Economic Growth of Nations: Total Output and Production Structure*, Cambridge: Harvard University Press.

Marelli, E., 1981 : "Optimal City Size, the Productivity of Cities, and Urban Production Functions", *Sistemi Urbani*, $^1/_2$, pp. 149-63.

Mazumdar, D., 1976 : "The Urban Informal Sector", *World Development*, 4, pp. 655-79.

McGee, T. G., 1967 : *The Southeast Asian City*, London: Bill & Son Ltd.

Mehta, S. K., 1961 : "A Comparative Analysis of the Industrial Structure of the Urban Labour Force of Burma and the United States", *Economic Development and Cultural Change*, 9, pp. 164-79.

Mouley, J. and E. Costa, 1974 : "Employment Policies in Developing Countries: A Comparative Analysis", *International Labour Office*, Geneva.

Nourse, H. O., 1978 : *Regional Economics*, New York: McGraw Hill.

Oshima, H. T., 1971 : "Labour-Force 'Explosion' and the Labour-Intensive Sector in Asian Growth", *Economic Development and Cultural Change*, 19, pp. 161-83.

Pratten, C. K., 1971 : *Economies of Scale in Manufacturing Industry*, Cambridge: Cambridge University Press.

Ramos, J. R., 1970 : *Labour and Development in Latin America*, New York: Columbia University Press.

Sabolo, Y. (assisted by J. Gaude and R. Wery), 1975 : *The Service Industries*, Geneva: International Labour Office.

Sawyer, M. C., 1981 : *The Economies of Industries and Firms: Theories, Evidence and Policy*, New York: St. Martin's Press.

Segal, D., 1976 : "Are There Returns to Scale in City Size?", *Review of Economics and Statistics*, 58, pp. 339 50.

Singer, H., 1964 : *International Development: Growth and Change*, New York: McGraw-Hill Book Company.

Sorrentino, C., 1969 : "Comparing Employment Shifts in 10 Industrialised Countries", *Monthly Labour Review*. 94, pp. 3-11.

Stigler, G. J., 1951 : "The Division of Labour is Limited by the Extent of the Market", *Journal of Political Economy*, 59, pp. 185-93.

Tinbergen J. and H. C. Bos, 1972 : *Mathematical Models of Economic Growth*, New York: McGraw Hill.

Turnham, D., 1971 : *The Employment Problem in Less Development Countries,* Paris: OECD.

United Nations Economic Commission for Europe, 1977 : *Structure and Change in European Industry*, New York: United Nations.

Wan, H. Y., 1971 : *Economic Growth,* New York: Harcourt Brace Jovanovich Inc.

15

# CHANGING LANDUSE AND CROPPING PATTERN IN BIHAR

RABINDRANATH BHATTACHARYA

IN THIS paper, first the changes in the landuse characteristics of Bihar from 1958-59 to 1970-71 are studied in terms of a five-fold classification both for the entire State and for the districts as they existed prior to their bifurcation. Subsequently, the changes in cropping patterns are discussed in a similar fashion against the backdrop of slowly increasing net sown area and lower level of multiple cropped area. The first part of the theme is dealt with on the basis of physical units, because spatial variations are the hall-mark of this aspect in Bihar as well and the State can be divided into three different physical units : the *Plateau* — consisting of the districts of Palamau, Hazaribagh, Ranchi, Singhbhum, Dhanbad and Santal Parganas; the *Sub-Himalayan tract* — confined to Champaran district; and the *Plains* — comprising the remaining districts of Saran, Muzaffarpur, Darbhanga, Saharsa, Purnea, Bhagalpur, Monghyr, Patna, Gaya and Shahabad. Among these districts Darbhanga, Saharsa and Monghyr have a special characteristic because they are affected by the river Kosi. Moreover, Monghyr lies astride both the banks of the river Ganga. The same physical basis of division has also been followed in the following.

## General Landuse Characteristics

Table 15.1 shows the changes in landuse from 1958-59 to 1970-71 in Bihar as a whole. The striking feature is the decline in forested land and less striking decrease in other uncultivated land (excluding current fallow), while the increase in the categories of fallow land, land not available for cultivation and net sown area (NSA) can be described as little (1.3), low (2.1) and slightly high (2.8) respectively. The decrease in forested land is striking for the reason that in all these years there has been a policy of increasing the forested land through new afforestation schemes, better management and the *vana mahotsava* programme. It was expected that land under forests will increase (Ahmad, 1965) but in reality it has decreased from 22.6 per cent of the total land to 16.9 per cent, a decline of 5.7 per cent which is the most striking feature of this aspect. Another point to note is the decrease in the area cultivated more than once, despite the progress of various irrigation projects and increase in the net area sown. The decrease in the area cultivated more than once below the level of 1958-59, and increasing in the NSA in absolute terms is a riddle. It indicates a lower level of intensification of agriculture.

*Forest*

Considering the spatial aspect of this category (Appendix-I) the Plateau region, on the whole, can be described as an area of severe deforestation. Only the district of Palamau shows a little decrease of only 1.1 per cent. The severest deforestation took place in Singhbhum (18.1), followed by Santal Parganas (14.0), Hazaribagh (9.4) and Dhanbad (6.7). Ranchi was not as severely affected, with only 4.7 per cent decline. The picture for the period 1951-52 to 1965-66 as noted by Jasbir Singh (1974) is very much similar. The only discordance is that the district of Palamau is also shown to have lost over five per cent of its forested land in common with other Plateau districts except Ranchi, which depicts a picture similar to the one just described. The Sub-Himalayan district of Champaran shows the second highest increase in the forested land in Bihar with an increase of 2.4 per cent. For the period 1951-52 to 1965-66, Singh (1974) noted an increase at the rate of 6-10 per cent. The picture for the Plains is not even. Monghyr shows the highest increase of 6.4 per cent and Patna a marginal increase of 0.1 per cent. For the other districts the picture is in tune with the State, i.e. one of decline, with Saharsa joining the districts of Saran, Muzaffarpur and Darbhanga in 1970-71 which had no forests even in 1958-59. This makes deforested land in the north Bihar plains not only continuous from the west but extending eastward. The natural vegetation from this tract of land has not only been completely annihilated but no new land has been brought under

TABLE 15.1: Land utilisation in Bihar

| Categories | Years | | | | | |
|---|---|---|---|---|---|---|
| | 1958-59 | 1966-67 | 1967-68 | 1968-69 | 1969-70 | 1970-71 |
| 1. Forests | 21.8 | 16.7 | 17.1 | 17.4 | 17.4 | 16.9 |
| 2. Land not available for cultivation | 12.9 | 15.1 | 14.4 | 14.7 | 15.0 | 15.0 |
| (a) Land put to non-agricultural use | 7.9 | 8.4 | 8.6 | 8.7 | 8.8 | 8.9 |
| (b) Barren & uncultivated land | 5.0 | 6.7 | 5.8 | 6.0 | 6.2 | 6.1 |
| 3. Fallow land | 12.9 | 19.7 | 15.3 | 14.6 | 14.0 | 14.2 |
| (a) Current fallow | 8.0 | 14.4 | 9.8 | 9.2 | 8.8 | 9.1 |
| (b) Other than 3(a) | 4.9 | 5.3 | 5.5 | 5.4 | 5.2 | 5.1 |
| 4. Other uncultivated land excluding 3 | 6.4 | 5.7 | 5.4 | 5.3 | 5.2 | 5.1 |
| (a) Permanent Pasture | 1.2 | 1.2 | 1.1 | 1.0 | 1.0 | 1.0 |
| (b) Cultivable Waste | 5.2 | 4.5 | 4.3 | 4.3 | 4.2 | 4.1 |
| 5. Net Sown Area | 46.0 | 42.8 | 47.8 | 48.0 | 48.4 | 48.8 |
| Total | 100.0 | 100.0 | 100.0 | 100.0 | 100.0 | 100.0 |
| Cultivated more than once (% of total area) | 17.9 | 10.4 | 15.1 | 14.8 | 15.4 | 14.8 |

*Note*: Figures shown as percentage of the total area.
*Source:* Same as for Appendix-I.

afforestation. Another feature to note in this connection is the highest percentage decline in Shahabad (8.1) in western South Bihar which also indicates a disturbing trend. In passing it may be mentioned that all these districts in the plain area are shown to have increased their forested land during the period 1951-52 to 1965-66 (Singh, 1974) at rates varying from less than two to 2.5 per cent. Except for the abovementioned cases the comparative picture from 1958-59 to 1970-71 is not much different from Singh's observation for the earlier period, 1951-52 to 1965-66.

*Other Uncultivated Land*

This category of land shows decline throughout the State, with very little to distinguish it in its spatial pattern (Appendix-I). The Plateau districts show a markedly lower decline, the plains greater, but the most notable decrease is shown by the districts of Champaran, Monghyr and Saharsa. In the matter of detail the culturable wasteland remains high over five per cent in the Plateau districts except in Palamau where it is only 2.2 per cent, while in the case of Saharsa it has decreased from 12.1 (in 1958-59) to 2.5 per cent (in 1970-71). Permanent pasture land as a whole remains very low. In 1970-71 it is 2.1 per cent in Hazaribagh, 4.5 per cent in Santal Parganas, 1.1 per cent in Dhanbad and 1.0 per cent in Gaya. All other districts have less than 1.0 per cent under pasture. The observation for the period 1951-52 to 1965-66 is one of decline (Singh, 1974).

*Fallow Land*

This category has shown a slight increase for the entire State from 1958-59 to 1970-71. In terms of spatial distribution (Appendix-I) the Plateau districts show an increase from 2.5 per cent (Dhanbad) to 4.2 per cent (Santal Parganas), but the district of Ranchi shows a decline of 0.7 per cent. The Sub-Himalayan district of Champaran shows a decline of 1.7 per cent. In the plains the Kosi districts of Saharsa and Darbhanga indicate a slight increase of 0.4 per cent in the former and static condition in the later. Among the other Plains districts, Bhagalpur shows the lowest decrease (0.1%) and Patna the highest decrease (4.4%), while the districts of Muzaffarpur and Monghyr show an increase of 0.7 per cent and 7.0 per cent respectively. Once again the spatial reflection of the changes is not uniform nor identical.

### *Land Not Available for Cultivation*

Theoretically this category should show an increase due to increasing demand on land resources necessitated by development of roads, factories, urban areas and also increasing mining activity. The trend from 1966-67 to 1970-71 is more or less a static one, since the fluctuations as indicated are very minor. The breakdown of this category (Appendix-I) shows a very slow increase in the category under *land put to non-agricultural uses*, only 1.0 per cent in 12 years. The *barren lands* also show a similar picture for the entire State. The Plateau districts display an extremely uneven growth in this regard. Singhbhum shows the highest increase with 24.9 per cent, followed by Hazaribagh with a moderate increase of 5.7 per cent, Santal Parganas 2.8 per cent, Dhanbad 2.1 per cent, Ranchi 1.2 per cent, and Palamau shows the least increase with 0.4 per cent. This also indicates the rapid development process working in Singhbhum and lesser intensive activities being carried out elsewhere in the Plateau. At the same time, it may also be assumed that Dhanbad had originally a high percentage of land under non-agricultural use (19.9%) and it increased to only 20.3 per cent in 1970-71, and it has little attraction at present for such activity. In the case of Singhbhum, the barren land has increased from 5.1 per cent in 1958-59 to 29.1 per cent in 1970-71 and this is the single largest increase in the entire State of Bihar. In the sub-Himalayan Champaran district barren land had decreased from 3.1 to 2.0 per cent in 1970-71, while the land under non-agricultural use had increased to 12.3 per cent from 8.1 per cent. The total increase is only 3 per cent for this category. As for the districts in the Plains, the picture is uneven. The Kosi districts of Saharsa and Darbhanga show an increase of 0.2 and 0.8 per cent respectively, while Monghyr shows a decrease of 2.1 per cent. The neighbouring Gaya district shows a decrease of 1.4 per cent. In the case of Gaya only the barren land has declined, while the other has increased, whereas in Monghyr both the sub-categories show a decline with land put to non-agricultural use decreasing more than the barren land. All other Plains districts show an increase of varying order from 6.0 per cent for Bhagalpur to 10.8 per cent for Patna. There is no contiguous spatial pattern or relationship. Singh's (1974) comparative figures for the period 1951-52 to 1965-66 are not in conformity in the case of Singhbhum (rate of increase indicated is lower than stated above) and Shahabad, Bhagalpur and Dhanbad (which are indicated as areas of decrease).

*Net Area Sown*

The figures for this category show a decline in 1966-67 and pretty sharp increase in the following year, which has remained more or less steady at 48.0 per cent level. Spatially (Appendix I) if the figures for the Plateau districts are consulted for 1958-59 and 1970-71, a distinct disparity between them can be observed. In the 1958-59 period Ranchi and Santal Parganas had more than 30.0 per cent of their land under net sown area, Dhanbad and Palamau came next with 23.2 and 22.1 per cent of their land under the NSA, respectively, and Singhbhum and Hazaribagh had only 19.4 and 19.3 per cent, respectively, under the NSA. Of these districts Singhbhum has the greatest increase of 10.5 per cent, followed by Santal Parganas (8.5%), Dhanbad (5.3%), Ranchi (1.7%) and Hazaribagh (0.5%), while Palamau records a decrease of 0.8 per cent. Singh's observations for the period 1951-52 to 1965-66 are in harmony with the above figures of increase for the districts of Singhbhum, Dhanbad and Hazaribagh. He indicates negative growth for Ranchi and Santal Parganas which is positive in the present case, and positive for Palamau (Singh, 1974) which is negative as noted earlier. For the Sub-Himalayan district of Champaran, the two sets of figures show a slight decrease in 1970-71 (0.3%). Singh's observation broadly for the district is in agreement though the actual rate of decrease is not. At this stage, it may be mentioned that Champaran is more akin to the neighbouring Plain districts. For the Kosi districts, while Saharsa shows an increase of 16.0 per cent over the 1958-59 figure and this is in agreement with Singh's observation (Singh, 1974). Darbhanga district shows a marginal decrease ranging from 11 to 15 per cent for the period 1951-52 to 1965-66. The other Kosi district of Monghyr also records a decrease of 9.6 per cent and records the low figure of 48.5 per cent under the NSA, the lowest for Kosi districts and for the plains. For the rest of the plains, it may be mentioned that the NSA figures for 1958-59 were between 53.5 (Bhagalpur) and 79.1 (Muzaffarpur) per cent. In 1970-71 also the lowest figure of 53.4 per cent still belonged to Bhagalpur among the Plains' districts. The highest increase is recorded by Shahabad district (9.0%), followed by Gaya (4.7%), Purnea (4.3%), Patna (4.0%) and Saran (0.2%). In fact the NSA in Saran district is being maintained above 71.0 per cent level. On the other hand, Muzaffarpur district records a decrease of 5.5 per cent though the net NSA is 75.6 per cent. The spatial pattern for the north Bihar plains in 1970-71 is one of over 60 per cent of the total land being net sown, which is an improvement not only over the 1958-59 pattern, but also over the 1965-66 pattern. In the later period unevenness in this respect was much more than the earlier period. The NSA distribution among the Kosi districts is uneven though not of the order of the Plateau districts. In the south Bihar plains as one moves westward the percentage of land under the NSA increases reaching its highest in Patna district and decreasing slightly westward. This pattern is not much different from the 1958-59 pattern nor from the 1965-66 pattern.

So far the five-fold division has been discussed, though their sub-divisions have also been touched upon. No area cultivated more than once is being discussed for an overview of the intensiveness of the cultivation process.

*Area Cultivated More Than Once*

A perusal of this data in Table 15.1 reveals a picture which cannot be described as one of progress. It is revealing that the actual area cultivated more than once was greater earlier than in any time subsequently. From Table 15.1 it is evident that only 18 per cent of the total land was ever cultivated more than once. Table 15.2 is prepared on the basis of index numbers, with figures of 1958-59 being taken as 100 for the NSA for Area More Than Once Cultivated, and for Total Cropped Area (TCA). In another exercise area cultivated more than once is expressed as Index Number for NSA, and the last column is the cumulative index number for the TCA. The picture that emerges indicates the fluctuations and the attainment of individual heads in their true perspective. From this it is quite evident that though the net area sown was increasing in absolute terms, the area cultivated more than once and the total cropped area remained below the 1958-59 level. A perusal of Appendix-II reveals the spatial picture of the same. The Plains districts register a scrappy improvement over the earlier figures, but the striking feature is the reduction in the area cultivated more than once in the Plateau districts. The later feature has a close correlation with the irrigated area. This fact only underlines the positive correlation that exists between the irrigated area and the more-than-once cultivated area. But the districts of Gaya and Shahabad register a sharp fall comparable to the Plateau districts, though the irrigated area remains high.

TABLE 15.2: Index numbers for NSA, TCA, and more-than-once cultivated area

| Years | NSA* | More than once cultivated* | TCA* | Area more than once cultivated as Index No. of Col. 1 | TCA (Total of Cols. 1 & 4) |
|---|---|---|---|---|---|
| | 1 | 2 | 3 | 4 | 5 |
| 1958-59 | 100.0 | 100.0 | 100.0 | 39.04 | 139.04 |
| 1966-67 | 93.1 | 57.9 | 83.3 | 22.60 | 115.70 |
| 1967-68 | 103.9 | 83.9 | 98.3 | 32.74 | 136.64 |
| 1968-69 | 104.5 | 82.7 | 98.4 | 32.30 | 136.80 |
| 1969-70 | 105.4 | 85.7 | 99.9 | 33.48 | 138.88 |
| 1970-71 | 106.1 | 82.7 | 99.5 | 32.27 | 138.37 |

* Base Year 1958-59 = 100.

*Source:* Same as for Appendix-I.

The area cultivated more than once in almost all the Plains districts is over 30 per cent of the NSA (except Gaya). The picture regarding the Plateau districts has already been described. The districts lying to the west of the river Kosi and north of the Ganga (Darbhanga, Muzaffarpur, Saran and Champaran) are the only notable areas having some orchards and it varies from 3.0 to 4.6 per cent of the NSA.

### *Area Under Seasonal Crops*

Now, the problem can be viewed from the point of view of the types of crops cultivated, i.e. *Bhadoi, Aghani, Rabi* and *Garma* and the percentage of the NSA devoted to them. This aspect of study clearly brings out the fact that most of the crops are dependent on rains, only a small percentage of the cropped area is devoted to Rabi or *garma* cultivation that is dependent on irrigation, an infra-structure that has been developed in parts of the State. These later crops generally occupy the areas that are cultivated more than once. The earlier statement that such areas are not more than one-third of the NSA are reflected in the cropping pattern. A study of Table 15.3 reveals that about 90 per cent of the NSA is dependent on rains. The Rabi crops are not at the level of 1958-59 but they account for slightly more than one-third of the NSA. A simple calculation shows that an *aghani* cultivation covered 2.5, 2.2, 2.3, 2.4 and 2.3 times of the area under *bhadoi* for the years 1958-59, 1966-67 to 1970-71 respectively.

Similarly *aghani* covered 1.5, 1.8, 1.6, 1.7, 1.7 and 1.6 times of the area under Rabi for the same order of the years as already stated above. With respect to *garma* crops, *aghani* covered 24 times of the area in 1958-59 and declined to 17.2 times in 1966-67 and remained very much at the same level. But this fact not only indicates a decline in the *aghani* area but also an increase in the *garma* area. It is revealing that the Rabi area is slightly higher than more-than-once cultivated area and appreciably higher than the irrigated area. It is suggestive of the fact that some of the Rabi crops are not dependent on irrigation and maybe all the Rabi cultivation areas do not belong to the area cultivated more than once. The *garma* (summer) crops are again quite insignificant. Moreover, the *bhadoi* and *garma* crop areas remain proportionately more or less the same, with *garma* having improved from the 1958-59 figure. As already noted, the *aghani* crop areas have declined from the 1958-59 level. The land under orchards has risen slightly to 1966-67 and from then onwards, it has declined marginally. The proportion of irrigated land has shown fluctuation and it has declined to below the 1958-59 level as indicated earlier.

TABLE 15.3: **Distribution of seasonal crop area, more-than-once cultivated area and irrigated land in Bihar**

*(Expressed as percentage of NSA)*

| Years | Bhadoi | Aghani | Rabi | Garma | Orchard | More-than-once-cultivated area | Irrigated area |
|---|---|---|---|---|---|---|---|
| 1958-59 | 26.90 | 66.49 | 41.62 | 2.54 | 1.01 | 39.04 | 25.9 |
| 1966-67 | 26.62 | 59.30 | 32.96 | 3.45 | 1.95 | 24.28 | 21.6 |
| 1967-68 | 27.35 | 61.56 | 37.43 | 3.45 | 1.73 | 31.52 | 24.3 |
| 1968-69 | 26.03 | 62.98 | 36.57 | 3.59 | 1.74 | 30.91 | 26.1 |
| 1969-70 | 26.48 | 62.96 | 37.10 | 3.55 | 1.68 | 31.77 | 27.1 |
| 1970-71 | 26.76 | 60.36 | 37.88 | 3.42 | 1.71 | 30.42 | 25.5 |

*Source:* Same as for Appendix-I.

Spatially (Appendix II), the Plateau districts of Singhbhum, Dhanbad and Santal Parganas have over 88 per cent of the NSA under *aghani*, followed by Hazaribagh (70.3%) and Ranchi (60.3%). At the same time Singhbhum has only 10.8 per cent of its NSA under *bhadoi*, while Santal Parganas has only 14.7 per cent and Dhanbad 15.6 per cent. On the other hand, Hazaribagh has 33.2 per cent and Ranchi and

Palamau have over 42 per cent of the NSA under *bhadoi.* It is interesting to note that Palamau is the only district in the Plateau area having almost equal areas under *bhadoi, aghani* and *Rabi* cultivation. A perusal of Appendix II also indicates though Palamau, Santal Parganas and Hazaribagh have more than 10 per cent of their NSA under the category cultivated more than once, Palamau district's figure is very much close to the irrigated area figure. All other Plateau districts do not have such a close relationship. The statement is true for both the years compared. This also indicates a decline in the irrigated area for all these districts. This at once suggests that much of the second or third crop is not dependent on irrigation, in other words they are hardier varieties with low water requirement. While Ranchi had more than one per cent of the NSA land under *garma* and about one per cent under orchards, the other Plateau districts had significantly lower areas under these crops, with Dhanbad and Singhbhum having none under orchards.

Only two north Bihar districts, i.e. Saran and Purnea (one in the extreme west and the other in the extreme east) have more than 40 per cent of their NSA under *bhadoi* crops. The Kosi area shows a variable figure for *bhadoi*, i.e. 26.8 per cent, (Darbhanga) to 38.7 per cent (Monghyr). Gaya and Shahabad have less than 10 per cent of the NSA under *bhadoi* and the other Plains districts do not exceed Saran's figure. The *aghani* cultivation accounts for more than 60 per cent of the NSA: Gaya (68.8%), Shahabad (64.2%), Darbhanga (60.2%), Purnea (60.2%), and Bhagalpur (60.1%), and they are not contiguous spatially except Gaya and Shahabad. The next group having more than 50 per cent of their NSA under *aghani* cultivation are Saharsa (56%), Patna (55.6%), Muzaffarpur (54.4%), and Champaran (54.3%). Saran (34.4%) and Monghyr (29.5%) have the lowest *aghani* cultivation area. The Kosi districts do not have any similarity in this case also. The proportion of Rabi cultivation in the Plains shows a wide fluctuation among the districts. The range is from 27.5 per cent of the NSA (Purnea) to 73.7 per cent (Patna). The highest values of over 70 per cent of the NSA are in the two contiguous districts of Patna and Shahabad with over 60 per cent of the net sown area being irrigated.

In the second row of districts are Monghyr (59.0%) and Saran (51.0%), followed by the districts of Gaya (49.9%), Champaran (48.5%), Bhagalpur (46.7%) and Muzaffarpur (41.9%). The districts of Darbhanga (34.7%) and Saharsa (31.9%) have a lower proportion of their area under Rabi crops. The relationship between irrigated area and Rabi cropped area is not very distinct. The district of Gaya has the highest area under irrigation (74.7%) with only about 50 per cent of its NSA being under the Rabi crops and about 29 per cent of the NSA being cultivated more than once. This indicates that even *aghani* crops are irrigated. The next highest proportion of the NSA irrigated is in Shahabad (65.0%) and Patna (61.5%) with the second highest (47.6%) and highest (50.3%) proportion of the NSA being cultivated more than once in these districts respectively. This suggests not only a high proportion of Rabi crops are irrigated but a sizeable proportion also are of the hardier variety, requiring less water. It may be noted in passing that the three contiguous districts of Shahabad, Gaya and Patna have the best irrigational facilities. Champaran and Saran, the two contiguous districts in north Bihar, have more than a quarter of their NSA irrigated, with the former having a greater proportion of the NSA cultivated more than once (47.6%) and the latter having a lower proportion (34.8%). This again can be partly correlated with Rabi cultivation.

The district of Monghyr (which lies astride both the banks of Ganga) has the most interesting combination of crops, with Rabi leading the proportion (in acreage) and having a very small *aghani* cultivated area. It has the sixth highest proportion of the NSA under irrigation among the Plains districts but with a lower proportion cultivated more than once (30.8%). This again indicates a larger proportion of Rabi crops may be of the hardier variety.

The *garma* cultivation is insignificant in the Plateau, for quite obvious reasons like paucity of water and poor moisture retention capacity of soils. Among the Plains districts it covers nearly a quarter of the NSA (23.8%) in Saharsa. The next significant area under *garma* is in Muzaffarpur district. Lying in between them, Darbhanga district has only 9.8 per cent of the NSA under *garma*. The only other district having a significant area under *garma* is Champaran. In the rest the percentage of the NSA under *garma* varies from two to six per cent. This indicates a sporadic nature of its cultivation and it is unrelated with the irrigation system as a whole.

### *Irrigated Area*

Before discussing the proportion of individual crops to total cropped area, it may be mentioned that the districts of Patna, Gaya and Shahabad in south Bihar had the highest percentage of their NSA under irrigation (over 60%) in 1970-71. The picture for 1958-59 and 1965-66 was same for these districts. The next lower rank is occupied by Saran and Champaran districts (over 26%) in north Bihar. The position in 1965-66 was very much the same, but in 1958-59 while Saran had 18.3 per cent of its NSA under irrigation, Champaran had only 5.9 per cent. Palamau, in the Plateau, and Monghyr spread over both the banks of the Ganga have 25.8 and 22.7 per cent of their NSA under irrigation. While in 1965-66 they are grouped under the same class (16.30%), in 1958-59 the picture was different, with 24.9 per cent in Palamau and 16.8 per cent in Monghyr. The contiguous districts of Saharsa and Darbhanga have 13.9 and 10.5 per cent of their NSA under irrigation respectively. The only other district having more than 10 per cent of its land irrigated is Bhagalpur. In 1965-66 while Saharsa and Darbhanga had only one to five per cent of their NSA under irrigation, Bhagalpur had between 31 and 45 per cent of its land under irrigation. In 1958-59, Bhagalpur had over 60 per cent of its NSA irrigated, Darbhanga had only 5.2 per cent and Saharsa none at all. All other districts have small areas under irrigation. The infrastructure of the irrigation system as available in 1970 indicates that the districts of Champaran, Saran, Muzaffarpur, Darbhanga, Saharsa, Shahabad, Patna, Gaya and Bhagalpur are well endowed. In 1965-66 less than one per cent of the NSA in Purnea was irrigated, while 1.5 per cent of the NSA was irrigated in the districts of Muzaffarpur, Hazaribagh and Ranchi, and 6.15 per cent in the districts of Santal Parganas, Hazaribagh and Singhbhum. In 1958-59 less than 1 per cent of the NSA was irrigated in Purnea (not the static nature), 1.2 per cent in Muzaffarpur, 2.2 per cent in Ranchi, 6.9 per cent in Dhanbad, 14.3 per cent in Singhbhum, 15.8 per cent in Hazaribagh and over 30 per cent in Santal Parganas.

J. Singh (1974) has noted that during the period 1951-52 to 1965-66, the irrigated area in Patna, Gaya and Ranchi districts had declined by one to three per cent; in Darbhanga the irrigated land had decreased by four to six per cent and the highest decrease was recorded by the districts of Santal Parganas, Dhanbad and Hazaribagh. On the other hand, the districts of Champaran, Shahabad, Saharsa and Singhbhum showed increases varying from one to five per cent; Saran, Muzaffarpur, Purnea and Palamau recorded less than one per cent increase; Monghyr recorded six to ten per cent increase and Bhagalpur over 15 per cent. These facts show that the infrastructure of the irrigation system though related to physical terrain, its development and utilisation in some parts is not related to the terrain.

## Crop Rank and Crop Combination

A cursory glance at Table 15.4 reveals that from 1950-51 to 1970-71, the leading crop for the entire State was *rice.* From the table it is apparent that the average figure (49.2 per cent of TCA) for 1950-51 to 1957-58 was exceeded only twice, in 1968-69 and 1969-70, and rice accounts for nearly 50 per cent of the TCA. The next crop of importance was pulses in 1950-51 to 1957-58 and in 1958-59 as it occupied more than 10 per cent of the TCA. This crop increased in 1958-59 from the average figure of previous years, but the figure for 1966-67 is very poor accounting for only seven per cent of the TCA and in subsequent years it improved but could not attain the position of 1958-59. It is evident from Table 15.5, showing the first six crops for the State of Bihar, that pulses lost its second position to maize in 1966-67 and itself was relegated to the fourth position, which it could retrieve only partially in 1968-69 and managed to retain it. This indicates a declining interest in this crop though some of them were not very demanding in respect of water. The decline of pulses also demonstrates the lesser emphasis on leguminous crops and lesser acreage for vegetable protein. Moreover, this indicates a moving away from the natural and in-built system of replenishment of soil fertility in the cropping pattern. Maize was the third crop in the 1950s and for one year (1966-67) it became the second crop with an appreciable increase in area, but only with a marginal decline it was relegated to third and in the subsequent year to fourth position with still more reduction in area. The only point to note is that in 1970-71 the area had increased but the area under other crops also increased correspondingly, and maize remained in the fourth position.

TABLE 15.4: **Area under different crops in the following years for Bihar**

*(Expressed as percentage of the TCA)*

| Years | Rice | Maize | Wheat | Barley | Gram | Pulses | S Cane | Tobacco | Jute |
|---|---|---|---|---|---|---|---|---|---|
| 1950-51 to 1957-58* | 49.2 | 6.4 | 5.8 | 3.7 | 5.1 | 14.9 | 1.4 | N.A. | 2.0 |
| 1958-59 | 47.2 | 7.0 | 5.5 | 3.7 | 5.3 | 15.5 | 1.5 | N.A. | 2.4 |
| 1966-67 | 47.7 | 9.3 | 9.0 | 3.8 | 3.2 | 6.8 | 1.6 | 0.15 | 1.7 |
| 1967-68 | 47.2 | 9.1 | 9.7 | 3.3 | 3.0 | 9.0 | 1.1 | 0.12 | 1.5 |
| 1968-69 | 49.6 | 8.7 | 10.1 | 2.7 | 2.2 | 9.3 | 1.4 | 0.13 | 0.8 |
| 1969-70 | 49.3 | 8.5 | 10.5 | 2.6 | 2.2 | 9.6 | 1.5 | 0.12 | 1.3 |
| 1970-71 | 48.0 | 9.0 | 11.9 | 2.3 | 2.4 | 9.2 | 1.5 | 0.12 | 1.3 |

* Average for the years 1950-51 to 1957-58.

N.A. = Not Available.

*Source:* Same as for Appendix I.

TABLE 15.5: **Ranking of crops in Bihar (1950-51 to 1970-71)**

| Years | Crops | | | | | |
|---|---|---|---|---|---|---|
| | First | Second | Third | Fourth | Fifth | Sixth |
| 1950-51 to 1957-58 | Rice | Pulses | Maize | Wheat | Gram | Barley |
| 1958-59 | Rice | Pulses | Maize | Wheat | Gram | Barley |
| 1966-67 | Rice | Maize | Wheat | Pulses | Barley | Gram |
| 1967-68 | Rice | Wheat | Maize | Pulses | Barley | Gram |
| 1968-69 | Rice | Wheat | Pulses | Maize | Barley | Gram |
| 1969-70 | Rice | Wheat | Pulses | Maize | Barley | Gram |
| 1970-71 | Rice | Wheat | Pulses | Maize | Gram | Barley |

*Source:* Same as for Appendix I.

Wheat's position has improved over the years, the area actually more than double from the 1950's. The emergence of wheat not only indicates an improvement in the irrigation facility, but also its encroachment of the hardier Rabi crop areas, e.g. barley, gram and pulses. The fifth and sixth crops were gram and barley. The position was interchanged in 1966-67 and remained so till 1969-70, and in 1970-71 the original status was restored. In terms of area, the picture is dismal, the area has shrunk continuously for barley and gram, the only exception being in 1970-71 when the gram-cultivated area increased marginally. The interchange of position was more due to a rapid decline in the area under gram than any increase in the area under barley. This underlines the observation already made earlier. It will be pertinent to point out that in the first two periods (Table 15.4) pulses was the only crop cultivated in more than 10 per cent of the TCA and ranking next to the very dominating rice. From 1968-69 onwards, area devoted to wheat cultivation was more than 10 per cent of the TCA. All other crops individually account for less than 10 per cent of the TCA in all the years reported. The area under maize and wheat in 1966-67, maize, wheat and pulses in 1967-68, and maize and pulses in the following years accounted for eight to nine per cent of the TCA.

At this stage it will be pertinent to point out that the major Kharif or rainy-season crops occupied only about 54 per cent of the TCA in the 1950s but in 1970-71 they occupied 57 per cent of the TCA, while the major Rabi crops, which occupied about 30 per cent of the TCA in 1950s, occupy only about 26 per cent of the TCA in 1970-71. This suggests a total decline in Rabi area in 1970-71. This fact is in consonance with the statement in Table 15.3. The cash crops present, on the whole, either a stagnant picture (sugarcane) or a decline (tobacco and jute).

Spatially, the emerging picture is very interesting. For a comparative picture the earlier studies of E. Ahmad (1965) (Fig. 15.1), Jasbir Singh (1974) (Fig. 15.2) and Tapeshwar Singh (1977) are compared with the one (Fig. 15.3) worked out on the basis of area of major crops as percentage of the TCA in each district (Appendix IV). In the latter case crops having less than 10 per cent of the TCA are not considered. A cursory perusal of Appendix IV shows that Darbhanga district has not experienced any change either in crops or in the relative importance. for Purnea district the first two crops have remained unchanged. The third crop has changed from maize to wheat in 1970-71 and it is found that wheat occupies less than 10 per cent of the TCA. Doi's method of crop combination does not indicate any third crop (T. Singh, 1977) in Purnea. In Saharsa district, in 1965-66, jute was replaced by wheat and subsequently it is found that wheat outranks maize.

Since in the present paper only three crop combinations or the minimum of 10 per cent of the TCA has been considered for crop combination, the only district that fulfils such requirements even for the fourth crop is Monghyr. This district has shown remarkable diversity as well as the decline of rice's predominance. This feature, however, could be indicated only in 1970-71. It is the lone case where maize has obtained the first rank. Gram has uniformly disappeared from ranking in all the districts by 1965-66 (Singh, 1974). A similar fate has overtaken barley (Singh, 1974), (Fig. 15.4). The decline of pulses as said earlier is much widespread (Fig. 15.4), though not very much apparent from the data (Appendix IV). Bhagalpur is the only district where the rice area declined during 1951-56 to 1961-66 (Singh, 1974), but that has not affected the paramountcy of rice even now (Singh, 1974).

The area under rice has increased in 11 out of the 17 districts of Bihar during 1951-56 to 1961-66 though the data for the State as a whole do not reflect the fact. Other crops which have shown increases in the same period are: pulses (Bhagalpur), wheat (Shahabad, Gaya) and oilseeds (Palamau, Ranchi and Hazaribagh) (Singh, 1974). A glance at Appendix IV reveals that pulses had overtaken gram in the districts of Patna, Gaya and Shahabad and maize in Bhagalpur. While the picture of crop-combination in 1970-71 for the districts of Patna and Gaya are identical with Jasbir Singh's observation, it appears to be different for Shahabad, though the result obtained by Doi's method concurs with Singh's observation. The district of Champaran indicates a gradual increase in land under wheat, decrease under barley and sugarcane so much so that they are below 10 per cent level. Singh's observation (1974) for the districts of Saran and Muzaffarpur holds good in 1970-71 too.

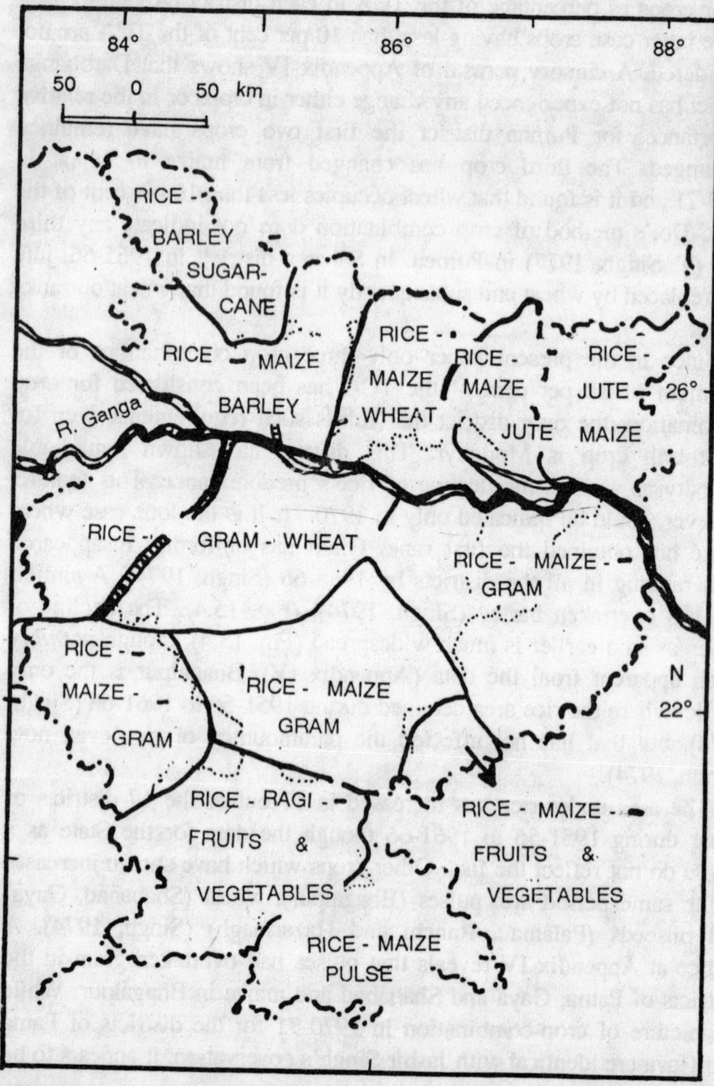


FIG. 15.1: Bihar : crop region 1950-51 to 1957-58. Three leading crops.

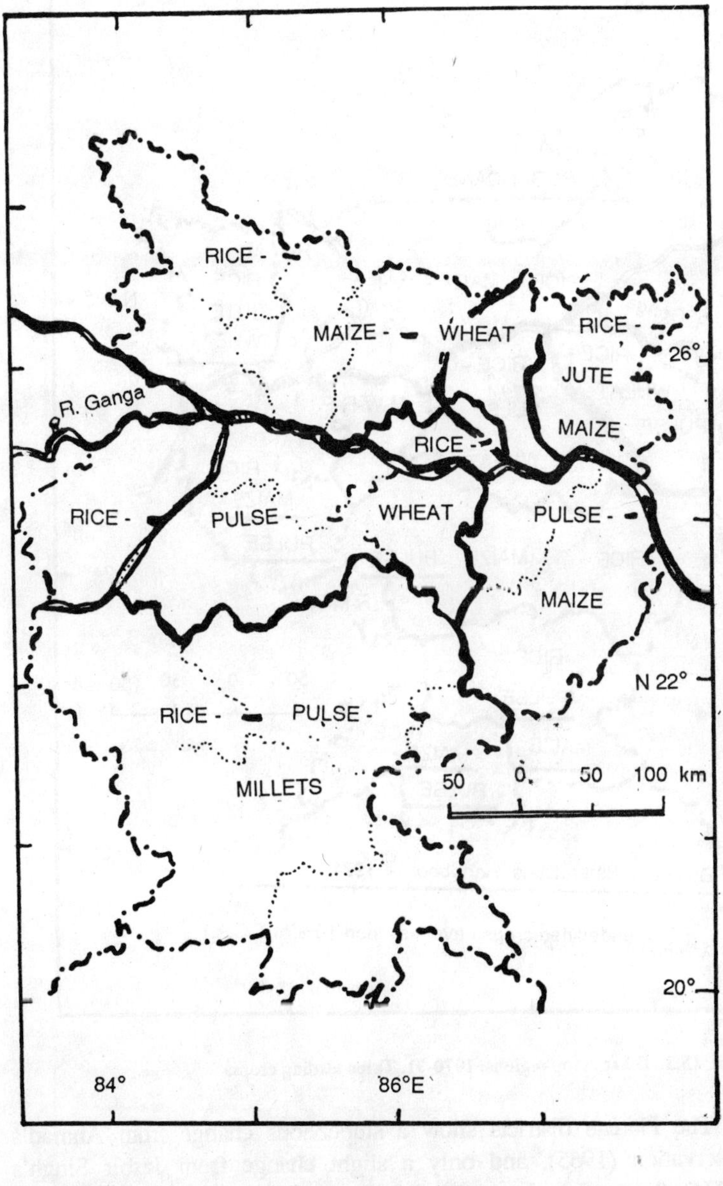


FIG. 15.2: Bihar : crop regions, 1961-65. Three leading crops.

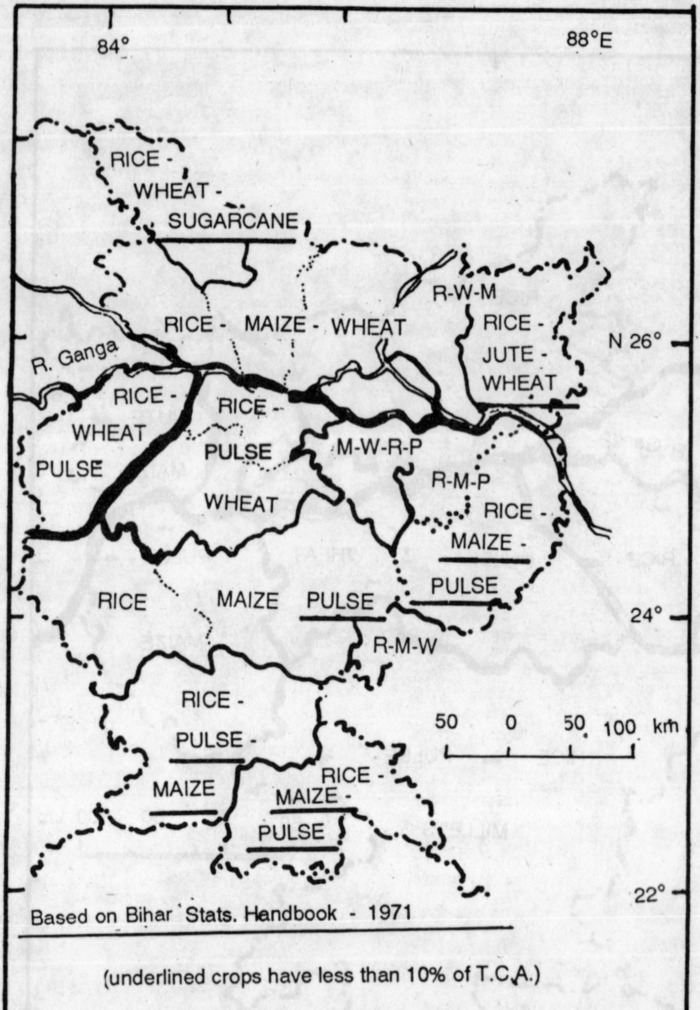


**FIG. 15.3: Bihar : crop regions, 1970-71. Three leading crops.**

The Plateau districts show a stupendous change from Ahmad's observation (1965), and only a slight change from Jasbir Singh's (1974) observation. Large areas have become mono-cultural. Only in

Hazaribagh and Palamau districts maize appears as second crop. This indicates a complete dependence on nature. The contiguity of the areas is revealed from Fig. 15.5 as well as the splendid isolation of Monghyr (Fig. 15.3).

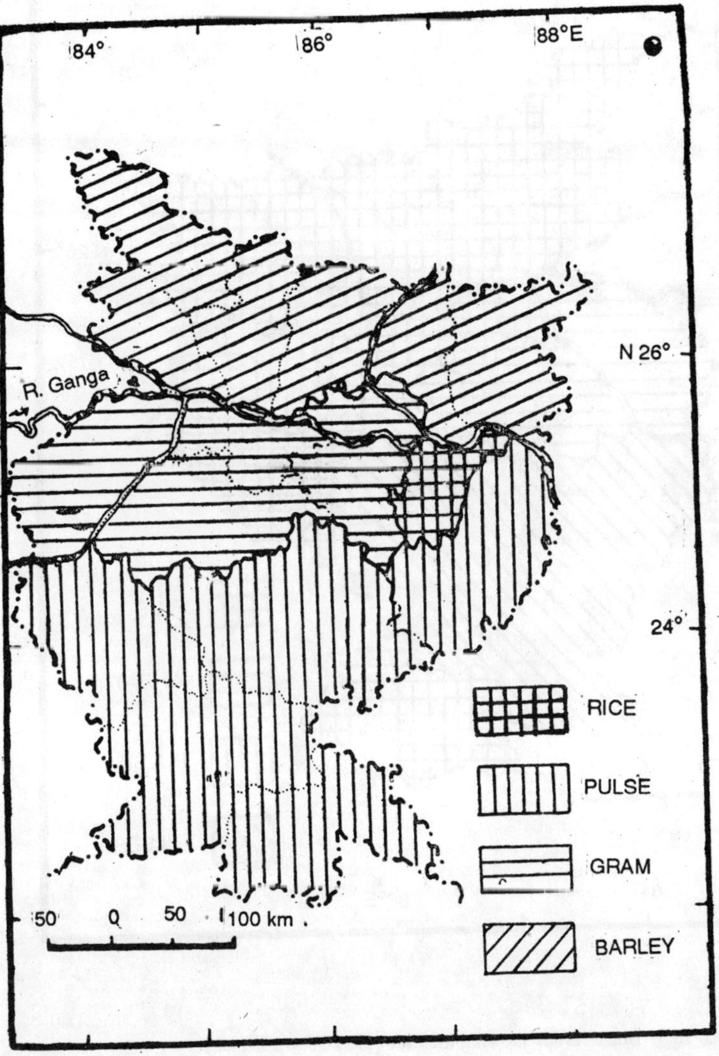


FIG. 15.4: Bihar : crops showing marked decrease (1951-55 to 1961-66).

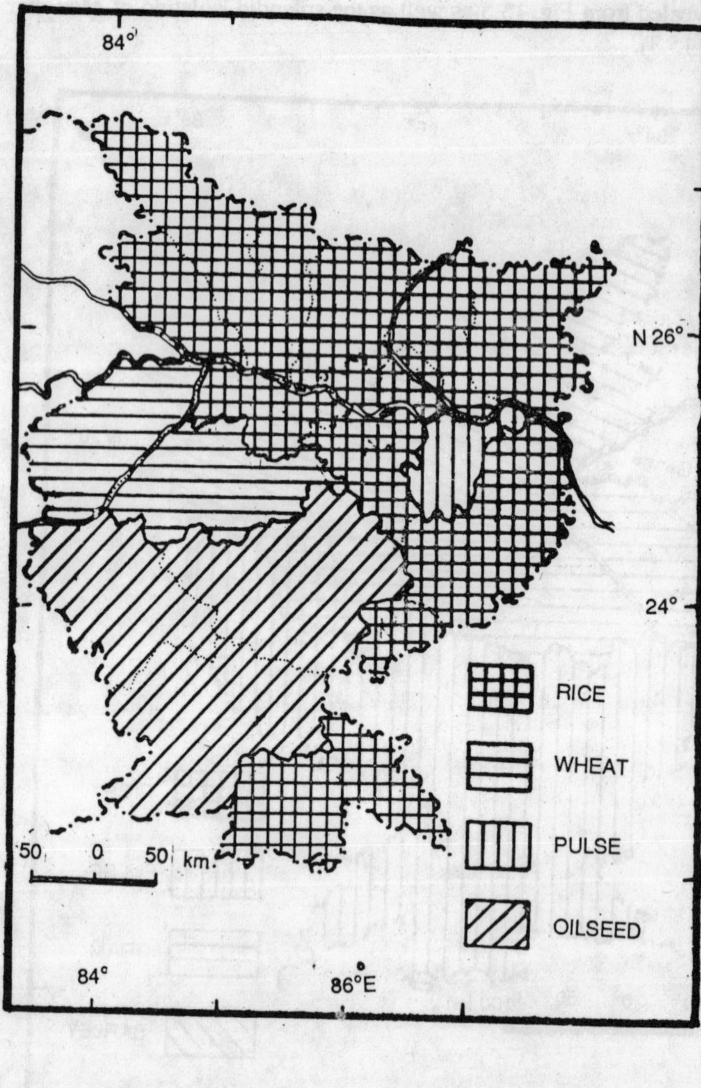


FIG. 15.5: Bihar : crops showing increase (1951-55 to 1961-66).

**Discussion**

The general decrease in the area under forests is really disturbing and, in the context of measures for afforestation and *vana mahotsava*, it is much more so. It reveals a lack of co-ordination in planning. Ahmad (1965) had calculated that the NSA could increase up to a maximum of 53.2 per cent. The excess land was estimated to be derived from culturable wastes and other fallows. In 1970-71 the increase in the NSA is only 2.8 per cent, of which only 1.1 per cent was from culturable waste, probably, and from permanent pasture another 0.2 per cent and the rest in all probability, came from the forests. Within the NSA the *aghani* crop areas have declined, so too have the Rabi areas. The decrease of *aghani* indicates a slight moving away from the dependence on nature but the subsequent modification of crop calendar or diversification of agriculture is not evident from the data. It appears there has been a lessening of the intensification process also. Even if the drought year 1966-67 is left out, the intensification process (as evidenced from more-than-once-cultivated area) and improvement in the use of irrigational facility has been of no great consequence. Moreover, the sources of minor or medium-size irrigational facilities have been neglected, as is evident from the decline in the irrigated area in the plateaus and neighbouring areas (e.g. Bhagalpur) and the spectacular and more capital-intensive large irrigation systems have been erected or improved upon or expanded in the South Ganga and North Ganga Plains. This has resulted not only in re-development but also in re-diversification of agriculture in the Plateau areas and its neighbourhood. The more costly and, for large sections of agriculturists, prohibitive agricultural infrastructure has contributed to the general decline in the intensive system of agriculture.

Consequently, the picture is one of a decline in agricultural diversity and growth of a new pattern of crop combination. The cash crops of sugarcane, tobacco and jute have lost ground or probably are so non-remunerative that they are increasingly being cultivated in less and less area. In their place, a new type of cash crop, the elite cereals, is finding favour. Side by side the efforts to raise cereal production has increased the importance of the elite cereals. The hardier and the less elite (coarse) staple cereal of general masses has lost its ground. The other effect, that is, a decrease in the area of cultivation of pulses (including gram) and oilseeds, has not only reduced their production but also disturbed the ecologically sound and less costly crop-rotation system of preserving the soil fertility. The picture of crop combination that has emerged makes Bihar a producer of food crops and, in that, of elite cereal crops. Bihar's position as a producer of coarse grains, pulses and oilseeds has declined. Probably some of the causes of dearth, in parts of the country, of pulses, oilseeds and coarse grains can be traced to this.

Tapeswar Singh's plea (1977) for quinquennial evaluation of crop combination in Bihar is worth consideration and perusal. In this context the picture obtained presently does not display a greater diversification, rather it is one of gradual reversion to mono-crop tendencies.

**Conclusion**

The picture presented so far is one of change, but only in the cropping and crop combination level. The primary landuses remain very much the same, though fluctuating at times with the monsoons. The last 30 years' planning efforts and the expected progress is not much evident from the landuse statistics. Table 15.5 shows the change in the individual crop's position very clearly as do Appendices III and IV. But the questions that beg to be answered are, the role of market prices, yield pattern of individual crops and governmental incentives given through minimum price support and the like, and the agricultural efficiency that tries to surface by an interaction of these; how far the socio-economic factors are responsible for the emergence of the current crop combination and the likely direction of its movement in future.

*APPENDIX-I*

**Districtwise distribution of landuse, 1958-59 and 1970-71**

*(shown as percentage of the total area)*

| Categories | Forests | Land put to Non-Agri. use | Barren Land | Current Fallow | Other than Current Fallow | Perma-nent Pasture | Culti-vable Waste | Net Sown Area |
|---|---|---|---|---|---|---|---|---|
| Years | 1958-59/ 1970-71 | 1958-59/ 1970-71 | 1958-59/ 1970-71 | 1958-59/ 1970-71 | 1958-59/ 1970-71 | 1958-59/ 1970-71 | 1958-59/ 1970-71 | 1958-59/ 1970-71 |
| | 1 | 2a | 2b | 3a | 3b | 4a | 4b | 5 |
| Districts: | | | | | | | | |
| *Plateau* | | | | | | | | |
| Santal Parganas | 24.2/10.2 | 7.9/7.9 | 3.9/6.7 | 9.8/13.3 | 8.5/9.2 | 3.5/4.4 | 10.3/6.8 | 31.9/40.4 |
| Hazaribagh | 49.6/40.2 | 4.2/5.0 | 5.1/10.0 | 6.8/10.1 | 6.4/6.7 | 2.3/2.1 | 6.3./4.8 | 19.3/19.8 |
| Ranchi | 28.2/23.5 | 4.2/4.7 | 5.0/5.7 | 10.7/10.6 | 11.3/10.7 | 0.7/0.4 | 5.2/4.7 | 34.7/39.4 |
| Palamau | 49.1/48.0 | 2.2/3.1 | 6.8/6.3 | 8.9/11.3 | 5.8/6.9 | 0.6/0.8 | 4.5/2.2 | 22.1/21.3 |
| Dhanbad | 15.1/8.4 | 19.9/20.3 | 10.2/11.9 | 9.9/12.3 | 8.7/8.8 | 0.9/1.1 | 12.1/7.9 | 23.2/28.5 |
| Singhbhum | 47.2/29.1 | 6.5/7.4 | 5.1/29.1 | 6.8/8.6 | 6.6/8.4 | 1.1/0.9 | 7.3/6.1 | 19.4/29.9 |

*(Contd.)*

(Contd.)

| | 1 | 2a | 2b | 3a | 3b | 4a | 4b | 5 |
|---|---|---|---|---|---|---|---|---|
| *Sub-Himalayan Tract* | | | | | | | | |
| Champaran | 7.3/9.7 | 8.1/2.3 | 3.1/2.0 | 5.3/4.2 | 2.1/1.5 | 1.4/0.5 | 4.8/0.4 | 67.9/67.6 |
| *Plain* | | | | | | | | |
| Saran | 0.0/0.0 | 10.9/10.6 | 1.8/4.6 | 10.8/8.3 | 2.1/1.7 | 0.6/0.3 | 3.8/0.9 | 71.0/71.2 |
| Muzaffarpur | 0.0/0.0 | 10.3/13.4 | 1.4/2.8 | 5.6/5.4 | 0.8/1.7 | 0.4/0.3 | 2.4/0.1 | 79.1/73.6 |
| Darbhanga | 0.0/0.0 | 12.4/17.1 | 5.2/1.7 | 6.8/6.9 | 1.2/1.1 | 1.1/0.4 | 3.2/0.3 | 70.1/69.9 |
| Saharsa | 0.1/0.0 | 9.1/10.8 | 7.5/6.0 | 11.6/11.8 | 4.6/4.8 | 0.7/0.6 | 12.1/2.5 | 54.3/60.3 |
| Monghyr | 7.4/13.8 | 10.3/9.1 | 9.2/8.3 | 6.5/12.0 | 2.7/4.2 | 0.5/0.6 | 5.3/2.6 | 58.1/48.5 |
| Purnea | 1.3/0.2 | 9.7/10.9 | 3.2/5.0 | 13.2/11.0 | 4.5/3.0 | 2.3/0.6 | 3.8/1.0 | 62.0/66.3 |
| Bhagalpur | 12.0/8.1 | 9.8/8.8 | 5.9/12.9 | 8.1/8.6 | 3.7/3.1 | 1.5/0.9 | 5.5/3.2 | 53.5/53.4 |
| Patna | 0.7/0.8 | 14.1/14.0 | 2.2/3.1 | 7.7/3.4 | 0.8/0.7 | */0.0 | 1.3/0.4 | 73.2/77.2 |
| Gaya | 18.2/12.1 | 8.3/9.3 | 7.7/5.3 | 6.0/4.8 | 2.2/2.1 | 0.9/1.0 | 2.9/1.4 | 53.8/58.5 |
| Shahabad | 23.7/15.6 | 8.8/8.8 | 2.3/3.8 | 2.9/2.4 | 1.7/0.7 | 0.1/0.1 | 1.7/0.6 | 58.8/67.8 |

* Negligible.

*Source:* 1. Enayat Ahmad (1965): *Bihar — Physical, Economic & Regional Geography*, Ranchi University, Ranchi, for 1958-59 Data.

2. (1975): *Bihar — Statistical Handbook*, 1971, Directorate of Statistics & Evaluation, Govt. of Bihar, Patna, for 1970-71 Data.

*APPENDIX-II*

**Districtwise distribution of area sown more than once and irrigated area (1958-59 & 1970-71) and different seasonal crops (1970-71)**

*(All are indicated as percentage of Net Area Sown)*

| Districts | Area sown more than once 1958-59/ 1970-71 | Net area irrigated 1958-59/ 1970-71 | Bhadoi | Aghani | Rabi | Garma | Orchard |
|---|---|---|---|---|---|---|---|
| 1 | 2 | 3 | 4 | 5 | 6 | 7 | 8 |
| *Plateau:* | | | | | | | |
| Santal Parganas | 62.3/14.5 | 30.8/9.5 | 14.9 | 88.3 | 10.9 | 0.4 | 0.1 |
| Hazaribagh | 35.7/11.7 | 15.8/5.6 | 33.2 | 70.3 | 7.7 | 0.5 | 0.1 |
| Ranchi | 15.7/9.1 | 2.2/3.3 | 42.8 | 60.3 | 4.8 | 1.2 | 0.8 |
| Palamau | 23.7/24.4 | 24.9/25.8 | 44.3 | 45.0 | 34.9 | 0.3 | 0.1 |
| Dhanbad | 51.7/6.6 | 6.9/3.8 | 15.6 | 89.1 | 1.4 | 0.4 | — |
| Singhbhum | 54.3/4.6 | 14.3/5.7 | 10.8 | 88.8 | 4.8 | 0.2 | — |

*(Contd.)*

*(Contd.)*

| 1 2 | 3 | 4 | 5 | 6 | 7 | 8 | |
|---|---|---|---|---|---|---|---|
| *Sub-Himalayan Tract:* | | | | | | | |
| Champaran | 26.5/47.6 | 5.9/26.5 | 35.7 | 54.3 | 48.5 | 9.1 | 2.9 |
| *Plains:* | | | | | | | |
| Saran | 38.7/34.8 | 18.3/26.7 | 44.0 | 34.4 | 51.0 | 5.4 | 3.4 |
| Muzaffarpur | 31.7/35.4 | 1.2/6.7 | 24.7 | 54.4 | 41.9 | 14.5 | 4.6 |
| Darbhanga | 31.5/31.5 | 5.2/10.5 | 26.8 | 60.2 | 34.7 | 9.8 | 4.3 |
| Saharsa | 40.2/42.6 | Nil/13.9 | 30.9 | 56.0 | 31.9 | 23.8 | 2.3 |
| Monghyr | 20.6/30.8 | 16.8/22.7 | 38.7 | 29.5 | 59.0 | 3.9 | 1.4 |
| Purnea | 33.9/33.3 | 0.4/9.1 | 40.5 | 60.2 | 27.5 | 5.0 | 1.7 |
| Bhagalpur | 48.9/36.4 | 64.2/15.4 | 27.6 | 60.1 | 46.7 | 2.2 | 1.2 |
| Patna | 53.1/50.3 | 67.2/61.5 | 15.2 | 55.6 | 73.7 | 5.7 | 0.8 |
| Gaya | 43.9/28.9 | 75.8/74.7 | 7.9 | 68.8 | 49.9 | 2.3 | 0.3 |
| Shahabad | 74.4/47.6 | 76.3/65.0 | 6.5 | 64.2 | 72.7 | 4.3 | 1.7 |

*Source:* Same as for Appendix-I.

*APPENDIX-III*

## Districtwise distribution of crops, 1970-71

*(shown as percentage of Total Cropped Area)*

| Districts | Rice | Wheat | Pulses | Maize | Gram | Barley | Sugarcane | Tobacco | Jute |
|---|---|---|---|---|---|---|---|---|---|
| *Plateau* | | | | | | | | | |
| Santal Parganas | 69.29 | 1.29 | 4.92 | 8.74 | 0.98 | 0.86 | 0.18 | — | 0.06 |
| Hazaribagh | 55.70 | 1.53 | 1.63 | 12.73 | 0.71 | 0.61 | 0.61 | — | — |
| Ranchi | 61.60 | 0.52 | 2.06 | 1.80 | 0.67 | 0.05 | — | — | — |
| Palamau | 30.06 | 4.31 | 9.22 | 14.73 | 5.99 | 4.55 | 0.84 | — | — |
| Dhanbad | 78.22 | 0.44 | — | 8.00 | — | — | — | — | — |
| Singhbhum | 83.60 | 0.38 | 1.52 | 2.86 | 0.95 | — | — | — | — |
| *Sub-Himalayan Tract* | | | | | | | | | |
| Champaran | 49.89 | 17.38 | 7.08 | 4.86 | 0.40 | 5.00 | 8.36 | 0.04 | 0.75 |
| *Plains* | | | | | | | | | |
| Saran | 30.23 | 18.18 | 7.11 | 19.34 | 5.05 | 6.87 | 4.50 | 0.06 | 0.06 |
| Muzaffarpur | 40.58 | 11.57 | 7.94 | 12.56 | 0.57 | 4.57 | 4.36 | 0.88 | 0.16 |
| Darbhanga | 46.94 | 11.44 | 6.43 | 11.75 | 1.01 | 4.56 | 1.82 | 4.56 | 0.35 |
| Saharsa | 40.95 | 12.76 | 5.40 | 12.36 | 0.16 | 1.78 | 0.16 | — | 4.93 |
| Monghyr | 20.62 | 21.27 | 11.03 | 26.92 | 7.16 | 2.17 | 0.72 | — | 0.07 |
| Purnea | 55.39 | 9.20 | 4.16 | 4.34 | 0.77 | 1.85 | 0.21 | 0.17 | 11.60 |
| Bhagalpur | 39.74 | 9.71 | 13.18 | 17.34 | 3.96 | 3.87 | 1.29 | — | — |
| Patna | 38.23 | 12.78 | 27.28 | 7.97 | 4.24 | 0.94 | 0.70 | — | — |
| Gaya | 53.08 | 13.75 | 18.24 | 2.44 | 3.10 | 1.09 | 1.35 | — | — |
| Shahabad | 44.33 | 25.22 | 15.23 | 1.88 | 4.55 | 1.38 | 0.82 | — | — |

*Source:* No. 2 of Appendix-I.

## APPENDIX-IV

### Districtwise crop combination in Bihar, 1957-58, 1961-65, 1970-71

| | E. Ahmad (1957-58) | J. Singh (1961-65) | T. Singh 1970-71 | | 1970-71* (as obtained by area Ranking) |
|---|---|---|---|---|---|
| | | | Ranking Method | Doi's Method | |
| 1 | 2 | 3 | 3 | 4 | 5 |
| Districts | | | | | |
| *Plateau:* | | | | | |
| Santal Parganas | R.M.G | R. | R.P.M. | R. | R. |
| Hazaribagh | R.M. Ragi | R. | R.M. Ragi | R.M. | R.M. |
| Ranchi | R. Ragi Fr. & V | R. | R.P. Ragi | R. | R. |
| Palamau | R.M.G. | R.P. Mi. Os. | R.M. Os. | R. | R.M. |
| Dhanbad | R.M. Fr. & V. | R. | R.P.M. | R. | R. |
| Singhbhum | R.M.G. | R. | R.P.M. | R. | R. |

*(Contd.)*

(Contd.)

| 1 | 2 | 3 | 3 | 4 | 5 |
|---|---|---|---|---|---|
| *Sub-Himalayan Tract:* | | | | | |
| Champaran | R.B.S. | R.M.W. | R.W.S. | R.W.S. | R.W. |
| *Plains:* | | | | | |
| Saran | R.M.B. | R.M.W. | R.M.W. | R.M.W. | R.M.W. |
| Muzaffarpur | R.M.B. | R.M.W. | R.M.W. | R.M.W.P. | R.M.W. |
| Darbhanga | R.M.W. | R.M.W. | R.M.W. | R.M.W. | R.M.W. |
| Saharsa | R.M.J. | R.M.W. | R.M.W. | R.W.M.P. | R.W.M. |
| Monghyr | R.M.G. | R.P.M.W. | M.P.W. | M.R.W.P. | M.W.R.P. |
| Purnea | R.J.M. | R.J.M. | R.J.W. | R.J. | R.J. |
| Bhagalpur | R.M.G. | R.P.M. | R.M.P. | R.M.P. | R.M.P. |
| Patna | R.G.W. | R.P.W. | R.P.W. | R.P.W. | R.P.W. |
| Gaya | R.G.W. | R.P.W. | R.P.W. | R.P.W. | R.P.W. |
| Shahabad | R.G.W. | R.P.W. | R.P.W. | R.P.W | R.W.P. |

* Crops having less than 10 per cent of TCA have been ignored.

*Source:* Same as for Appendix I, and Jasbir Singh (1974): *An Agricultural Atlas of India : A Geographical Analysis*, Vishal Publ., Kurukshetra; Tapeshwar Singh (1977): *A Study of Crop Combination Regions, Ind. Jour. Reg. Sc.*, Vol. IX, No. 2.
R = rice; M = Maize; W = Wheat; B = Barley; G = Gram; S = Sugarcane; J = Jute; Mi = Millets; Fr & V = Fruits and Vegetables; Os = Oil Seeds; P = Pulses.

## REFERENCES


Ahmad, Enayat, 1956 : *Bihar, Physical, Economic, Regional Geography*, Ranchi University, Ranchi, pp. vii+420.

Anonymous, 1970 : *India : Irrigation and Power Projects (Five Year Plan)*, (1978) Govt. of India, Central Water & Power Commission, New Delhi, pp. xxv+407.

———, 1970 : *Bihar : Statistical Hand Book, 1971*, Directorate of Statistics & Evaluation, Govt. of Bihar, Patna, pp. xv+206.

Singh, Jashir, 1974 : *An Agricultural Atlas of India : A Geographical Analysis*, Vishal Publication, Kurukshetra, pp. xxxvi+356.

Singh, Tapesar, 1977 : "Bihar : A Study of Crop Combination Regions", *Ind. Jour. of Reg. Sc.*, Vol. IX(ii), pp. 145-156.

16

# AGRICULTURAL PRACTICES AND AGRICULTURAL CHANGE IN BUNDELKHAND

M. FAROOQ SIDDIQI

BUNDELKHAND SPREADS over a greater part of trans-Yamuna region of Uttar Pradesh and administratively includes twenty tahsils and forty-seven agricultural development blocks in the districts of Lalitpur, Jhansi, Jalaun, Hamirpur and Banda of Jhansi Division. It stretches over 29,463.84 sq. kilometres and is located between 24°11′ N and 26°27′ N latitudes and 78°10′ E and 81°34′ E longitudes. It forms a geographical unit of its own lying in a transitional zone between the Ganga Plain in the north and the Peninsular Plateau in the south. The southern portion, composed of Vindhyan formations, is an upland area which gradually slopes northward into a lowland tract of Ganga alluvium. The upland region has an increased gradient resulting into over-drained eroded soils whereas the lowland region has a gentle slope leading to occasional waterlogging of clayey and clayloam soils.

The region is traversed by both the branches of the southwest monsoon which give copious rainfall but the fall is uncertain and variable both in space and time. The variability of rainfall and the typical sticky nature of soils, when wet, over major portion have given rise to such an agriculture which is characterised by extensive land occupancy during the rabi season, resulting in two crops of different water requirements mixed together and rarely getting two crops from the same plot in a year.

Agriculture is a major activity in Bundelkhand engaging directly or indirectly 78 per cent of the total population of the region. The practice of agriculture, however, is comprised of a few simple enterprises which are not well-developed and lead to a semi-subsistence type of farming. The chief crops of the region are: big and small millets, pulses, wheat, gram and oilseeds. Jalaun district has the largest cultivated area, being 76 per cent of the total area while district Lalitpur with 36 per cent under cultivation is the least in Bundelkhand. Hamirpur, Banda and Jhansi have a cultivated area of 70, 61, and 60 per cent respectively.

**Agricultural Development**

Five Year Plans of the government after independence of the country have given great impetus to agriculture in Bundelkhand. The farmer in these parts has increased the cultivated area from around 48 per cent in 1950-51 or a year before the implementation of First Year Plan to about 71 per cent of the total area in 1985-86. The increase is largely the result of reclamation of culturable waste which has gradually declined from 22 per cent to 9 per cent in the same period. The reclamation has been possible in eroded wastelands or in land lying uncultivated owing to an overgrowth of a weed known as *kans*. The reclamation or eroded lands has been highest in areas of upland portions of the region, mainly in Jhansi and Lalitpur districts. Some area has also been reclaimed in lowlying Jalaun district but here the total increase in 35 years is only 10 per cent of the total area. The cultivators of Banda and Hamirpur districts have nevertheless reclaimed about 25 per cent of the total area which has been brought under plough between 1950-51 and 1985-86.

**General Landuse**

Relief and drainage play an important part in determining the extent of agricultural landuse as it becomes obvious from the fact that fallow lands and wastelands occupy larger area in upland region which, owing to high surface gradient, has undergone considerable soil loss and soil exhaustion through surface erosion and low absorption of water. The lowland area has gentle slope which, coupled with compact clayey (*mar*) and clayey-loam (*kabar*) soils, has remained largely safe from erosion and is mostly devoted to cultivation.

The total area available for cultivation in Bundelkhand is 71 per cent but only 64 per cent of the total area was put under cultivation in 1985-86 and the remaining 7 per cent was left fallow for the whole year. Wastelands on the whole occupy 13 per cent of the area of which

9 per cent can still be reclaimed for cultivation. The district-wise breaking reveals that the districts falling completely in the lowlying zone, namely Jalaun and Hamirpur, have a greater cultivated area (80% or more) and a nominal wasteland area (7 to 10%) in comparison to district like Lalitpur lying entirely in the upland zone where only half of the total area is actually available for cultivation. In the districts which have both upland and lowland areas like Banda and Jhansi, about 70 per cent of the total area is available for cultivation while 10 to 17 per cent remains culturable and unculturable wasteland.

**Irrigation**

The role of irrigation is not important in the agricultural scene of Bundelkhand. In fact *mar* and *kabar* soils occupying a greater part of the region inhibit farmers from practising irrigation. These are dark coloured soils wherein clay and silt constitute 60 to 70 per cent of the soil material. Owing to large concentration of clay particles the soil is very susceptible to moisture variations and gets quickly saturated with water. Any saturation at the time of rabi sowing creates much difficulty in preparing the seed-bed and has even adverse effects on the growth of plants. This soil however has moisture retaining capacity and crops can grow in the rabi season without the application of water on the basis of moisture retained in the rainy season. During the kharif season monsoon rains make the land sticky owing to which cultivation is not possible. If the monsoons fail the soil becomes too hard to plough without copious irrigation which economically is not feasible. The practice of irrigation in areas of black soils is therefore not a common sight. In other soil types, wherever permitted by terrain and a reasonable water table, irrigation is gradually gaining ground.

Only one-seventh of the net cultivated area in 1950-51 was irrigated in Bundelkhand. The percentage ranged from a minimum of 8 in district Hamirpur to a maximum of 22 in district Jalaun. But with the guidance and monetary help provided by the government, the poor farmers of the area are now adopting the practice of irrigation. The total area in the region increased to 25 per cent of the total cultivated area in 1985-86. Baring Lalitpur the irrigated area in other districts ranges from 19 per cent of the net cultivated area in Hamirpur district to 29 per cent in Jhansi district. The net cultivated area in Lalitpur district is only 40 per cent of the total area but the absence of clayey and clayey-loam soils coupled with the presence of canals and wells has put 44 per cent of the net cultivated area under irrigation.

The traditional sources of irrigation, i.e., canals and wells, are also the main source of irrigation in Bundelkhand. Canals hold a dominating position and irrigate about three-fourth of the total irrigated area of the region. In Banda, Jalaun and Hamirpur districts 90 per cent, 88 per cent and 69 per cent of the total irrigated area receives its water from canals. In the district of Lalitpur canal shares only 41 per cent while 38 per cent of the total irrigated area receives its water from the wells. In Jhansi district canals provide water to 60 per cent of the total irrigated area and 34 per cent is shared by wells. Wells in Lalitpur and Jhansi districts are thus second most important source of irrigation. Some noticeable area, being about 14 per cent of the total irrigated area, is irrigated by wells in Hamirpur district but in the remaining two districts of Jalaun and Banda well irrigation with 2 to 3 per cent of the total irrigated area is insignificant. The practice of tube-well irrigation is also gaining ground but still it is not common everywhere. Only 5 per cent of the total irrigated area in Bundelkhand is served by tube-wells. It is only in Hamirpur and Jalaun districts where tube-well irrigation is worth mentioning because here about 12 and 9 per cent of the total irrigated area respectively receives its water from tube-wells.

## Cropland Use in the Kharif Season

All crops requiring relatively high temperature throughout their vegetative growth till ripening are cultivated during the kharif season. The whole of Bundelkhand is well suited in respect of temperature for the growth of kharif crops but the actual areal spread of such crops is restricted to only about 30 per cent of the total cultivated area because of the typical character of soil. The onset of kharif season corresponds with the monsoonal rains which make the *mar* and *kabar* soils sticky under waterlogged conditions making ploughing difficult and unprofitable. It is only on slightly uplying areas of these soils or on lands with other types of soil where kharif crops are cultivated. If, however, there are depressions in the land, the accumulated water makes ideal conditions for the growth of rice irrespective of the fact whether the soil is clayey, loam or sandy-loam. Such areas are nevertheless found in only a few places. In view of these limitations kharif crops are not very important and about 70 per cent of the area available for cultivation is left fallow during this season. The importance is further minimised by the fact that kharif crops occupy such areas which are less productive and do not give good return to the farmer. These crops are cultivated in undulating areas of loamy or sandy-loam soils which owing to sloping surface have lost their fertility through erosion and have turned into areas of marginal productivity. Kharif crops however give better output in those areas of loamy soil where irrigation facilities are available. Grain crops dominate the cropping pattern in the kharif season. They together occupy 92 per cent of the net area sown while other crops like oilseeds, groundnut, sugarcane and other miscellaneous crops extend over the remaining cropped area of the season. Jowar (big millet: *Sorghum vulgare*), rice and arhar (a pulse crop: *Cajanus indicus*) in order of importance are main crops of the region together occupying three-fourth of the net cropped area in the season.

Jowar is the leading kharif crop accounting for 43 per cent of the total area devoted to crops in this season. The area under this crop in the two decades of fifties and sixties recorded a downward trend. From a figure of about 32 per cent of the total kharif sown area in 1950-51 it gradually declined to an all time low of 16 per cent in 1970-71. Since then the farmer with the rise of new high yielding varieties of seeds, increased irrigation facilities and the resultant higher returns has found it profitable to devote larger area to jowar, so that in the eighties the jowar area in Bundelkhand shot up to around 45 per cent of the total sown area of the kharif season.

The cultivation of jowar is common throughout Bundelkhand but its concentration is greater in Hamirpur and Banda districts where together it accounts for 59 per cent of the total jowar area of Bundelkhand. Jhansi stands at third place with 18 per cent, followed by Lalitpur 19 per cent and Jalaun 10 per cent of the total jowar area of the region.

Rice is second in order of areal importance in the kharif season but its areal occupancy has been fluctuating from year to year depending on the variation of rainfall. It occupied 12 per cent of the total cropped area of kharif season in 1950-51 but rose up to 15 per cent in 1960-61 and then declined to almost 7 per cent in 1970-71. It has since then shown a rising trend with its area reaching 16 per cent in 1985-86.

Banda district is famous for its rice as it commands three-fourth of the total area devoted to this crop in Bundelkhand. It is followed by district Lalitpur which shares one-eighth of the rice area. Districts Hamirpur, Jhansi and Jalaun are not important with respect to rice and they three together account for about one-tenth of the total rice area of the region.

Arhar sown as a sole crop or mixed with millets like jowar or bajra ranks third amongst the grain crops of the kharif season. Its yearly progress in Bundelkhand shows as if it is no more regarded as attractive to the farmers. It is constantly loosing ground from 28 per cent in 1950-51 to only 14 per cent of the net kharif sown area of Bundelkhand.

District Banda with 35 per cent and Hamirpur with 32 per cent of the total arhar area of Bundelkhand are the leading producers of arhar, followed by Jhansi (18%) and Jalaun (15%). Lalitpur district is conspicuous for the absence of arhar.

Bajra (bulrush millet) is not important too in the overall cropping picture of Bundelkhand. It occupies only 5 per cent of the net sown area of the kharif season. Jalaun followed by Banda are the only districts that produce bajra as compared to other districts of Bundelkhand where its production is insignificant.

*Urd* (*Phaseolus radiatus*), *moong* (*Phaseolus mungo*), maize, small millets are other grain crops which together occupy 15 per cent of the seasonal cropped area. *Moong* is seen in all the districts but Jhansi has a significant area devoted to this crop. It is sown mixed with big millets and other kharif pulses. *Urd* is also sown mixed, though in small quantities, with big millets and other kharif pulses. Maize requires well-drained soils and therefore Lalitpur district with a large undulating area has a three-fourth share of maize area in Bundelkhand. Jhansi ranks second in areal extent, while in Banda, Jalaun and Hamirpur districts it is almost absent. Maize however is not an outstanding crop of Bundelkhand and occupies only 3 per cent of the kharif season cropped area. Small millets namely *kodon, sawan, kakun, kutki* also extend over a small area with a share of 2 per cent in the seasonal cropped area of Bundelkhand.

Some non-grain crops are also grown in the kharif season but they together occupy only about 7 per cent of the net sown area. With 3 per cent area, *til* is an important oilseed and is often grown mixed with millets and pulses.

## Cropland Use in the Rabi Season

Rabi is the main agricultural season of Bundelkhand where about three-fourth of the total cultivated area remains occupied by crops. The highest cropped area of around 85 per cent of the total cultivated area was recorded in Jalaun and Jhansi districts, followed by Banda and Hamirpur with about 70 per cent of the area under rabi crops. District Lalitpur with extensive marginal soils reported only 60 per cent of its cultivated area under rabi crops in 1985-86.

Grain crops, like kharif season, dominate the agricultural scene in the rabi season and occupy about 89 per cent of the seasonal cropped area. Wheat and gram with 50 per cent and 37 per cent are the leading grain crops. The other crops in order of importance are masur, a pulse crop, and alsi, an oilseed but their areal spread with less than 10 per cent of land occupancy is insignificant.

Wheat as the chief rabi crop, with half of the seasonal cropped area, is generally concentrated in the lowland part of Bundelkhand particularly in the districts of Banda and Hamirpur which together share 55 per cent of the total wheat area of the region. The rest is accounted for by Jhansi (17%), Jalaun (15%) and Lalitpur (13%). An analysis of the area under wheat reveals a rising trend in its land occupancy. In the year 1950-51 it occupied only one-fourth of the total cultivated area of Bundelkhand but has since then been increasing so that by 1985-86 it occupied one-third of the sown area of the region.

Gram is another important grain with a share of 37 per cent of the net sown area in the rabi season. It is sown as a sole crop and is also grown mixed with wheat or barley. The practice of sowing gram is however loosing ground to wheat. In 1950-51 it was the leading crop of Bundelkhand when it commanded more than 55 per cent of the net cropped area. Since then its cultivation has declined constantly so that after a lapse of thirty-five years in 1985-86 it covered only 30 per cent of the net sown area. Presently it holds second rank in grain land occupancy and its cultivation is widespread in all the districts of the region.

Masoor (*Lens esculenta*), barley and peas are other grain crops which together account for only 12 per cent of the season's cropped area. Masoor amongst them with a share of 9 per cent area is somewhat important. It has increased its importance in due course of time because thirty-five years ago in 1950-51 it occupied only one per cent of the season's cultivated area in the region. About half of the total masoor area of Bundelkhand lies in Jalaun district, followed by Jhansi and Hamirpur with 18 per cent each and Banda district with 11 per cent. Lalitpur district has only 3 per cent of the region's masoor area.

Barley, though grown all over Bundelkhand, occupies an insignificant area of only 2 per cent in the rabi season.

Alsi (linseed), mustard, potato, some vegetables and a little fodder are other non-grain crops of Bundelkhand but they together occupy only 5 per cent of the total cropped area of the rabi season.

## Mixed Cropping

An important phenomenon of agriculture in Bundelkhand is growing two or three crops of differing water requirements mixed together. The farmer is compelled to adopt this practice owing to dependence on rainfall which is variable both in amount and in its incidence and to typical soils which in the presence of large percentage of silt and clay particles are much responsive to rainfall variations. Bundelkhand lies at the junction of the two currents of south-west monsoon and hence the rainfall, though heavy, is most variable in its incidence. The result of a variable rainfall is that when it is deficient in the kharif season, kharif crops suffer and there is much difficulty in preparing the rabi seed-bed by breaking hard clayey or clayey-loam soil, and if the rainfall is excessive, the kharif crops suffer to an equal degree and the preparation of seed-beds becomes difficult owing to the soil becoming too sticky and untractable. If there is no rainfall in the cold weather season corresponding to rabi crops, the crops on light soils are severely affected while heavy rain causes the spread of fungus in wheat and linseed. As rainfall is responsible for the type of crops grown and also determines the areal distribution of each crop, any variation in its occurrence has great agricultural consequences. The influence of rainfall variation is further increased with the total dependence of agriculture on rainfall almost everywhere except in a few areas where irrigation facilities are available. In order to save the crops from the vagaries of rainfall the cultivators resort to sow those crops mixed together which differ in their water requirements. Important crops involved in this practice are millets mixed with pulses mainly *arhar* or sometimes in a limited way with other pulses like *urd* and *moong* in the kharif season and wheat mixed with gram or with both barley and gram in the rabi season.

The practice of sowing millets with pulses or wheat with gram or barley or with both provides security to the cultivators against the adverse rainfall conditions. If the rainfall is excessive in September or October the millet crop is seriously affected and the crop is damaged or the yield is poor but under these conditions the pulse crop (arhar) grows well and the outturn is good. If however the rainfall is deficient in the kharif season, millet crop yields well but arhar does not perform well. Similarly a deficient rainfall in the rabi season is responsible for forming lesser amount of grain in the wheat crop than in the season of favourable rainfall, but it affects gram favourably which gives fairly good yields. Any occurrence of rainfall in the end of September or the beginning of October hinders the process of fertilisation in millet crop but instead is very beneficial for the vegetative growth of arhar. Likewise, if damp weather is extended to January and February, there is a great chance of the spread of rust which may destroy the whole wheat crop, but gram remains safe. In this way by resorting to the practice of mixed cropping, the cultivator ensures at least one successful crop from his land.

The mixture provides some other advantages, the most important being an unconscious improvement of soil fertility. Since the clayey soil can be cultivated only in the rabi season and remains fallow in the kharif season owing to its stickiness when wet, the cultivator puts it under crop year after year because he has no other season suitable to grow crops other than rabi crops. This continuous practice of cultivation, without of course any addition of manures, leads to soil exhaustion. But the practice of sowing a leguminous crop like gram mixed with other grain crop helps in restoring the soil fertility. The cultivation of wheat requires many ploughings for a better tilth but gram can be grown with little land preparation. The cultivator therefore mixes gram with wheat after only one or two ploughings and saves himself the cost and labour of better tilth preparation. Another benefit which accrues to the cultivator from the practice of mixed cropping is the availability of two types of grains, a grain for bread and a pulse to be taken with bread and thus providing his total food requirements from the same field.

### Double Cropping

Cultivation of land in both kharif and rabi seasons is a rare practice in Bundelkhand. In the whole of the region only about 8 per cent of the net area available for cultivation was put to double cropping. Farmers may not be blamed for it but in fact the nature of soil inhibits double cropping. *Mar* and *kabar* soils owing to their compactness and stickiness when wet cannot be ploughed during the kharif season and are therefore left fallow. They are thus devoted to only one crop in the whole year. The other soils lying in upland Bundelkhand have witnessed large scale erosion and have lost their fertility. They cannot thus support two crops in a year. It is only in a few fields of loamy soil given to millets, oilseeds and vegetables in the kharif season where irrigation facilities are available that another crop is grown in the rabi season also. Some water-logged low lying fields in the clayey or clayey loam areas devoted to rice crop are also clear by the time of rabi sowing and are thus cultivated again. Such areas are very few and are localised to certain favourable places with respect to physical character of the land and double cropping is not a common feature.

## REFERENCES


Beck, K.J. and Bennema, J., 1972 : "Land Evaluation for Agricultural Land Use Planning — An Ecological Methodology", Agricultural University, Wagningen, The Netherlands.

Coleman, A., 1961 : The Second Land Use Survey : Reform and Prospect, *Geographical Journal.*

Dagli, V., (Ed), 1974 : *A Regional Profile of Indian Agriculture.*

F.A.O., 1978 : A Framework of Land Evaluation, Soil Bulletin, 32.

Kalamkar, R. J., 1950 : "Agronomy and Land Utilisation", Proceedings of the National Institute of Science of India, Nov-Dec.

Mohammad N., 1978 : Agricultural Land Use in India, Concept, New Delhi.

Mohammad N. 1981 : *Perspectives in Agricultural Geography*, Vol. I-V, Concept Publishing Co., Delhi.

Shafi, M., 1960 : Land Utilisation in Eastern Uttar Pradesh.

Shafi, M., 1967 : Land Use Planning, Land Classification and Land Capability : Methods and Techniques, *The Geographer*, Vol. XVI, Aligarh.

Singh, G. B., 1979 : *Transformation of Agriculture*, Kurukeshtra.

Singh, R. P., 1967 : Concept of Land Use, *Patna University Journal*, Vol. 23, No. 1, Jan. 67.

Stamp, L. D., 1960 : *The Land of Britain : Its Uses and Misuses*, Longmans, London.

# 17

# WASTELANDS IN INDIA
## Some Issues and Problems

NOOR MOHAMMAD

THE CONCEPT and definition of wasteland is highly controversial and debatable. Inspite of the continuous efforts of the individual scholars and research institutions together with several voluntary and government organisations, a precise definition of wasteland has not yet emerged. Similar is the case with the areal strength of the wasteland and their spatial distribution. In the present paper an attempt has been made to describe and explain some of these issues and problems.

The common understanding of wasteland is that of a vast, dreary, uncultivated, uninhabited expanse, the mining devastated area, deforested mountain slope, the land turned into desert like condition, and even the areas subject to nuclear test explosion or chemical warfare (Tolba, 1982). The Chambers Dictionary explains wastelands as "uncultivated and almost sparsely inhabited, lying unused, unproductive, devastated, and ruinous", while Webster International Dictionary describes it as "uncultivated barren land". The Longman Dictionary of Geography explains it as "any wild, uncultivated, and uninhabited land . . . ". In these descriptions and explanations the major emphasis has been given on the suitability or otherwise of land for agricultural purposes which speaks the partial truth. In several other definitions, being attempted by the economists, the economic potential and the actual returns from the land seems to have formed the basis of identifying the wasteland. According to it any land which gives less than twenty five per cent of its potential may be considered as wasteland. But this concept has its own limitations and problems because the productive potential of land is mainly the function of technological input, and with increasing level of technological input the productive potential of land also increases. Hence, it is a dynamic concept and the estimation of wasteland on the basis of this definition would be difficult as the status of land will keep on changing with change in the status of the technology of the region. Moreover, the concept has not taken into account the ecological factors which do play a major role, and their balance must be maintained (Bhumla and Khare, 1987).

In the category model of landuse proposed by the Technical Committee on Co-ordination of Agricultural Statistics (TCCAS) in 1950 wasteland has been given as a separate category entitled "Culturable wasteland". Besides this, it has also some amount of wasteland in other categories of landuse. Mohammad (1978) and Singh (1974) have simplified this model and have given adequate coverage to it. In another study wastelands have been defined as those lands which are "unculturable or presently lying unutilised due to different constraints but have been used previously which is giving very low actual return of its economic potential, which are ecologically unstable or whose top soil has completely lost its fertility status, which has developed toxicity for growth of crops and trees due to environmental or anthropogenic problems, have been abandoned and no further use has been found" (Yadav, 1987). Of course it is a very lengthy definition trying to incorporate loosely everything said about wastelands, nevertheless it has some weight in it and it needs to be sharpened. In order to make it more specific the author has developed the category model of TCCAS and has identified culturable wasteland and fallow other than current fallow as wasteland. Finally he has identified 13 categories of culturable wastelands and two categories of fallow other than current fallow. Bhumla and Khare (1987) in their article "Estimates of Wastelands in India" have defined wastelands as the lands which are ecologically unstable, have nearly completely lost their top soil, and have developed toxicity in the root zones for growth of most plants, both annual crops and trees. This definition covers the lands affected by water erosion, wind erosion, floods, waterlogging, soil salinisation, and soil alkalinisation but it does not cover certain other categories of lands which are wastelands. Moreover, it is highly biased towards the suitability of land for growing the crops under certain given level of technology.

As a result of all these conflicting and controversial definitions and view points, and the severe threat being imposed by expanding wastelands demanding their control and reclaimation, in 1984, a

National Wastelands Development Board under the Ministry of Environment and Forests was constituted to look into the matter. The Board, as a first step, constituted a "Technical Task Group" to define and classify the wastelands so that effective measures could be taken to control and reclaim them. The Group published its report on Definition and Classification of Wastelands in 1986 which defines wastelands as "that land which is degraded and is presently lying unutilised except as current fallow due to different constraints". It has divided the wastelands into two levels. In the first level wastelands are classified into Culturable and Unculturable wastelands which is based on its potentiality and suitability for agricultural purposes, while level II is based on major limitations. The culturable wastelands have been classified into 11 categories such as gullied and/or ravinous land; undulating upland with or without scrub; surface water-logged land and marsh; salt affected land; shifting cultivation area; degraded forest land; degraded pastures/grazing land; degraded non-forest plantation land; strip land; sands; and mining/industrial/wastelands. Unculturable wastelands have been divided into 3 categories which are barren rocky/stonywastes/sheet rock area; steep sloping area; and snow covered and/or glacial area. The Group has arrived at this nomenclature and classification of wastelands after considering several major points such as maximum possible compatibility to existing data terminology, easier understanding of the technical terms at different levels of the administrative hierarchy and scientific circles to enable identification and delineation of wastelands without confusion or error, feedback collected from various research and development organisations and field extension units based on a tentative classification of wastelands circulated by National Remote Sensing Agency, specific requirements of wasteland development and so on. The Group has also taken into cognizance some other points putforth by its members such as utilisation of land below optimum productivity level, spatial cannotation of wasteland distribution, cannotations and specificity of terminology together with its exhaustiveness, compatibility with existing data base and possibility of further categorisation.

The definition of the Technical Task Group seems to be comparatively better but the details included into it are highly debatable. The marshy lands and snow covered and glacial areas are not wastelands everywhere. They perform several useful economic and ecological functions for the economic development of the area. For example, the marshy lands of Kashmir valley functions as a floodlung during the rainy season. It absorbs the rain water and consequently controls the floods during the rainy season, whereas during dry months it gradually releases the water which is utilised for irrigation purposes. Likewise the snowy caps and the glaciated areas are the perennial source of water in our rivers which is utilised for irrigation purposes. Such lands, therefore, may be wastelands in certain areas or may be performing very useful functions. This definition, therefore, does not have universal applicability.

Since the report of the Technical Task Group was widely circulated among eminent scholars of related discipline, academic institutions and organisations and the concerned ministries of the Government of India, a lot of comments and criticism were received. In the light of these comments, therefore, the Group reconsidered the entire issue and came up with a new booklet entitled "Description and Classification of Wastelands" in 1987 which describes wastelands as "degraded land which can be brought under vegetative cover with reasonable efforts, and which is currently underutilised and land which is deteriorating for lack of appropriate water and soil management or on account of natural causes". Wasteland can result from inherent/imposed disabilities such as by location, environment, chemical and physical properties of the soil or financial and management constraints. The land degradation connotes loss of land productivity, qualitative and quantitative through a variety of processes. Many interesting results may be inferred from it. Firstly, it is difficult to give a precise and universally applicable definition of wastelands as it is a comparative term and its definition keeps on changing from region to region and from time to time; secondly, it refers to degraded lands which have lost their productivity due to variety of reasons; thirdly, it has excluded many types of lands earlier classified as wastelands from wasteland category; and finally, the number of categories of wastelands has been reduced to only 7 as against 14 in earlier classifications.

From the above discussion, therefore, it may be concluded, that firstly, it is difficult to frame a concise definition of wastelands which could be applied to international or even at national level; secondly, wastelands should be defined on the basis of their productive and functional utility; and finally, wastelands should be defined with reference to its area of recurrence because a particular type of land may be waste in one region but productive in other region.

The actual areal strength of wastelands is even more controversial and debatable. It varies from 175 million hectares to 51 million hectares. However, the first comprehensive review of the wastelands was done by the National Commission on Agriculture (Ministry of Agriculture, 1976) which estimated different types of wastelands and suggested different measures and guidelines for their reclaimation and development. The Soil and Water Commission Division (Ministry of Agriculture) revised these estimates on the basis of 1980-82 statistics. The official estimate of wastelands in India was 175 million hectares which is more than 50 per cent of geographical area of the country (Mukherji et. al., 1985). Realising the estimate to be on higher side the Soil and Water Commission Division modified it to 146 million hectares excluding 27 million hectares which have been claimed to be treated for its improvement (Das, 1985). In these estimates firstly, certain percentage of area under each category of landuse has been assumed as problem area, and secondly, the area under certain specific problem category has been estimated and added to the area calculated in the first step. Hence, it has inflated the estimate by double counting of some areas under wastelands.

Bhumla and Khare (1987) have made a more systematic and scientific approach in estimating the extent of wastelands in India. Out of the landuse categories proposed by the Directorate of Economics and Statistics, Ministry of Agriculture only seven categories viz. Forests, Barren and uncultivable land, Permanent pastures and other grazing lands, Culturable wastelands, Fallow land other than current fallows, Current fallows, and Net area sown, have been considered for estimating wastelands. The two categories of landuse viz. area put to non-agricultural uses, and land under miscellaneous tree crops and groves have been left out because they are considered to have lesser or inconsiderable amount of wastelands. Nine major possible problems of varying nature have been identified which have finally been grouped into following three broad categories:

(a) Water erosion which includes sheet erosion, ravine, riverine lands, water logging, gully erosion, shifting cultivation.
(b) Wind erosion which includes shifting sand dunes, wind erosion, extreme moisture stress, coastal sand dunes.
(c) Salinity and alkalinity.

The relationship between the three problem categories and seven chosen landuse categories is ascertained. The forests have been considered as having all the three group of problems while the barren and uncultivated land having the only problem of salinity and alkalinity. The remaining five categories are liable to have only two type of problems viz. water erosion and wind erosion. However, due to several definitional problems and coverage, the extent of wastelands in the forest category of landuse has not been done. Finally on the basis of a large number of documents, literature, government records, geographical conditions prevalent in the country and with their own perception of the situation, the authors have made the estimate of the area affected by salinity alkalinity, water erosion and wind erosion. The area affected by salinity and alkalinity works out to be 7.36 million hectares, while the estimate of wind eroded area has turned out to be 12.93 million hectares. The total water eroded area is estimated to be 73.6 million hectares of which 35.38 million hectares are affected by water erosion in uncultivated area, while the same is 38.21 million hectares in cultivated area. As such the estimate of total wastelands in India turns out to be 93.67 million hectares. A statewise status of these types of wastelands is given in Table 17.1.

TABLE 17.1: **A statewise status of wastelands in cultivated area**

*(Figures in 10,000 hectares)*

| States/U.Ts. | Saline and Alkaline Lands | Wind Eroded Area | Water Eroded Area | Total Waste Lands | % age to Total Area |
|---|---|---|---|---|---|
| Andhra Pradesh | 2.40 | — | 74.42 | 76.82 | 27.39 |
| Assam | — | — | 9.35 | 9.35 | 11.90 |
| Bihar | 0.04 | — | 38.92 | 38.96 | 22.41 |
| Gujarat | 12.14 | 7.04 | 32.35 | 71.53 | 36.39 |
| Haryana | 5.26 | 15.59 | 2.76 | 24.01 | 54.29 |
| Himachal Pradesh | — | — | 14.24 | 14.24 | 25.74 |
| Jammu and Kashmir | — | — | 5.31 | 5.31 | 2.38 |
| Karnataka | 4.04 | — | 67.18 | 71.22 | 37.13 |
| Kerala | 0.16 | — | 10.37 | 10.53 | 27.09 |
| Madhya Pradesh | 2.42 | — | 127.05 | 129.47 | 29.23 |
| Maharashtra | 5.34 | — | 110.26 | 115.60 | 37.56 |
| Manipur | — | — | 0.14 | 0.14 | 0.62 |
| Meghalaya | — | — | 8.15 | 8.15 | 36.25 |
| Nagaland | — | — | 5.08 | 5.08 | 30.79 |
| Orissa | 4.04 | — | 27.53 | 31.57 | 20.26 |
| Punjab | 6.88 | — | 4.63 | 11.51 | 22.63 |
| Rajasthan | 7.28 | 106.23 | 66.59 | 180.08 | 52.60 |
| Sikkim | — | — | 1.31 | 1.31 | 17.93 |
| Tamil Nadu | 0.04 | — | 33.88 | 33.92 | 26.08 |
| Tripura | — | — | 1.08 | 1.08 | 9.96 |
| Uttar Pradesh | 12.95 | — | 53.40 | 66.35 | 22.53 |
| West Bengal | 8.50 | — | 13.27 | 21.77 | 24.77 |
| Union Territories | 0.16 | — | 8.73 | 8.89 | 7.40 |
| Total | 71.65 | 129.26 | 736.00 | 936.82 | 28.46 |

The lower estimates of wastelands in this study has come basically due to two reasons: firstly, it has not included the degraded forests land; and secondly, lower estimate of wastelands in the cultivated category. There are different estimates regarding the degraded forests and it varies from 40 million hectares (Chowdhary, 1985) to 19.5 million hectares (Ministry of Agriculture). However, if was take an average figure of 30 million hectares of degraded forests and add it up to Bhumla and Khare's estimate of 93.67 million hectares, the total wastelands in India come to 123.67 million hectares which is far less than the estimate of the Soil and Water Conservation Division.

Since 1970s nationwide aerial mapping of wastelands have been carried out and the estimates of its areal extent have been made which are far below the earlier estimates. The figures based on these surveys and aerial maps come out to be about 55 million hectares. According to the Institute of Coastal and Offshore Research the estimate of wastelands is given in Table 17.2 which is based on the satellite imageries of National Remote Sensing Agency. A statewise break-up of wastelands as worked out from satellite imageries is given in Table 17.3.

TABLE 17.2: Aerial estimates of wastelands

| Sl. No. | Description | Area in million hectares |
|---|---|---|
| 1. | Salt affected land | 3.99 |
| 2. | Gullied or ravinous land | 6.73 |
| 3. | Undulating upland with or with out scrub | 11.74 |
| 4. | Jhum or forest blank | 6.24 |
| 5. | Sandy area | 13.94 |
| 6. | Barren hill ridge or rock outcrop | 2.70 |
| 7. | Snow covered or glacial area | 10.07 |
| | Total | 55.41 |

*Source*: Institute of Coastal and Offshore Research.

The spatial distribution and areal delineation of wastelands is equally complicated because of the definitional problem and lack of authentic mapping and surveying (Fig. 17.1). As a result it has not been properly done, although for the development of wastelands, complete information about the location of such lands on the surface of the country must be known. The planning Commission had appointed the

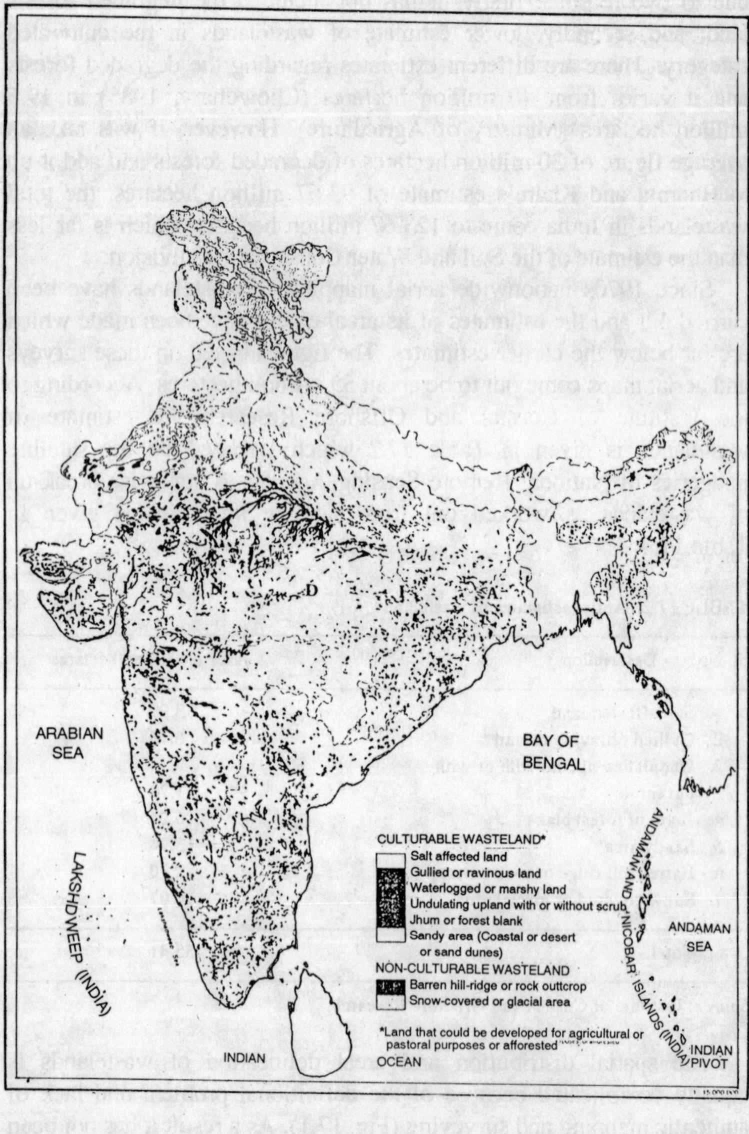


**FIG. 17.1: India : Spatial distribution of wasteland.**

Wasteland Survey and Reclaimation Committee in 1963 which has included in its report the sketch maps of different states showing the location of wastelands covering 250 acres or more. On the basis of the report and the included sketch maps, National Atlas and Thematic Mapping (NATMO) has prepared landuse maps on a scale of 1" = 4 miles on which the demarcation of wastelands has been done. Besides this, many states have also produced maps showing wastelands of various types on the basis of field surveys conducted by various agencies. A generalised picture of wastelands has been given in the Landuse Map prepared by NATMO.

TABLE 17.3: Statewise distribution of wastelands

*(Figures in 000' hectares)*

| State/U.Ts. | Total Area | Culturable Wasteland | % age | Non-Cult. Wasteland | % age | Total Wastelands | % age to Total |
|---|---|---|---|---|---|---|---|
| Andhra Pradesh | 27682 | 2063.3 | 7.45 | 404.0 | 1.56 | 2467.3 | 8.91 |
| Assam | 7852 | 191.2 | 2.44 | — | — | 191.2 | 2.44 |
| Bihar | 17388 | 902.6 | 5.19 | 105.1 | 0.60 | 1007.7 | 5.80 |
| Gujarat | 19598 | 3251.9 | 16.59 | 1530.7 | 7.81 | 4732.6 | 24.40 |
| Haryana | 4422 | 873.8 | 6.19 | 59.6 | 1.35 | 333.4 | 7.54 |
| Himachal Pradesh | 5567 | 84.3 | 1.51 | 1957.4 | 35.16 | 2041.7 | 36.68 |
| Jammu & Kashmir | 22224 | 85.8 | 0.39 | 13260.1 | 59.76 | 13365.9 | 60.14 |
| Karnataka | 19177 | 1374.9 | 7.17 | 268.8 | 1.40 | 1643.7 | 8.57 |
| Kerala | 3887 | 186.5 | 4.80 | 39.9 | 1.09 | 226.4 | 5.82 |
| Madhya Pradesh | 44284 | 2600.5 | 5.87 | 311.4 | 0.70 | 2911.9 | 6.58 |
| Maharashtra | 30776 | 2890.8 | 9.39 | 90.7 | 0.29 | 2981.5 | 9.69 |
| Manipur | 2236 | 320.9 | 14.35 | — | — | 158.6 | 9.59 |
| Meghalaya | 2249 | 158.9 | 7.07 | 9.0 | 0.40 | 167.9 | 7.47 |
| Nagaland | 1653 | 158.6 | 9.59 | — | — | 158.6 | 9.59 |
| Orissa | 15578 | 1145.7 | 7.35 | 8.4 | 0.05 | 1154.1 | 7.40 |
| Punjab | 5036 | 352.1 | 6.99 | — | — | 352.1 | 6.99 |
| Rajasthan | 34221 | 11878.5 | 34.71 | 1061.6 | 3.10 | 12940.1 | 37.81 |
| Sikkim | 730 | 69.3 | 9.49 | 200.4 | 37.45 | 269.7 | 36.94 |
| Tamil Nadu | 13007 | 899.8 | 6.92 | 111.5 | 0.86 | 1011.3 | 7.76 |
| Tripura | 1048 | 93.7 | 8.94 | — | — | 93.7 | 8.94 |
| Uttar Pradesh | 29441 | 2806.3 | 9.53 | 1511.7 | 5.13 | 4318.0 | 14.67 |
| West Bengal | 8785 | 545.9 | 6.21 | 18.0 | 0.20 | 563.9 | 6.42 |
| Union Territories | 11874 | 507.7 | 4.28 | 960.9 | 8.09 | 1468.6 | 12.37 |
| Total | 328715 | 32843 | 9.99 | 21929.2 | 6.67 | 54772.2 | 16.66 |

Recently National Remote Sensing Agency has published the maps of almost all the states on a scale of 1:3500000 showing the spatial distribution of wastelands. It has also produced a map on wastelands in India on a scale of 1:1,000,000 which shows the pattern of distribution of all the six categories of culturable wastelands, and the two categories of non-culturable wastelands. A critical view at these maps indicates that there are several deficiencies in them. Some of them are technical while some are procedurial and methodological. For example, the distribution of wastelands in U.P. shows that the entire Ganges plain mainly consists of saline and alkaline wastelands, while in all the hilly states specially in Jammu and Kashmir the coverage of wastelands is far less than their actual areal strength. Same is true regarding the distribution of wastelands in other states. The deficiencies have occurred because of the scale problem in satellite imageries. Since the imageries have been taken on a very large scale, the small patches of wastelands covering 500 hectares have not emerged on the imageries, while small patches of cultivated land being surrounded by patches of wastelands have emerged as a single patch of wasteland. The former is more true in case of hill states, while the later usually holds good in case of the states of the plain regions. The deficiency, perhaps, may be reduced if satellite imageries are taken at small scales so that minor details may also emerge.

From the foregoing discussion it becomes quite clear that there are several issues and problems related to wastelands which should be solved with highest degree of precision. It could be done at various levels by individual scholars and voluntary and government organisations.

## REFERENCES

Bhumla, D. R. and Khare, A. K. : "Estimate of Wasteland in India", Published by Society for the Promotion of Wasteland Development, N. Delhi.

Clark, A. N. : *Longman Dictionary of Geography*, London.

Das, D. C., 1982 : "Problems of Soil Erosion and Land Degradation in India", Quoted in *Development India's Wasteland : A Brief Report* by Centre for Science and Environment, New Delhi.

Mohammad, N., 1978 : *Agricultural Landuse in India*, Delhi.

———, 1981 : "Patterns of Landuse in Ghaghara-Rapti Doab", in N. Mohammed (Ed.), *Perspectives in Agricultural Geography*, Vol. III, New Delhi.

Shankeranarayan, K.A., 1988 : *Wasteland Development and Utilisation*, Jodhpur.

Singh, J., 1974 : *An Agricultural Atlas of India : A Geographical Analysis*, Kurukshetra.

Singh, P., 1989 : *Problems of Wasteland and Forest Ecology in India*, New Delhi.

Tolba, M. K., 1987 : "Rolling Back the Wastelands", in H.R. Yadav (Ed.), *Wastelands: Diagonis and Treatment*, Concept Publishing Co., New Delhi.

Yadav, H. R., 1987 : *Wasteland Diagonis and Treatment*, Concept Publishing Co., New Delhi.

———, 1976 : Report of the National Commission on Agriculture, Ministry of Agriculture and Irrigation, Part-V, New Delhi.

———, 1986 : *Developing India's Wastelands: A Brief Paper*, Centre for Science and Environment, New Delhi.

———, 1986 : Technical Task Group Report on 'Definition and Classification of Wastelands, National Wasteland Development Board, New Delhi.

———, 1987 : Report on 'Description and Classification of Wasteland', National Wasteland Development Board, New Delhi.

# 18

# TURNING THE TIDE OF AGRICULTURAL LAND LOSS

ALICE COLEMAN

OUR DEEPEST concern for British agriculture should be whether it will exist at all in the future. After a period of controversy there is an emerging consensus that agricultural land losses are very serious indeed. The Ministry of Agriculture, on the basis of a main survey in 1963, has recently introduced a new category of loss, "land unaccounted for", which add 80 per cent to the previous figure for loss to urban use, and if project into the future, suggests that all our improved farmland will have gone within 400 years. Since this new category is still growing, the period will probably shorten. The maps of the Second Land Utilisation Survey, which have no "land unaccounted for", show a great deal of abandoned farmland which has not been taken for any specific purpose but is simply the result of urban pressures. The main survey referred to above had suggested that if the contemporary rate of loss continued, as already said, all improved land would have gone within 400 years, while sample re-surveys in the 1970s show an acceleration which makes 200 years a more reasonable estimate. The rate varies locally: Surrey, for example, would lose 70,900 hectares[1] in 132 years, northern Kent and southern Essex would lose 32,500 hectares in 87 years, and Merseyside would lose 20,425 hectares in only 41 years.

These figures are projections, not predictions. Critics, who mistakenly treat them as predictions discount them on the ground that any trend will bend. This trend is indeed bending at the moment, but in the wrong direction, and there are indications that it will accelerate still more if left to its own devices. For example, the figure quoted is a net rate, in which the true gross rate of loss is concealed by compensatory improvements of previously unimproved land. But the supply of improvable land is limited, and once it is exhausted the gross rate could continue unmasked as an apparently sudden spurt. Another line of evidence is the fact that highly productive land is taken faster than land of lower quality. R.H. Best has shown that 12 per cent of all land taken for urban purposes in Grade I, and since this grade occupies only 2.3 per cent of the national area, it is clearly being lost at a rate which is proportionately five times as fast as the other grades. In fact, there is a high positive correlation between land grade and land take, holding good through all five grades, which means that the loss in productivity is even greater than the loss in area. At one time it was believed that "We shall always be able to produce more food from less land", but as the increase in total production has faltered and gone into reverse in some recent years, this belief is more than suspect.

**A Chain — Reaction of Outward Pressure**

Diminishing agricultural acres are part of an overall national pattern of landuse change which can only be described as obsessed with spaciousness. This has begun in inner city areas and spread outward like ripples on a pond, in shockwaves that have dislocated one zone of landuse after another.

From the best of intentions, but with disastrous results, high-density in inner city areas have been emptied of their population and converted into low-density areas with only one-quarter or one-third of the number of people who lived there early in the century. London alone lost million-and-a-half people in 30 years, and the G.L.C. structure plan called for a further decanting of 1.7 million in only 20 years, a 70 per cent increase in the annual rate. With policies like this, a continued upsurge in the rate of farmland loss can be expected even if the total population remains stable.

The displaced inhabitants of *townscape* have largely been provided for by new development on greenfield sites in new towns, expanded towns, out-of-town housing estates, and the fringes of existing cities. Some developments have been designed as compact new townscape, taking the minimum of farmland necessitated by their generously low densities. Many, however, have been scattered as new sprawl which has fragmented the farmland and rendered it under-productive, often to the point of abandonment as idle wasteland. The resulting mixture of settlement sprawl, fragmented farmland and wasteland vegetation make up a patchy pattern which is referred to as rural-urban or ***rurban fringe.*** In south Essex and north Kent between 1962 and 1972, no less than 42 km$^2$ (4,200 hectares or 10,380 acres) went out of farming, but only one-third of these were used for specific purposes. No less than 2,800 hectares (6,920 acres) went down to waste vegetation because the growth of urban pressures had made them unfarmable. This type of situation is very common. Although the Second Land Utilisation Survey does not yet have final figures, the indications are that the urban fringe, which began some 50 years ago, covers about twice as much land as all the townscape created over the last millennium. It is all the more ironic that for 33 of the 50 years we have had planning authorities which were expected to prevent sprawl and conserve farmland.

The dilution of townscape and the growth of rurban fringe has inevitably been at the expense of *farmscape*, i.e. the areas where improved farmland is dominant and uninterrupted, and where productivity is greatest. Farmscape now contains only about three-quarters of the total improved farmland, since the remainder is scattered in discontinuous patches in either the rurban fringe or the marginal fringe, and hence the farmscape use pattern is diminishing even more rapidly than the farmland use category.

Simultaneously with this severe squeeze upon its resource land, the farming profession has been urged to produce "More Food From Our Own Resources", and hence the chain of landuse pressures has been carried further, as farmers have impinged upon wildlife habitats: grubbing up hedges and copses in the farmscape, upgrading rough land in the ***marginal fringe***, and even reclaiming parts of the ***wildscape***, as in Exmoor. Marginal fringe is the zone of poor quality land where improved and unimproved patches are interspersed, and where the improved land expands or contracts according to the rise or fall in farm prices. A sustained boom may permit a complete upgrading to form new farmscape, while a sustained depression may cause complete downgrading to form new wildscape. Wildscape is the zone of continuous unimproved rough grazings, together with any non-pasturable kinds of semi-natural vegetation.

Thus, the chain reaction of outward pressures extends from the inner city to wildlife habitats, with farmland caught in the middle. It is destroyed by expanding rurban sprawl but also attacked by a backlash from wildlife interests, who object to its attempts to compensate for its losses. It is undoubtedly true that wildlife habitats have been far more vociferously defended than the productive farmland resource. For example, great publicity has been given to the possible loss of all our hedgerows within 140 years with no thought for simultaneous disappearance of most of our existing farmland, so that any successfully conserved hedges would no longer have a rural setting.

Urban interests, farmland interests and wildlife interests are all fighting each other, yet none of them are really benefiting from it. Wildlife habitats are being lost, farmland is being destroyed, and the inner city where the chain reaction starts is so much worse off than before that it is said to be dying. Landuse studies suggest that this mutually destructive impasse is totally unnecessary; looking at the whole range of landuse in a unified context suggests that a mutually satisfying alternative is possible. Whichever of the three major interests provides a starting point, the solution is the same, and in the context of this paper the farming viewpoint will be adopted.

## Landuses Compatible with Agriculture

During the past few decades rural areas have been redefined in the public consciousness as "the countryside", which everyone should have the right to enjoy. Wildlife habitats, landscape beauty, historical and archaeological sites and recreational amenities must be conserved. The farmer has been brainwashed into believing that it is his duty to cater for all these interests, and also to surrender his land for urban and industrial expansion when required. Almost everything is regarded as important except farming and forestry, which are the basic working machinery of rural land.

Increasingly, however, it is recognised that we live in a world of shrinking resources, and that the future must give increasing priority to renewable resource land, i.e. to precisely those landuses which have been regarded as poor relations in the recent past. It is essential, therefore, that we should ask ourselves which other landuses are compatible with agriculture, and which are incompatible.

Wildlife habitats are not generally incompatible with farming. The two have always co-existed and there is evidence that strong, varied and complex ecosystems can help to keep pests and diseases in check. But potentially productive land given over to wildlife will not be compatible with hungry populations in a future situation in which oil revenues have run out and Britain can no longer afford to import half her food, or when world food surpluses no longer exist to be purchased. Wildlife enthusiasts may themselves be among the hungry, and it would take very dedicated people indeed to accept starvation in the cause of protecting birds. Indeed, protection may be impossible; birds may have to be eaten.

Conflict between naturalists and farmers is not conducive to maximising wildlife habitats. The correct policy for both sets of rural interests would be to combine against urban and rurban consumption of rural land. If land is protected and farmers enjoy security of production, they are far more likely to cooperate in conservation than if they are constantly under threat. They themselves are countrymen par excellence, and wildlife is of intrinsic interest to them. It is an interest that needs encouragement rather than bullying.

Landscape beauty, too, is not incompatible with agriculture. Indeed, it has largely been created in the first place by the efforts of farmers, foresters and rural landowners, and it soon decays when the unremitting labour of these people is withdrawn. This is seen in the rurban fringe where a great deal of rural land remains unbuilt on, but unfarmable. It soon becomes weed-infested and depressing in appearance, with vandalised trees, and fields despoiled by litter and rubbish dumps. No one looks after farmland like a farmer, and only a prosperous farmer can afford to do so well. One of the royal roads to landscape beauty is the maintenance of prosperous agriculture. Another is by finding ways in which landscape beauty is in the farmer's self-interest. If hedgerow timber could be recognised as a legitimate source of farm income which could be harvested and replanted in due course, it would do far more to conserve tall and attractive hedgerows than fossilising tree preservation orders or taxes on timber wealth.

Archaeological and historic sites take up only little improved farmland. Prehistoric populations were small, and occupied uplands which are now largely wildscape rather than improved land. Urban and industrial archaeology is often on sites that have been disturbed and sterilised for agricultural purposes. But some archaeological and historic sites do exist where they are in direct competition with agricultural production. These might well be conserved by means of some sort of compensation for lost production, which should not be unduly expensive as long as the objective is purely scientific. Once these sites are opened to the general public, however, they have to be classed as recreational amenities, and a negligible degree of incompatibility may be transformed into a very serious one.

It takes a certain amount of courage to suggest that rural recreation is not necessarily a good thing, since many planners seem to regard it as the leading function of the countryside. Gordon Cherry's book on *Rural Planning Problems*, published in 1976, two years before he became President of the Royal Town Planning Institute, does not contain even a single chapter on farming or forestry, only a few scattered references where these subjects happened to be briefly incidental to something else. Recreation, by contrast, is described as vital in terms of the quality of life, and enriching to the community.

The damage to agriculture is considered a necessary price to pay for the supreme benefit of countryside enjoyment by all. The Strutt Report on "Agriculture and the Countryside" suggests financial compensation to farmers, but it seems doubtful whether the scale of the loss is appreciated. The present writer is preparing a questionnaire on financial losses due to trespass and damage, and is coming to believe that farmers themselves do not realise how serious a measure of social injustice is perpetrated against them. The Countryside Commission is currently committing several million pounds in an attempt to redeem a tiny area of rurban fringe which may contain perhaps one-thousandth of the farmland in England and Wales.

It seems unlikely that the country can carry the huge financial burden of genuine compensation to farmers, and perhaps, instead, we should face the fact that recreation and agriculture are rarely compatible. If one has to be sacrificed upon the altar of the other, should it always be agriculture that suffers? And if farmers are robbed of prosperity in this way, can they afford to support the conservation of wildlife, landscape beauty, and sites of archaeological or historic importance in the countryside? Once again, we conclude that the potentially compatible interest should join forces to protest against the incompatible intrusions, rather than fight each other.

**Can Present Trends be Reversed?**

We have recognised two very strong and destructive trends in British landuse: the consumption of farmland for urban purposes, and the damage to production on the remaining land by rural recreation. Some critics argue that these trends are powerful natural tides which only unrealistic people like Canute's courtiers would attempt to turn. They believe we must wait for the natural turn, which will come when the consequences of land loss are sufficiently dire to be obvious to all.

They thus combine inactivity at present with optimism for the future.

Unfortunately, it will then be too late. Agricultural land losses are largely irreversible, and the fragmented areas that can be reclaimed are unlikely to yield as efficiently as the uninterrupted farmscape which we are steadily destroying. We need to look ahead and take action now.

The first step is to stop thinking of these trends as inexorable tides and to recognise that they are merely fashions. A fashion is something that has been created, and that can also be superseded if a new fashion is created in its place. So if we can demonstrate that today's drift to the country is an artificially stimulated fashion, we can overcome the first great psychological barrier. This should present no problem, as in many parts of the world, as well as in our own past, a drift towards the towns has been a far stronger trend than a drift towards rural areas.

There are at least five artificial factors that have stimulated today's urban exodus. In the first place, the decanting of people from the inner city was not a natural trend but a planned compulsion. Because few dwellings were rebuilt where many were demolished, most people had little choice but to move out. Secondly, the rebuilding has been largely in the form of flats, which have continued to be imposed by housing authorities regardless of the misgivings of the residents. Families who were in a position to choose houses had to go elsewhere. A third factor in promoting an exodus has been the increase of urban crime and vandalism, and this has now been scientifically demonstrated as having a high correlation with eight design and layout features commonly found in post-war council estates, as well as with a high density of children, which has also resulted from public policy in housing allocations (O. Newman 1974; S. Wilson and A. Sturman, 1976). A fourth factor has been the large-rate increases to be borne by a declining number of residents, while last but not least has been the enormous inflation in city house prices, exacerbated by the land shortage due to the hoarding of derelict sites owned by public authorities. All these factors make it quite clear that people are not fleeing the inner city location as such; they are fleeing the sort of place it has become as a result of artificial policy decisions. But policy decisions can be reversed, and must be, if a new fashion for inner city living is to be created to relieve the pressure on our renewable resource land.

An essential ingredient for a successful new fashion is to show that it does something positive for the multitudes who adopt it, i.e. it is in their own self-interest. Fortunately, there are some very real factors of self-interest that can be harnessed to arrest the urbanisational and recreational invasion of farmland. One is the inflation of commuter fares which has made many people wish to live nearer their work, and there is, in fact, plenty of unused city land available for the kinds of homes and environments they would like. In Tower Hamlets, for example, no less than 15 per cent of the borough is unused land, and London as a whole has 80 $km^2$ in this category. It would seem unethical to take even another acre of farmland until all this dereliction has been brought into urban use.

The building of homes with gardens would also simultaneously relieve recreational pressures on the countryside to some extent. Gardening is the most popular form of outdoor recreation with both men and women, and yet millions of families have been deliberately denied the opportunity for it by the flat-building policy.

The implications of turning the tide are enormous, and cannot be more than hinted at in the space of this article. But one thing seems certain. The inner city is the source of the great chain reaction of landuse pressures, and it is in delimiting the inner city that the battle for the countryside should begin.

## NOTE

1. The figures quoted here refer (unless otherwise stated) to improved farmland, which produces about 98 per cent of our farm output, and to England and Wales.

## REFERENCES

Best, R. H., 1978 : "Myth and Reality in the Growth of Urban Land", in A.W. Rogers (ed.), *Urban Growth Farmland Losses and Planning*, Institute of British Geographers.

Coleman, A. M., 1978 : "Agricultural Land Losses: The Evidence from Maps", in A.W. Rogers (ed.), *Urban Growth, Farmland Losses and Planning*, Institute of British Geographers.

———, 1966 : "Is Planning Really Necessary?", *Geographical Journal*, Vol. 142, Part 3, 1978; pp. 411-37.

———, 1978 : "Urban Regeneration and the Inner City: Planning and Land Use", *Chartered Surveyor*, Vol. 111, No. 5, pp. 158-63.

Ministry of Agriculture, Fisheries and Food, 1974. *Agricultural Statistics, England and Wales.*

Newman, O., 1974 : *Defensible Space*, Architectural Press.

Wilson, S. and Sturman, A., 1976 : "Vandalism Research aimed at Specific Remedies", *Municipal Engineering*, pp. 705-7.

# 19

# LANDUSE AND NUTRITION LEVEL IN HASPURA ANCHAL, BIHAR

A. KUMAR AND G. P. SHARMA

THE POTENTIALITY of arable land is revealed through the production of food crops and their caloric output. Haspura Anchal of Aurangabad district, located in the Southern Gangetic Plain of Bihar, provides a distinct agricultural unit to measure the agricultural efficiency of the region. The study is relevant in the context as the entire region still preserves the traditional dependence over food-crop economy with little potential to absorb more population. As the extension of arable land is limited, the only way to meet the increasing burden of population and reduce nutritional deficiency seems to be the improvement of existing landuse, cropping pattern and per-hectare yield. A detailed survey of availability of cereals, calories, food habits of the population and their nutritional requirements of Haspura Anchal is an attempt towards a proper planning.

Haspura Anchal or Development Blocks with a population of 66,806 spans over an area of 135.4 sq.km. in Aurangabad district of Bihar. The Anchal stretches in south-west to north-east direction over the interfluvial part of the Sone and Punpun rivers. The central part of the Anchal still registers the abandoned course of the river Sone in the form of "Chaur" or "Buh". The soil is old alluvium and varies in texture from north to south. Good soil, moderate climate with seasonal rhythm have provided conditions for a diversified cropping pattern.

The capability of soil and the natural facilities available together created conditions for the growth of population in this region. The average growth of population during 1961-71 was 1.99 per cent per annum but it was not uniform for all the 72 villages of the Anchal. The growth rate of one-third villages of this Anchal was less than the growth rate of the Anchal. The density of population per hectare varies from 14.65 to 4 persons. Area with low density, with less than four persons per hectare is confined to the eastern and southern drier parts of this Anchal. Population is mainly composed of primary population. In 47 villages over 90 per cent of their population is composed of primary population. Similarly, cultivators dominate over the agricultural labourers in 40 villages. Scheduled caste population comprises the bulk of agricultural labourers. It shares about 23 per cent of the Anchal's population.

## Landuse Pattern

Table 19.1 shows the eight broad landuse categories for the year 1975-76.

TABLE 19.1: Landuse categories, 1975-76

| Landuse Categories | Total Area in Hectares | Per cent |
|---|---|---|
| 1. Net Sown Area | 10732.31 | 78.92 |
| 2. Current Fallows | 638.62 | 4.65 |
| 3. Other Fallows | 105.49 | 0.72 |
| 4. Cultivable Waste | 723.70 | 5.35 |
| 5. Unculturable Barren | 48.98 | 0.33 |
| 6. Land under Miscellaneous Tree Groves and Crops | 49.31 | 0.34 |
| 7. Permanent Pastures | 101.62 | 0.81 |
| 8. Land put to Non-agricultural Use | 1198.63 | 8.88 |
| Total | 13598.66 | 100.00 |

An analysis of Table 19.1 reveals that net sown area alone shares 78.92 per cent of the total area of the Anchal. Land put to non-agricultural use sharing only 8.88 per cent takes second place. Cultivation is the main source of economy of the people and the heavy pressure on land and complete absence of other employment opportunities have resulted in high percentage of net sown area. Current and other fallows together share 5.37 per cent of the total area of the Anchal. The share of this category of landuse is higher in the eastern and central villages with a limited amount of canal irrigation. In contrast to these villages, entire cultivable land is put under plough in the northern and central canal-irrigated tracts of this Anchal. Cultivable waste and barren lands are mainly confined to the vast *chaur* tract, which suffers from waterlogging.

The change in the pattern of landuse is apparent since 1951-52. It may be partly due to the increase in the net area irrigated and improved land management over marginal lands of the *chaur* tract. Land under non-agricultural use has shown a gradual decrease as pressure on land has increased during these years. Percentage of area under net sown area as a whole increased to 12.61 per cent in 1976 over 1951-52. Wastelands still preserve some potentiality for future increase in cultivation but such potential lands are limited to a few villages only. Smaller villages have reached their saturation point where no future increase in cultivated land is possible. Demand for more land under cultivation has similarly reclaimed even those lands which were under miscellaneous tree crops and groves.

Though the Anchal has a greater share of its land under cultivation, the per capita share of net sown area is only 0.16 hectare. The increase in population has resulted in a net decrease in the per capita net sown area. The pressure of population on arable land is high. The economic density, which is related with the number of persons and intensity of cultivation, is 4.73 persons per hectare of arable land. Most of the villages of this Anchal record even higher pressure than the desired limit. Similarly, the caloric density and the optimum carrying capacity of land indicate that the Anchal as a whole has a moderate caloric density (4.6 persons/hectare of food crops) and the carrying capacity of land is almost equal (4.26 persons/hectare of food crops).* Only few villages have surplus food when food shortage is acute, in the area.

Under these pressing conditions the Anchal requires rational use of its available resources so that the needs of various uses of land can be adequately and scientifically fulfilled beside improving the nutritional standard of the population. It requires manifold achievement in land utilisation, crop combination and food production. The present paper attempts to analyse the present position of food production and nutritional standard of this Anchal under the prevailing conditions of landuse.

---

* Mukherjee has recommended a desired limit of 2.5 persons per hectare of arable land.

## Food Crop Production and Average Yield

The regional differences in the cropping pattern in this Anchal reveal that Aghani and Rabi crops occupy the same place in the crop-structure of the villages and in every village foodgrains occupy a dominating position among the various crops. The farmers in the villages do not make much distinction between food and cash crops, and to some extent the food crops are also regarded as low-value cash crops.

"Food-crop production is a very important topological characteristic of agriculture. It reflects the inclination of a particular village to produce specific agricultural goods". In the present study the average yield of major foodgrains has been determined by the detailed villagewise survey and then the gross food crop production is calculated for each village of this Anchal. The present calculation reveals great variation in the total production of foodgrains. Bigger villages with a high percentage of area under food crops produces more than the smaller villages. Production reflects apparent relation with the available facilities of cultivation such as irrigation, etc. This is more clear when villages with almost equal average under food crops are compared as in the case of Bagharewa, Haibaspur and Barewa. Haibaspur avails more irrigation facilities than Barewa and Bagharewa and its production is similarly high among them. The average per hectare production of foodgrains ranges from 26.5 quintal (Kamalpur) to 1.5 quintal (Bagharewa) per hectare of the total cropped area. The villages can be grouped into various productive categories (Table 19.2).

TABLE 19.2: Per hectare of total cropped area in various productive categories

| Quintal per hectare | No. of Villages | Percentage of Villages | Category Productive |
|---|---|---|---|
| 0.0–4.0 | 6 | 8.3 | Very low |
| 4.1–8.0 | 31 | 43.1 | Low |
| 8.1–12.0 | 24 | 33.3 | Fair |
| 12.1–16.0 | 8 | 11.1 | High |
| Above 16.0 | 3 | 4.2 | Mod. high |
| Total | 72 | 100.0 | |

High-potential productive arable lands are confined to 11 villages, sharing roughly 15.3 per cent of the total arable land of this Anchal.

Kamalpur leads in production of food crops. Other villages under this category are Jaitpur, Narsan, Bakhtiarpur, Udaibhanchak, Jalpura, Haspura, Paranpura, Nagauli, Shihari-Kalan and Koilvan. Their production varies between 12.3 and 26.5 quintals per hectare.

There are 24 fair productive villages of Haspura Anchal, which occupy 33.3 per cent of its total arable land. The production of food-crops per hectare of the total cropped area varies between 8.3 quintals and 11.7 quintals per hectare. Dihuri with only 54.42 hectares of cropped area produces 640.40 quintals of foodgrains. On the other hand, Paharpura produces only 1461.60 quintals of foodgrains from its 176.42 hectares of cropped land. Within this category certain villages, like Itwan, Ahiyapur, Haibaspur, Dhamani, Alpa, Raghunathpur, Makhbulpur, Jakhaura, Dindir, Chaurahi and Tal have fairly good production capacity as their foodgrain production varies between 10 and 11.7 quintals per hectare of the total cropped area.

Poor production of food crops per hectare is a common aspect of the arable use of land in the 31 villages of this Anchal. Altogether about 43.1 per cent of the Anchal area produces four to eight quintals per hectare of the cropped area. Excepting the villages located in the southern part of the Anchal, namely, Chanhat, Jhinguri, Itwan-Wazirpur, Kaithi, Sonhathu, Fatehpur, Tilakpura, Tilauti, Bihta and Munjahar most of the villages have low production, although these are canal-irrigated villages. This reveals the inadequate traditional method of agriculture and the lack of incentives among the cultivators to invest in manure, good seeds, etc., for the increase in output.

The physical potentiality of the cropped area of the Anchal is high, because proper investment in agriculture brings high production. If per hectare input in agriculture is increased and irrigation facilities are provided, an increased output can easily be obtained even from the moderate cropped land.

Very low productive villages occupy 8.3 per cent of the total villages of the Anchal. The six villages, namely, Barewa, Bagharewa, Mahuar, Nirakhpur, Shihari-Khurd and Suratbigha, produce even less than four quintals of foodgrains per hectare of the cropped area. The main concentration is on the cultivation of paddy and little attention is paid to the cultivation of other food crops in villages Barewa, Bagharewa, Shihari-Khurd and Suratbigha. Cultivators reside in neighbouring village, so little attention is paid to increasing the production of food crops. In Bagharewa, physical conditions also hamper cultivation as it lies in the "buh" tract of the Anchal. In most of the villages potential is high, only it requires proper management associated with adequate irrigation facilities.

**Availability of Cereals**

Cereals account for a major share of the total food requirement of the population of this Anchal. The details of production and consumption pattern in terms of basic agricultural produce, mainly cereals, reveal the pattern of availability of cereals per capita per annum/per day besides per head daily surplus of deficit in grams. In the present calculation of cereals production, only rice, wheat, maize, barley, marua and gram have been included and other crops like pulses oilseeds, sweet potato, vegetable and sugarcane do not cover a wide area of the Anchal. The actual production of each crop and other products were calculated on the basis of average yield in the year 1975-76. With the help of actual production obtained and total population per capita availability has been worked out.

It has been estimated that in Indian conditions per capita daily intake of 2,000 calories available from 580 grams of food should be treated as adequate and a reasonable minimum requirement (Shafi, 1960). Out of 580 grams of foodcrops approximately 480 grams are obtained from cereals and the remaining 130 grams from pulses, oil, animal products and other sources.

Considering 450 grams of cereals per head as measuring requirement of a man (Stamp, 1964) for maintaining subsistence level, which provides roughly 1,800 calories, the daily surplus and deficit per head in grams have been calculated for each village.

Figure 19.1 clearly reveals that only in 19 out of 72 villages of the Anchal, per capita available cereal is less than 450 grams while in four villages it is almost adequate. In the remaining 49 villages per-capita-per-day availability of cereals is more than the estimated requirement of 450 grams.

The villagewise per-day availability of cereals further reveals that in 22 villages it is above 700 grams. Such villages are intensively cultivated with high diversification of crops. But in villages the availability of cereal is actually much lower than the calculated figure because cereals are also cash crops. Besides, the produce of two uninhabited villages and of Barewa is consumed in the nearby villages which further increases the per capita availability of cereals. Surplus cereals mainly concentrate in small villages with less mouths to feed.

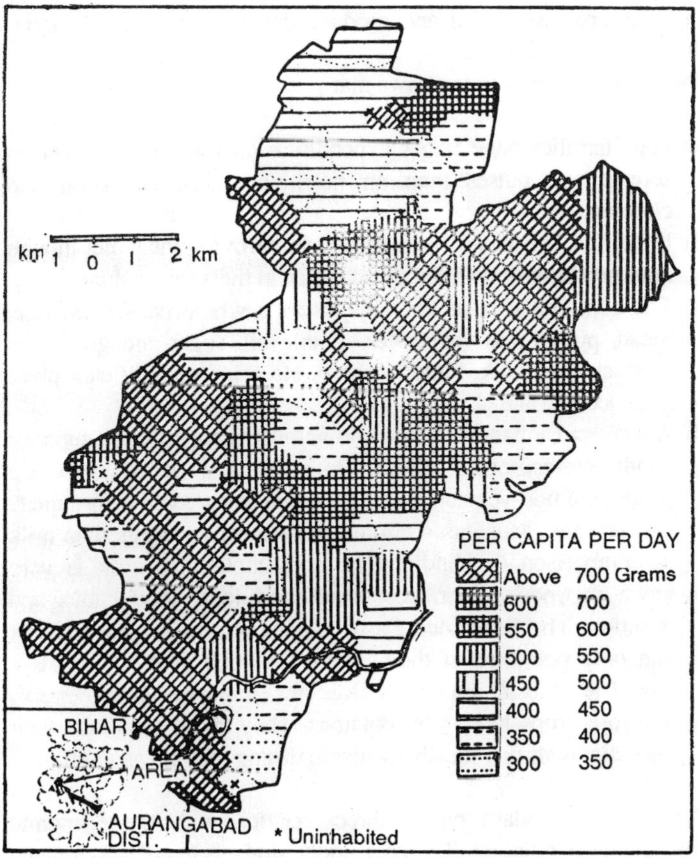


FIG. 19.1: **Haspura Anchal — availability of cereals, 1975-76.**

## Food and Dietary Standard

In examining the food requirement of population, not only the production of cereals and its per capita distribution is essential to estimate the level of nutrition but it is also required to examine the dietary habit and standard of population. Nutrition and diet are not synonymous. "Diet consists of various articles of food which are ingested and converted for use in the body for building up and maintaining it in vital condition. Nutrition is the end of the process by which these are assimilated and produce the desired result" (Borkar, 1957).

The personal observation shows that :

(1) Poor families have larger expenditure on food, particularly on cereals and pulses than on nutritious diets, education and clothing.
(2) Even rich families seldom think of improving their diet though milk products find an important place in their food habits.
(3) The common diet of the villages as a whole constitutes rice, bread, pulses and vegetables. Edible oil, sugar and gur, dairy products like milk, butter, cheese, etc., find insignificant place in the common diet.
(4) Apart from a handful of people belonging to different superior Hindu castes, other Hindus, Muslims and Scheduled Caste people are non-vegetarian but their budgets seldom afford meat, fish or eggs. Muslims consume more meat and eggs than milk in comparison to Hindus. The consumption of milk is very much dependent upon economic conditions, food habits and tradition. The Scheduled Caste population has the poorest diet and only one-third of their total population gets two meals a day. They subsist mainly on rice, stra (*sattoo*) of gram, *khesari* and root crops like sweet potatoes. They try to substitute their poor diet with fish locally available during the rainy season.

The dietary standard has a direct relation with the economic standard of the population. The area under study is mainly dominated by middle class agrarian communities and about half of its population is constituted by the poorer sections of society, especially the Scheduled Castes. Dietary standard varies from average peasants to agricultural labourers. Average high middle class Hindus and Muslims have sufficient caloric intake to keep them in good health. But dietary budget of middle class cultivators keep them below the standard balanced diet. The lower class families are badly suffering from malnutrition. In this Anchal there is too much use of cereals in the diet and not much stress is laid on its nutrient content. Though gur is a chief source of calories, yet, its consumption is limited to middle-class population. The majority of people subsist on a diet which lacks adequate protein, sufficient carbohydrates and is deficient in total calories.

**Availability of Calories from Cereals**

To determine whether a person suffers from undernourishment or not, we have to find out the caloric requirements of his body and the caloric value of the food he takes (calory being a measure of energy supplied by the food articles we consume) (Mishra, 1970). The villagewise study of the production and consumption of cereals and the food habits of the people together can be worked out to establish an indirect relationship between the use of the land and the nutritional state of the people.

The details of consumption and production pattern in terms of basic agricultural produce including purchases from Government fair price shops have been worked out for each village of this Anchal. The need for milk, milk products, fish, meat, eggs was estimated for each village on the basis of the investigation carried out during field work. On the basis of the survey, Food Balance Sheets for each village have been prepared, after making on allowances for seeds and wastage in cooking, storage, transport, etc. The net grains available for consumption have been converted into calories (Aykroyd, 1962). Calories per day available for 72 villages of the Anchal range between 654 and 6,281 and the average for the whole Anchal works out at about 2,500.

However, to achieve a proper and true caloric position in all the villages it is essential to analyse the per capita daily available calories from the Anchal average, i.e. 2,500. Altogether 42 villages reveal a deficit in per capita calories from the Anchal average. The deficit or surplus varies considerably and they can be classified into different groups. Thirty villages of this Anchal fall under surplus category. They are grouped in three categories: low, medium and high (Fig. 19.2).

*Low-to-Medium Surplus*

Villages having low surplus calories (less than 20%) are 15. They are situated in between the western canal irrigated tract and mid-eastern tract. Sonhathu in the south, Raghunathpur in the north and Dindir in the east are a few other scattered villages of this category.

Medium surplus calories (20-50%) are confined to only five villages, widely scattered over the Anchal. Four villages reveal a high surplus of calories by more than 50 per cent. In Habaspur, Tika, Kaithi and Alpa the higher surplus is very much related to the high regional coverage and good production per hectare. Barewa village has been recently inhabited by a few agricultural labourers, consisting of 23 persons in all. The actual calories available to them are far below the calculated average of the Anchal.

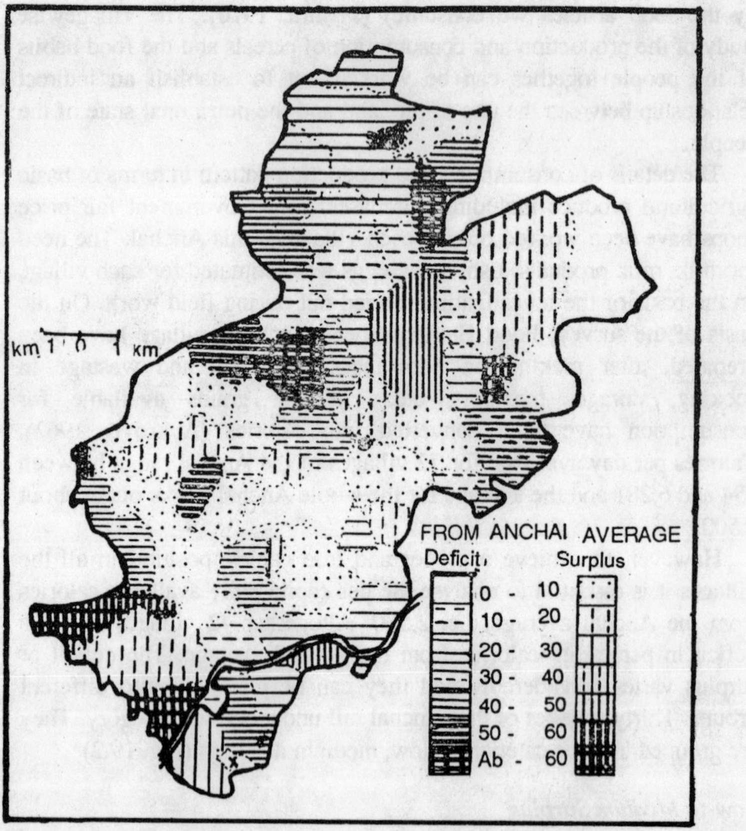


FIG. 19.2: Surplus and deficit calories (in per cent) calculated on daily per capita basis.

*Low-to-Medium Deficit*

Villages with low deficit (below 20%) are 15. Such villages are seen scattered in the outer margin of the irrigated tract of the Anchal mainly in the east, south and north. Twentytwo villages are medium deficit in calories by 20 to 40 per cent. They are distributed throughout the Anchal but are mainly clustered in the northern and southern sections of the Anchal. Villages having a very high deficit calories are only five and show a deficit in calories by more than 50 per cent. This is due to various causes like low per capita available land, low yield and high density of population.

The analysis, thus, reveals that the lowest per capita consumption of calories is in Itwan whereas, in surplus group, as high as 22,231 in Barewa. This figure is, however, misleading as the 23 persons residing in Barewa belong to the *Nut* community (a sauntering tribe) and they receive a fraction of the total produce of the village. After a painstaking field work and thorough investigation of the dietary condition of the various economic groups and communities, it is found that the average computed per capita/per day calories of this Anchal are more than the actual calory intake of this Anchal. As already mentioned, the cereals produced in the Anchal are not all available for local consumption, because they are also used as low-value cash crops. Farmers sell their surplus produce to meet their other requirements like clothes, medicine, social obligation, marriage, etc. Thus, the actual figure of cereals available for consumption is considerably low.

**Standard Nutrition Unit**

The per-capita-per-day caloric requirements have been, thus, reckoned for Haspura Anchal on the basis of scale recommended by the Nutrition Expert Group (1962, Indian Council of Medical Research). In calculation the requisite weight is given for the proportion of different age and sex groups.

The calculation reveals an average requirement of about 2,151 calories per day to keep the population in a healthy condition. This figure may, however, be well compared with Aykroyd's (1962) recommendation of 2,600 calories and Sukhatme's (1961) 2,300 calories for average Indian. Stamp (1963) recommended 2,460 calories (British) for the average which is equivalent to 9,00,000 calories a year and termed it a Standard Nutritional Unit (SNU). One SNU is estimated to be adequate for a person. The requirement of calories in India should be much less as children constitute the major portion of population; besides, the climate is warm here and also because of the physical metabolism of short stature of Indians. Shafi (1960), on the basis of his studies, considers intake of 2,000 calories adequate and reasonable minimum requirement under Indian conditions. Mohammad (1978) and Das (1972) have also supported this contention while establishing the nutritional levels in their region of study.

In the present study the average caloric requirement on the basis of population structure is though worked out on an average of 2,151. An intake of 2,000 calories per capita daily has been considered reasonable minimum requirement. Thus, the weighted average standard nutrition unit for indication in calories/person/annum for this region as a whole amounts to 730,000 for consumption.

The villagewise distribution of SNU reveals that out of the total 72 villages of this Anchal, 22 villages are deficit in SNU. In these villages the deficit of SNU ranges from 0.5 per cent to 44.4 per cent. Most of the deficit villages are situated in the northern part of the Anchal while some are scattered in the south of the Anchal also. Surplus of SNU ranges from 0.5 per cent to 1082.6 per cent.

Surplus and deficit per capita calories as discussed earlier seem to determine the availability of SNU in this Anchal. As a whole most of the villages are either adequate in SNU or they are surplus in SNU which may be available to nearby villages as cultivators' own land is widely scattered in the other villages. Thinly populated villages and uninhabited villages provide the surplus SNU to be shared in nearby villages. The availability of SNU ranges from .5 per head of population in Itwan, Tetarahi and Habaspur to 13.1 per head in Mohiuddinpur (Fig. 19.3). In four villages, namely, Barewa, Bagharewa, Nirakhpur and Kaithi-Bankat it is more than five per head of the total population. In 22 villages it is less than 1.0. The availability of adequate per capita SNU in most of the villages does not mean that the level of nutrition in such villages is perfect as the proportion of total calories derived from cereals is high. The quality of diet becomes lower as the proportion of calories from cereals increases. This produces higher incidence of malnutrition in the area under study.

### Total Sown Area and SNU

Standard Nutritional Unit is the produce of agriculture and it is related to area under cereals, yield and cropping intensity. It is measured in land productivity. SNU can be thus directly correlated to total sown area as it will give a better picture of the nutritional level of the people in terms of carrying capacity of land (Stamp, 1963).

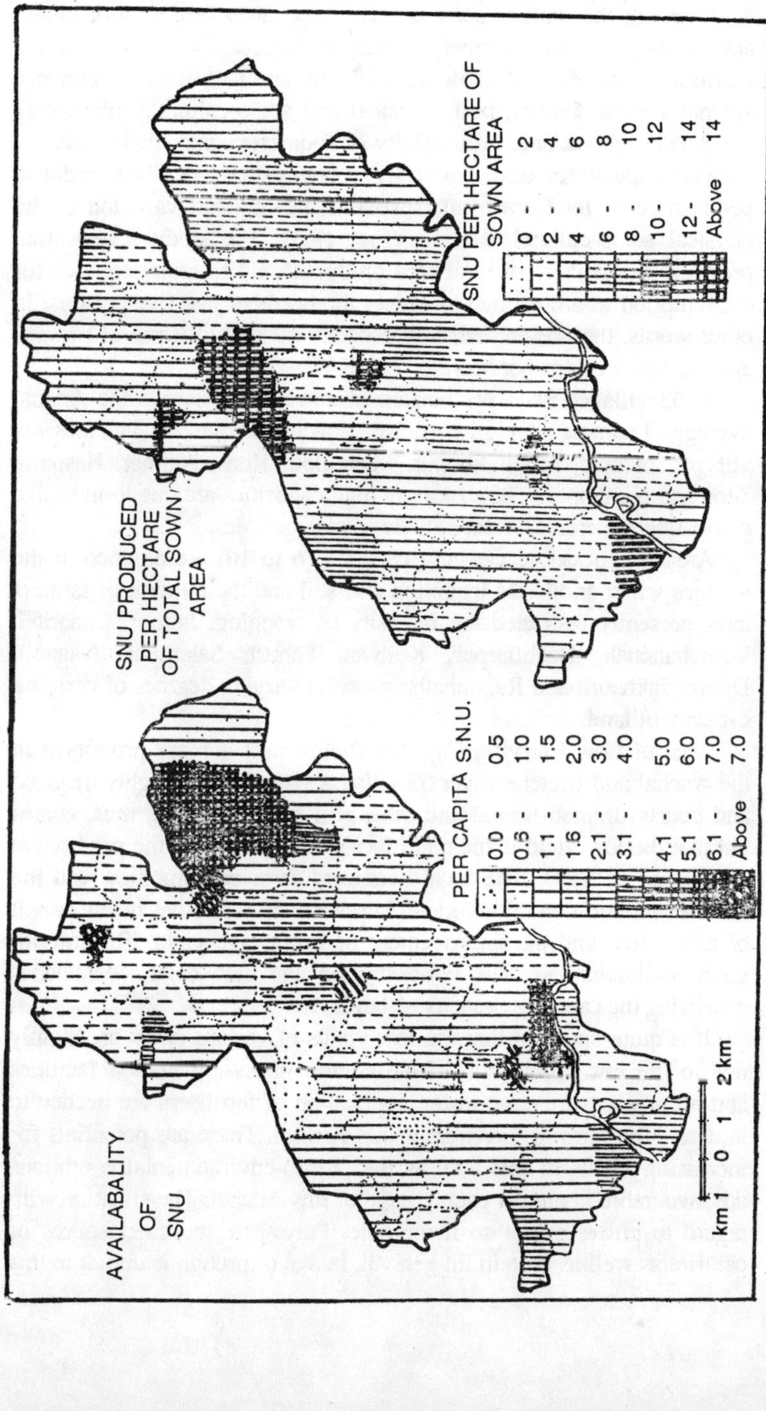


FIG. 19.3: Haspura Anchal.

It reveals the limit of persons it can adequately feed to good health and activity. "If the number of people exceeds this limit then the nutrition standards are liable to fall" (Singh, J., 1971). In Haspura Anchal various factors, both physical and socio-cultural, collectively resulted in the variable productivity of food crops and the differential carrying capacity of land. The distributional patterns of SNU produced per hectare of total sown area explains the regional variation in the physical socio-cultural controls (Fig. 19.3). The study reveals that presently per hectare of total area produces only as many calories for consumption as are enough to satisfy the needs of just five persons. In other words, the present carrying capacity of land in terms of persons is about five persons per hectare.

In 33 villages SNU per hectare of sown area exceeds the Anchal average. The area of high carrying capacity (above 10) includes six villages, namely, Mohiuddinpur, Makbulpur Khas, Amjhar, Haspura, Alpa and Nirakhpur. The strikingly high capacities are due to intensive cultivation of cereals, sweet potatoes and vegetables.

Areas of moderate carrying capacity (6 to 10) are attained in the villages where moderate irrigation and soil and the outlook of farmers have presently restricted the intensity of cropping. Jaitpur, Chaurahi, Udaibhanchak, Bakhtiarpur, Koilvan, Dihuri, Salempur, Nagauli, Dindir, Jakhaura and Raghunathpur reveal varying degrees of carrying capacity of land.

Area of Low Carrying Capacity (below six) is more prominent in the Anchal and stretches over 53 villages lying in both highly irrigated and poorly drained Kewal and other soil belts. The fact, thus, comes out that the low yield is the main factor which controls the per hectare production of SNU. The local backward farm communities, and the socio-economic conditions have hampered the effective development of agriculture and the improvement of hectareage yield. The inferior grain production, i.e. less concentration and high caloric cereals, is restricting the carrying capacity of land.

It is quite apparent that the low yields of various crops are mainly due to the low application of agricultural inputs. Irrigation facilities and better seeds increase in the application of fertilisers are needed to intensify the farming practice of this Anchal. There are potentials for increasing yields in this way as the physio-environmental conditions are favourable in almost each village of this Anchal. The situation with regard to pulses is not so favourable. Pulses are the chief source of calories as well as protein. In general, intake of protein is deficit in this

Anchal as protein deficiency diseases are very common. The pulses contribute only a part of the required protein and calories, and require additional areas to be brought under them to increase their output. They are not very responsive to fertilisers. This can be done only if sufficient hectareage under cereal crops is double-cropped. The production of protective foods like vegetables, fruits and animal products should be planned in order to reduce nutritional deficiencies. This requires a considerable change in the attitude of the people. Moreover, the improvement in handling, storage, cooking and uses of food is also required. It is true that habits die slow and it will take time to change the food habits in a short time but immediately still a lot can be done to avoid wastage of foodgrains by proper storing of grains.

There is no doubt that Haspura Anchal has adequate agricultural potentials to ensure a minimum level of nutrition to her people. At present a proper planning is required to change the traditional outlook of farmers with a proper combination and intensity of cropping with a view to increasing the yield of nutrition, cereals and cultivation of high-yielding cash crops. This is the prevailing requirement of the South Ganga Plain of Bihar. As it has been pointed out by various workers in this field of research an all-India norm is unsuitable for judging nutritional deficiencies in view of the regional variations in nutritional needs. Norms should be determined by population groups and their dietary habits. Haspura Anchal provides a typical example which requires an integrated development of crop production, livestock and poultry-fisheries beside improving the nutritional conditions of its population.

## REFERENCES

Ahmad, A. and Siddiqui, M. F., 1967 : "Crop Association Patterns in the Luni Basin", *The Geographer*, Vol. 14, pp. 69-80.

Aykroyd, W. R., Gopalan, C. and Subramaniyam, S. C., 1962 : *The Nutritive Value of Indian Food and the Planning of Satisfactory Diet*, I.C.M.R., New Delhi, pp. 49-55.

Borkar, G., 1957 : *Health in Independent India*, National Book Trust, New Delhi, p. 152.

Bose, P. K., 1977 : "Population-Food Nutrition Equation in India", *Everyman's Science*, Vol. XII, No. 1, pp. 17-34.

Das, K. N., 1972 : "Landuse and the Level of Nutrition in the Kosi Region, Bihar", *The Geographer*, Vol. XIX, pp. 56-85.

Kostrowicki, J., 1972 : "Some Methods and Techniques to Determine Crop and Other Landuse Combination as used in the Polish Landuse Studies", *Symposium on Landuse in Developing Countries*, (ed.), Shafi, M., 21st I.G.U., Aligarh, p. 90.

Misra, R. P., 1970 : *Medical Geography of India*, Ministry of Health, New Delhi, p. 26.

Mohammed, N., 1973 : "Nutrition and Nutritional Deficiency Diseases in Ghaghara-Rapti Doab", *The Geographer*.

———, 1978 : *Agricultural Landuse in India*, New Delhi.

Mukherjee, R. K., 1938 : *Food Planning for 400 Millions*, London, p. 1.

Nityanand, 1972 : "Crop Combination in Rajasthan", *Geog. Rev. of India*, Vol. 34, No. 1. pp. 46-60.

Sen, S. N., 1976 : "Nutrition and Rural Development : Some Economic Aspects', *Everyman's Science*, Vol. XI, No. 1, pp. 33-43.

Shafi, M., 1960 : *Land Utilisation in Eastern Uttar Pradesh*, Aligarh, p. 222.

Sharma, T. C., 1972 : "Patterns of Crop Landuse in Uttar Pradesh", *Deccan Geog.*, Vol. 1, pp. 1-17.

Singh, Jasbir, 1971 : "Optimum Carrying Capacity of Land, Caloric Density and Intensity of Population Pressure Changes in Punjab, 1951 — 61", *The Nat. Geog. Jr. of India*, Vol. XVII, pp. 1-35.

Stamp, L. D., 1963 : *Applied Geography*, Penguin Book, p. 115.

———, 1964 : *The Geography of Life and Death*, London, p. 96.

Sukhatme, P. V., 1961 : *Feeding India's Growing Millions*, Bombay, pp. 18-20.

# INDEX